math in 100 key breakthroughs

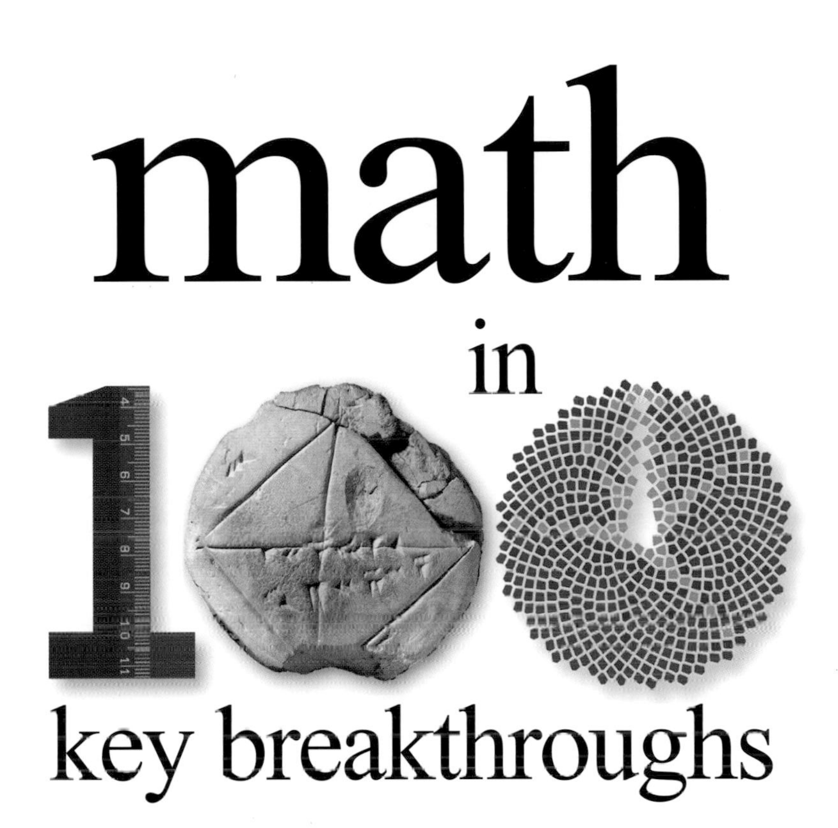

math in 100 key breakthroughs

richard elwes

Quercus

Contents

Introduction

Mathematics is a timeless subject. While historians study the peculiarities of place and era, and artistic tastes vary from culture to culture and person to person, no matter whether you are an ancient Babylonian shepherd or a 21st century computer programmer, one plus one is always equal to two. The same can be said of many branches of science, of course. After all, human anatomy has changed little within the last few thousand years, and gravity is (nearly) the same at every point on the surface of the Earth. Yet the fixedness of mathematical truths runs even deeper. If extra-terrestrial life exists, its biology will surely differ from that on Earth. We can even imagine other universes in which the laws of physics are fundamentally different, yet internally consistent. But it is harder to conceive of a world where $1 + 1 = 3$. Mathematics is not only true, but seems inevitably, necessarily true.

Of course, our ancestors did not emerge from the primeval swamp with a mastery of numbers. Discoveries are made at certain historical junctures; new techniques are invented by specific people. This is even true for the starting point of the whole subject: counting. That ability, too, emerged at a particular stage in our evolutionary history.

The nature of progress

So how does mathematics progress? The stereotypical picture is of a solitary scholar, whose outrageous genius concocts some dramatic discovery out of the blue. But this caricature overlooks the collaborative and incremental nature of the subject. As even the notoriously self-absorbed Isaac Newton admitted, "If I have seen further it is by standing on the shoulders of Giants."

Many of the breakthroughs described in this book do indeed involve the hard work and insight of a few dazzling individuals. Even so, very few emerged fully formed from nowhere, but instead built on the ideas of earlier thinkers. I believe it is better to see every development as a milestone on a longer road. With this in mind, I have tried to place each breakthrough in its proper context, telling the story of where it came from, the questions that researchers were seeking to address, and the consequences that followed once the answers were found.

Mathematics' golden ages

There have been several periods when mathematics has flourished: the Pythagorean cult of ancient Greece imbued the subject with a mystical importance. The Indian school of astronomy laid the foundation for the numerical system we know today. The Arabic translators of the House of

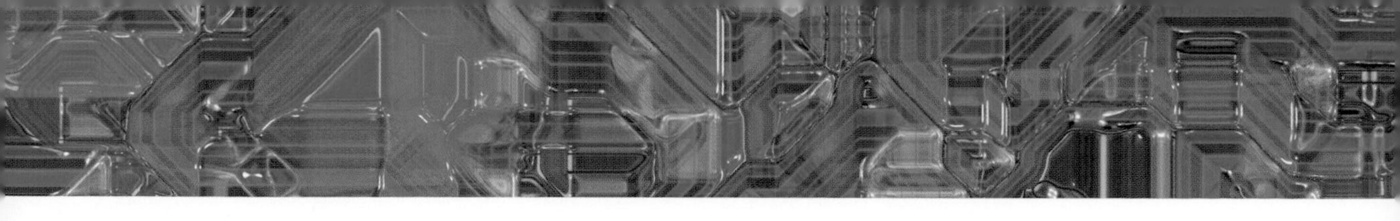

Wisdom gathered the world's mathematical knowledge into one supreme collection. The European enlightenment opened up new avenues of research, and led to a panoply of practical applications. All of these have claims to be golden ages of mathematics. But so too does the era in which we currently live.

The expansion of schools and universities around the world, the invention of the computer and the subsequent growth of the internet have all played a role in revolutionizing the subject's culture. Today's mathematicians are armed with sophisticated tools for research, as well as for teaching and disseminating their work. What is more, the subject is now truly global, and the mathematical community larger than ever, allowing people with common interests to communicate and collaborate more efficiently than ever before.

At the same time, the need for mathematics is becoming ever greater. With the development of relativity and quantum theory in the early 20th century, our understanding of the physical Universe reached a point where fluency in the language of advanced mathematics became an essential prerequisite to probe the deeper levels of reality. The same is true in other walks of life. With so much data gathered by businesses and governments, experts in probability, statistics and risk are constantly in high demand. Another burgeoning industry is computer science, a subject which emerged from Alan Turing and others' work in mathematical logic in the early 20th century. The deepest questions here are still mathematical in nature: ultimately what are computers capable of achieving? And what will they never do?

Tomorrow's world

So, the golden age of mathematics is today. This book tells some of the stories of how we arrived at this point. But what of the future? Let me finish with some predictions: ever more aspects of science and society will be illuminated by mathematics, as more and more nominally "pure" branches of the subject find unexpected practical applications, further blurring the boundaries between mathematics, physics, computer science and other areas of enquiry. Meanwhile, a stack of problems previously judged impossibly hard will quickly be proved by techniques as yet undreamt of. Yet, through all this progress, an embarrassing number of easy to state, seemingly obvious conjectures will still defy all attempts at solution, luring in new generations of thinkers to grapple with them.

RICHARD ELWES

1 The evolution of counting

BREAKTHROUGH SIMPLE ARITHMETICAL SKILLS ARE PRESENT IN A VARIETY OF ANIMAL SPECIES, FROM BIRDS AND BEES TO RHESUS MONKEYS AND CHIMPANZEES.

DISCOVERER COMPARISON WITH HUMANS' CLOSEST RELATIVES SUGGESTS THAT OUR ANCESTORS HAVE BEEN COUNTING AND CALCULATING FOR MILLIONS OF YEARS.

LEGACY WHILE HUMANITY'S MATHEMATICIANS PURSUE THEIR SUBJECT, EXPERTS IN ANIMAL COGNITION EXPLORE WHETHER MATHEMATICAL ABILITIES ARE SHARED WITH OTHER SPECIES.

Mathematics is part of humans' evolutionary heritage. Just as jellyfish, giraffes and jackdaws each found strategies to help them flourish in their own ecosystems, so humans came to occupy an intellectual niche, our enlarged brains providing us with powerful weapons to outmaneuver predator and prey. Part of this cognitive armory was an ability to reason abstractly with shapes and numbers. Over the millennia, these skills would grow into the subjects of geometry and number theory, and provide foundations for a scientific investigation of the world.

It is impossible to say when humans first became numerate. But we can get some fascinating hints from the mathematical abilities of our cousins across the animal kingdom. In 2010, Jessica Cantlon and Elizabeth Brannon trained two Rhesus monkeys named Boxer and Feinstein to add together single-digit numbers. The monkeys performed their arithmetic with arrays of dots. They were shown two screens displaying different numbers of dots, then challenged to add the two numbers together and select the correct total from two possibilities. Boxer and Feinstein achieved an accuracy of 76%, far above what pure guesswork would have achieved, if a little lower than human college students (94%). Interestingly, both humans and monkeys took longer to answer when the choices were nearby numbers, such as 11 and 12. The extra time and higher likelihood of error followed similar patterns in the two species. So it is not fanciful to imagine that the monkeys needed the extra time for thinking and counting.

OPPOSITE Experiments show that honeybees can reason abstractly about small numbers, mentally associating different patterns which contain the same number of elements, up to four. This is a useful skill for remembering the routes to sources of food.

Mathematical symbols

Higher mathematical ability requires some way of representing numbers using language or symbols. It might be thought that humans alone have this capacity, but in 1993, a chimpanzee named Sheba revealed that this is not so. Chimpanzees are humans' closest relatives, the two species dividing around 4 million years ago. Over the course of several years, primatologist Sarah Boysen trained Sheba to associate quantities of food with the symbols 0 to 9. The result was that Sheba successfully learned to interpret mathematical symbols, much as a human child does. Not only could Sheba move back and forth between a symbol and the corresponding amount of food with reasonable fluency: in time she could even master arithmetic presented purely symbolically, understanding that $4 + 2$ is equal to 6, for instance. Boysen's experiments with chimpanzees suggest the animals have an ability to recognize and manipulate numerical symbols, more so than preschool children.

Chimpanzees have an ability to recognize and manipulate numerical symbols, more so than preschool children.

Counting in birds and bees

It is not only primates who can count. In 2009, Hans Goss and colleagues tested the ability of honeybees to count the number of elements in a visual pattern. The bees performed impressively well on patterns up to four elements, a skill they surely use to remember sources of food. Meanwhile, one of the most famous experiments in animal cognition was carried out by psychologist Irene Pepperburg, who spent 30 years training an African grey parrot named Alex (standing for Avian Learning EXperiment). Alex became something of a celebrity due to his intelligence and mastery of the English language. In addition to a vocabulary of around 150 words, he was able to count to six reliably and perform simple arithmetical tasks as quickly as a human. In a significant step toward abstract mathematical thinking, Alex made the connection between numbers presented in three different ways: symbolically (such as the figure 6), a collection of six objects, and the sound of the word "six."

Extraordinary though Alex's achievements were, more recent studies have suggested that he was not quite as unique as first imagined. Several species of bird are capable of comparing numbers and making judgments about which is larger. In 2007, Kevin Burns and Jason Low performed an experiment in which robins of a New Zealand species were shown mealworms being lowered into holes. The challenge for the birds was to count the worms, and then choose the hole containing the greater number. It is not surprising that they were able to manage this when the numbers were low—between 0 and 2. Unexpectedly, though, the birds had the memory and intelligence to discriminate between numbers as high as 11 and 12. The researchers suggested that this skill derives from their hoarding behavior. When food is scarce, the robins try to hide food from each other—and indeed to raid each others' food hoards. In this battle of wits, an ability to keep count of pieces of food is a considerable advantage.

ABOVE In experiments, New Zealand robins have shown the ability to distinguish between numbers as high as 12.

Nature and nurture

An interesting question is whether numerical reasoning is an innate ability in animals, or, like Alex's use of language, a purely learned behavior. A telling observation is that even baby birds exhibit a basic grasp of number. In 2009, Rosa Rugani performed a study in which baby chickens were reared for three or four days with five chick-sized yellow balls in their nests, to which they formed emotional attachments. These balls were then divided between two screens, while the chick observed from behind a glass wall. When they were released, the chicks invariably preferred to join the larger groups of balls. But in order to know which screen concealed more, the chicks had to watch carefully as the balls vanished behind the screens. They needed to remember the results, and then have the intelligence to compare these two numbers, even when the collections were out of sight.

In a second experiment, some of the balls were moved after their initial distribution. This time, the chicks watched as balls traveled back and forth between the two screens. To keep track of which screen concealed a larger collection of balls, the chicks needed to perform addition and subtraction in their heads. Despite having no training in problem-solving, the chicks were able to pass this test of mathematical thinking.

2 Tallies

BREAKTHROUGH IN TOOLS USED FOR COUNTING, TALLIES REPRESENT THE FIRST TIME THAT HUMANS DEPICTED NUMBERS SYMBOLICALLY.

DISCOVERER THE USE OF TALLIES BEGAN WITH THE HUNTER-GATHERERS OF THE PALEOLITHIC ERA. THE FIRST DIRECT EVIDENCE WE HAVE OF THEIR USE DATES FROM AROUND 35000 BC.

LEGACY THESE SIMPLE, PREHISTORIC DEVICES MARK THE MOMENT THAT MATHEMATICS BEGAN.

While several species of animal have some ability to reason about numbers, only one has made the crucial transition from counting mentally to representing numbers symbolically. Evidence of prehistoric mathematics largely consists of tallies: sticks or bones with notches carved in them to help people keep count.

The earliest archaeological piece of evidence of mathematics is the Lebombo bone. Found in the mountains of Swaziland in southern Africa, this is a fragment of a baboon's leg bone into which someone has carefully carved 29 notches. This number suggests that the bone may have served as a simple lunar calendar; indeed, its design is similar to calendar sticks still used by Namibian bushmen today. Possibly it was a tool for keeping track of a lunar month, or counting the days of female menstrual cycle.

The Lebombo bone

Whatever its exact purpose, the bone was clearly intended as an aid for counting. The bone's owner had made the first great leap necessary for true mathematics: representing numbers in a fixed physical form, rather than storing them temporarily in the mind. The Lebombo bone has been dated to around 35000 BC, making its creator a modern human in evolutionary terms, but predating the first true civilizations, with their systems of agriculture and larger settlements. This later revolution did not begin until the Neolithic period, around 10,000 years ago. This gave human societies the stability and organizational structure that allowed

OPPOSITE Around 22,000 years old, the Ishango bone is the most striking evidence of Paleolithic mathematics. Discovered in the Democratic Republic of Congo, it is a baboon's leg bone which has been used as a tally.

later innovations to emerge such as writing, and technology including pottery and the wheel. The Lebombo bone's Paleolithic owner would have known none of this; he or she would have been part of a small band of hunter-gatherers, subsisting off the local wildlife. They were highly mobile, able to move locations depending on the season and the local animals, armed with tools made of stone, bone and wood.

The Ishango bone

The most famous evidence of prehistoric mathematics is the Ishango bone. This was discovered in 1960 in Ishango, in what is now Virunga National Park in the Democratic Republic of Congo. This bone is dated to around 22,000 years ago, also making it the property of a Paleolithic hunter-gatherer. Again, the tool is essentially a tally, but its configuration of notches is altogether more elaborate than that on the Lebombo bone. Grooves are grouped into three columns: the first reading 11, 13, 17, 19; the second 3, 6, 4, 8, 10, 5, 5, 7; and the third 11, 21, 19, 9. The first column has been interpreted as showing evidence of understanding of prime numbers (see page 49), but this seems highly speculative. Again, a lunar calendar has been suggested as an explanation, for tracking periods of time up to six months.

The bone's owner had made the first great leap necessary for true mathematics.

One-two-many

Besides tools unearthed by archaeologists, other evidence of our hunter-gatherer ancestors comes to us through people whose remote locations have isolated them from the outside world and whose ways of life have changed little in the intervening time. It is striking that some such people have survived over the millennia with amazingly limited vocabularies when it comes to numbers. The Warlpiri are an aboriginal people from Australia whose lifestyle has remained fairly constant over the last 30,000 years. Counting in the Warlpiri language begins with the word *jinta* (meaning "one"), followed by *jirrama* ("two"). But the Warlpiri have no word for "three" or "four." For any number larger than *jirrama*, the catch-all term *panu* is used, which is usually translated as "many." Other Australian aboriginal languages exhibit similar characteristics, some with additional numbers for three and perhaps four. It seems astonishing that at no stage have terms for larger numbers been invented, but the plain fact is that their particular mode of desert life has little need of them.

Languages like Warlpiri prompt deep questions about human cognition. Can people raised in a language without numbers have any concept of

LEFT A stone circle at Nabta Playa in Egypt. Dating from around 6000 years ago, it is believed to have been built as a calendar by nomadic cattle-herders.

arithmetic? To put the question at its crudest level: if a traditional Warlpiri bushman is faced with a choice between five and six pieces of food, can he even tell the difference? Even to ask the question is to answer it, and of course the answer is yes. While Warlpiri may lack the vocabulary to distinguish verbally between large numbers, they are just as adept as anyone else at making the mental distinction when required to do so. In 2009, the neuroscientist Brian Butterworth investigated the numerical abilities of Warlpiri children by asking them to lay out counters on the floor to match the number of sounds created when two sticks were hit together. The Warlpiri children fared just as well at this task as the English-speaking children.

Butterworth's experiments tell us that language is not the sole determinant of numerical ability. It is, however, a necessary prerequisite for developing mathematics of any sophistication.

Art and geometry

The Neolithic revolution, which began around 10,000 years ago, saw the emergence of art and technology, and with them the beginnings of geometry. There can be no clear distinction between geometry as a branch of mathematics and as a source of design. Yet the adornments on early pieces of pottery show clear evidence of geometrical thinking, as do megalithic sites such as Stonehenge in Britain and Nabta Playa in Egypt. A fascination with symmetry is already evident in these artifacts, and it is not entirely fanciful to see these as early investigations into group theory.

3 Place-value notation

BREAKTHROUGH BY ARRANGING NUMERALS IN COLUMNS, EARLY MATHEMATICIANS ALLOWED A NUMBER'S MEANING TO DEPEND ON ITS POSITION AS WELL AS ITS SYMBOL.

DISCOVERER BABYLONIAN MATHEMATICIANS, 3000–2000 BC.

LEGACY FAR MORE FLEXIBLE AND EXPRESSIVE THAN ANY PRIOR SYSTEM, PLACE-VALUE NOTATION REMAINS THE STANDARD WAY OF WRITING NUMBERS TODAY.

The scholars of ancient Babylon, whose golden age began around 5,000 years ago, have left us several cultural legacies. Among them is the hour, which we divide into 60 minutes, each of which is divided into 60 seconds. This comes from the Babylonian sexagesimal system of representing numbers, meaning that it is based on 60. Today we prefer 10 as the base of our numeral system. Whichever base is chosen, the Babylonian innovation of arranging numbers in columns, where the position carries as much information as the symbol itself, was a crucial moment in the history of human thought.

For thousands of years, humans have used numbers to describe and understand the world. But how could numbers best be represented physically? The tallies of prehistoric peoples (see page 13) were adequate for basic purposes, but as people settled into static communities and civilizations began to grow, more sophisticated numerical systems started to evolve. There was one particular innovation which would become critically important for humans. It first appeared on the clay tablets of ancient Babylon, the largest and most advanced city the world had ever seen.

OPPOSITE A Sumerian tablet, dating from around 2300 BC. Discovered in Tello in modern Iraq, it lists numbers of sheep and goats in cuneiform script.

Babylon was a city state in Mesopotamia, in what is now Iraq. Its power came from its advanced agricultural system. Over its millennium-spanning history as the capital of the region, Babylon changed hands several times between the Sumerian and Semitic peoples, while science,

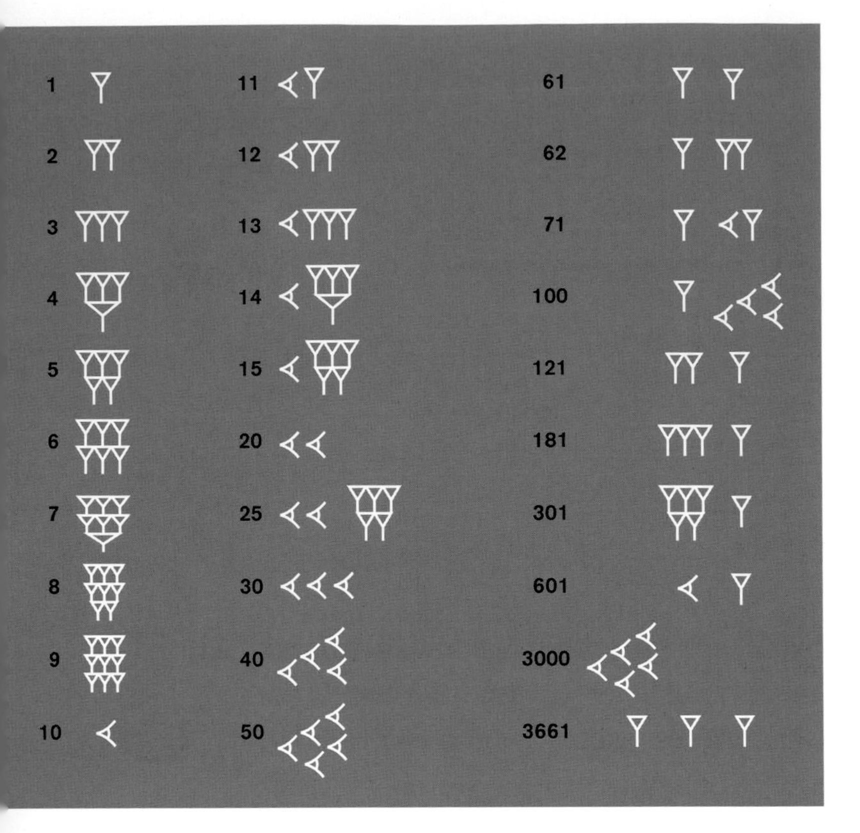

literature and culture flourished to unprecedented heights. The Babylonians put together a calendar of the year, which was divided into 12 months. It was they who first formulated the week, consisting of seven days, of which the last was a holiday. Babylonian scientists studied the stars, the local plants and animals, medicine and mathematics.

Babylonian mathematics

In cuneiform, a single notch represents the number 1. Notches were then overlaid to form symbols for the numbers 2 to 9, and a new wedge-shaped symbol was introduced to represent 10. Copies of this symbol for 10 were then combined to give the characters for 10, 20, 30, 40 and 50. Using these characters, the numbers 1 to 59 could all be written down.

ABOVE The Babylonian numeral system was the first to employ place-value notation, where a numeral's position conveyed as much information as its design.

For the numbers 1 to 59, the Babylonian system is unremarkable, and comparable to the numerals of many other cultures. Its true significance is revealed when it reaches the number 60. Here, instead of assembling a composite of six copies of the numeral for 10, the Babylonians started a new column to the left, where they entered the symbol for 1.

This is exactly analogous to how we write the number 10. It does not have its own character, as numbers 1–9 do. Instead, it is represented by a 1. But the position of that 1, in a new column to the left, indicates that it stands for "one ten."

Carrying and borrowing

It was only in time that the importance of place-value notation was revealed; but we can recognize in retrospect what a huge moment this was in the history of science. Place-value notation is an efficient way to represent numbers and makes them much easier to manipulate.

Other systems, such as Roman numerals, may be easy enough to read (with practice), but they make simple arithmetical procedures, such as multiplication and division, awkward and unnatural. With place-value notation, techniques of "carrying" and "borrowing" from one column to the next make arithmetical procedures more transparent, paving the way for more advanced theories of number and algebra.

The advantage of place-value notation is particularly noticeable when dealing with large numbers. As civilization and science have become ever more sophisticated, the numbers we need to use have grown ever larger. While a band of hunter-gatherers can manage using small numbers and a numeral system based on tallies, a city of 50,000 people with scientists taking on the skies demands something more. Thanks to their place-value notation, Babylonian scribes could express all numbers below 216,000 using just three symbols.

With place-value notation, techniques of "carrying" and "borrowing" from one column to the next make arithmetical procedures more transparent, paving the way for more advanced theories of number and algebra.

Babylonian clay tablets

Archaeologists have found several hundred clay tablets in modern Iraq, giving us some idea of the mathematical sophistication of the ancient Babylonians. The most famous of these is Plimpton 322, dating from around 1800 BC. For many years it was believed to be a table of Pythagorean triples, such as (3, 4, 5) and (5, 12, 13) (see page 27). But it is now believed to be a set of mathematical exercises for students studying to be scribes. It suggests evidence of methods for solving quadratic equations (see page 87), employing techniques that would not be fully formalized until the work of Al Khwarizmi. In geometry, meanwhile, the Babylonians had mastered what would come to be called Pythagoras' theorem (see illustration on page 26).

Murmurings of zero

Place-value notation automatically leads to the need for a concept of zero. To distinguish between "twenty-one" and "two hundred and one," it somehow needs to be conveyed that the middle column for tens is empty, rather than non-existent. The archaeological record clearly shows the evolution of this idea. Early Babylonian tablets simply left the middle column literally empty, as we might write "2 1." But of course this is open to misreading. By 700 BC, the Babylonians had taken to using a punctuation symbol to indicate an empty column. Although it is unlikely that they regarded this as a true number, it was an important precursor to the concept of zero, which came into its own in India many centuries later (see page 82).

4 Area and volume

BREAKTHROUGH THE DISCOVERY OF TECHNIQUES FOR CALCULATING THE AREAS AND VOLUMES OF A VARIETY OF SHAPES.

DISCOVERER EGYPTIAN MATHEMATICIANS HAD ASSEMBLED AN ARRAY OF METHODS BY 1850 BC.

LEGACY THE ABILITY TO PERFORM MEASUREMENTS IN DIFFERENT DIMENSIONS HAS BEEN ESSENTIAL TO SCIENTISTS FOR THOUSANDS OF YEARS.

One of the earliest uses of numbers was for measuring distance. A greater challenge was to extend this technique to quantify the size of 2- or 3-dimensional objects. This involved the analysis of area and volume. Several scrolls of ancient Egyptian mathematics tell us that scholars of the era were very interested in this problem and developed impressive methods for calculating answers.

There are several ways to measure distance, depending on the scale involved. A traditional measure for the height of a person or building is the foot, while it is customary in several parts of the world to measure horses in hands. The origins of these units are evident in their names. In fact, hands (and their subunits of fingers and palms) were standard in ancient Egypt, where the science of area and volume first took off. Of course, if you wish to measure the distance from one town to the next, feet or hands are not very practical. Longer measures such as miles or kilometers are used today. The Egyptians measured such distances in river units, which correspond to around 10 kilometers (6.2 miles). In between a hand and a river unit, the standard Egyptian unit was the cubit, which was 7 hands long. Surviving cubit rods tell us that a cubit was around 52.5 centimeters (21 in).

OPPOSITE A sphere has the minimum ratio of surface area to volume of any 3-dimensional shape. Bubbles naturally take this shape as it minimizes surface tension.

Questions of area

How can one quantify the area of a 2-dimensional shape? The method pioneered by the Egyptians, and still used today, is to use a square whose side is 1 unit long. So today we talk about square meters or square miles, while the Egyptians contemplated square cubits. Imagining one of these squares as a tile, the question is: how many tiles would be needed to cover the area being measured?

Questions of area were very important to the ancient Egyptian civilization, which flourished from around 3000 BC. When parents died, it was standard for their land to be divided up equally between all their children. Land was also taxed, so it was important both to the authorities and to individual citizens that its area could be calculated exactly. This is easy when the shape is a perfect rectangle. A patch of land 3 meters long and 2 meters wide needs 6 tiles to cover it, if each tile is a 1×1 meter square. Nowadays, we might write its area as $6m^2$ (where m^2 is short for "square meters").

Questions of area were very important to the ancient Egyptian civilization, which flourished from around 3000 BC. When parents died, it was standard for their land to be divided up equally between all their children.

The same principle applies to volume, in which 3-dimensional space is measured. Here, the fundamental unit is a $1 \times 1 \times 1$ cube. To measure the volume of a space is to ask how many of these cubes are required to fill it. If a room is 2 meters wide, 3 meters long and 4 meters high, then it requires 4 layers of 6 cubes to fill, giving a total volume of $2 \times 3 \times 4 = 24m^3$.

Area and volume become far trickier to measure when the shape cannot be filled by an exact number of tiles or cubes. Even with that simplest of shapes, the triangle, the answer is not perfectly obvious. If a triangle is w units wide and h units high, then its area is $\frac{w \times h}{2}$. This was one of many facts mastered by the geometers of ancient Egypt.

The Ahmes Papyrus

The first mathematician known by name is Ahmes, a scribe who worked around 1650 BC, and wrote the Ahmes Papyrus (sometimes also known as the Rhind Papyrus after the 19th-century antiquarian Alexander Rhind, who brought it to the UK). It is a list of 87 mathematical problems with their solutions, which Ahmes tells us are copied from an older work that hasn't survived. (He even hints that its contents ultimately derive from the work of Imhotep, the revered architect, medic and scientist who worked around 2600 BC, designing one of the earliest of the pharaonic pyramidal tombs.)

Several of the problems on the Ahmes Papyrus are geometrical in nature and entail calculating the areas of an assortment of shapes. The 50th problem asks for the area of a circular field whose diameter is

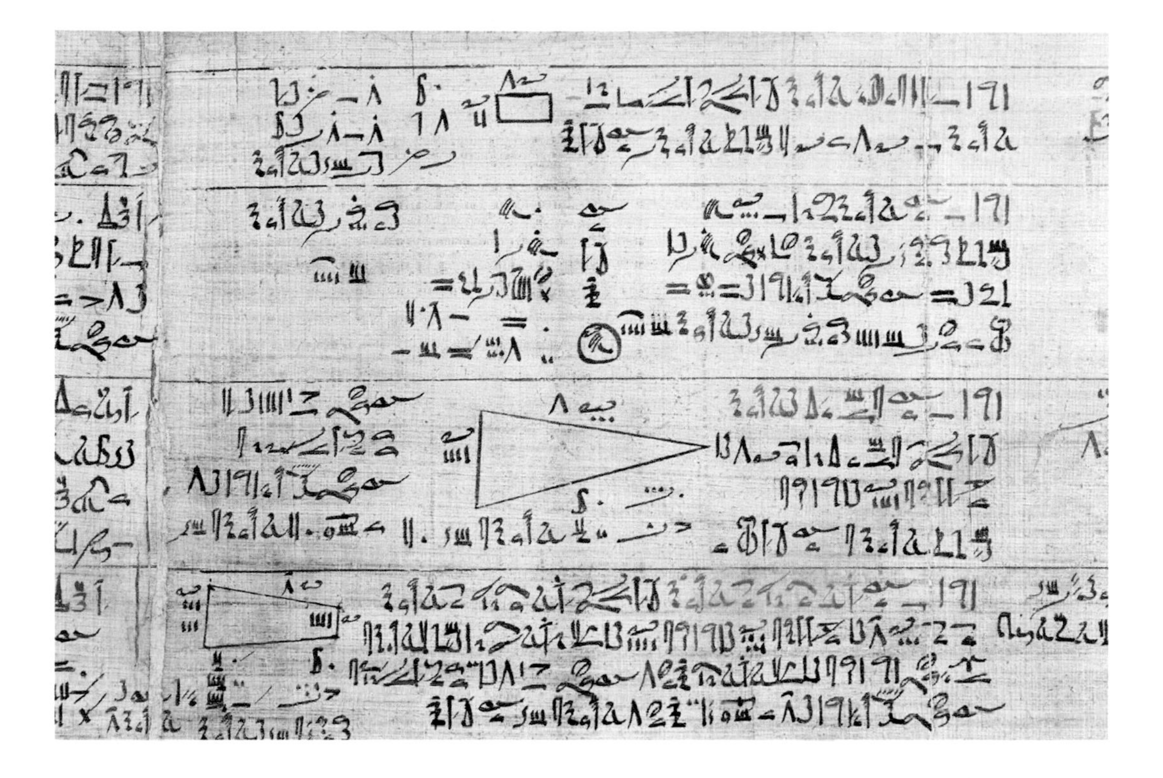

9 *khet* (1 *khet* being 100 cubits). The answer given is 64 square *khet*. This suggests that the approximation to π in use at the time was $\frac{256}{81}$ (see page 53). Other problems relate to finding the area of fields in the shape of triangles, trapezia and rectangles.

ABOVE A section from the Ahmes Papyrus of 1650 BC. Among its 84 problems are several concerning the areas of triangles, circles and other figures.

Pyramids and the Moscow Papyrus

Even earlier than the Ahmes Papyrus is the so-called Moscow Papyrus, dating from around 1850 BC. This consists of 25 mathematical problems and their solutions. The 14th of these is perhaps the most impressive surviving feat of ancient Egyptian mathematics. Appropriately enough, it involves reasoning about a pyramid. In fact, the pyramid is truncated, meaning that it has had its top sliced off. We are told that the pyramid's base is a 4 cubit by 4 cubit square, while its top is a 2 × 2 square and its height is 6 cubits. The challenge is to calculate the volume of the resulting shape. The right way to proceed here is by no means obvious. In fact, a truncated pyramid with a square base of width a, height h and a square top of width b has volume

$$\frac{1}{3}h(a^2 + ab + b^2)$$

The fact that the anonymous author is able to deduce the correct answer (56 cubic cubits) tells us that Egyptian geometers were in command of some fairly sophisticated geometrical formulae.

5 Pythagoras' theorem

BREAKTHROUGH A FUNDAMENTAL RELATIONSHIP BETWEEN THE SIDES OF A RIGHT-ANGLED TRIANGLE.

DISCOVERER THOUGH TRADITIONALLY ATTRIBUTED TO PYTHAGORAS (570–475 BC), THIS THEOREM IS LIKELY TO HAVE BEEN KNOWN TO EARLIER GEOMETERS.

LEGACY AN EARLY EXAMPLE OF THE ESSENTIAL ALGEBRAIC RULES BEING EXTRACTED FROM A GEOMETRICAL SCENARIO, IT IS THE PRINCIPAL WAY WE CALCULATE LENGTH EVEN TODAY, AND REMAINS THE CORNERSTONE OF ELEMENTARY GEOMETRY.

One of the first true mathematical theorems, and still perhaps the best known, Pythagoras' theorem is a fundamental fact about the geometry of triangles. If we know the length of two sides of a triangle, it tells us how to calculate the third. But Pythagoras' theorem is not valid for all triangles, only for a special subclass: right-angled triangles.

It is not known who was the first geometer to discover the relationship between the sides of a right-angled triangle. It was certainly familiar to Babylonian mathematicians, as long ago as 1700 BC. See the picture of the Yale tablet on page 31. There is also evidence that it may have been known to Indian and Chinese mathematicians, centuries before the Greeks took an interest. What is not clear is whether any of these early thinkers elevated its status from an observation into a true theorem by providing a proof valid for all right-angled triangles. The earliest surviving proof of this theorem appears in Euclid's *Elements* (see page 46). But even in the ancient world, the result was commonly attributed to the man whose name will forever be associated with it.

OPPOSITE Right-angled triangles being used for decorative tiles. Every shape built from straight lines can be broken down into right-angled triangles, making them a topic of interest not only to geometers but engineers and graphic designers.

The man and the myth

Although Pythagoras was unquestionably an influential figure in Greek culture, his life is shrouded in myth. He was hailed by one poet as the son of the god Apollo, while others honored him as an emissary from Zeus. Many stories were told about his miracle-working, including his ability to appear in two places at once. He was even said to have spent 207 years in the underworld before returning to the mortal realm.

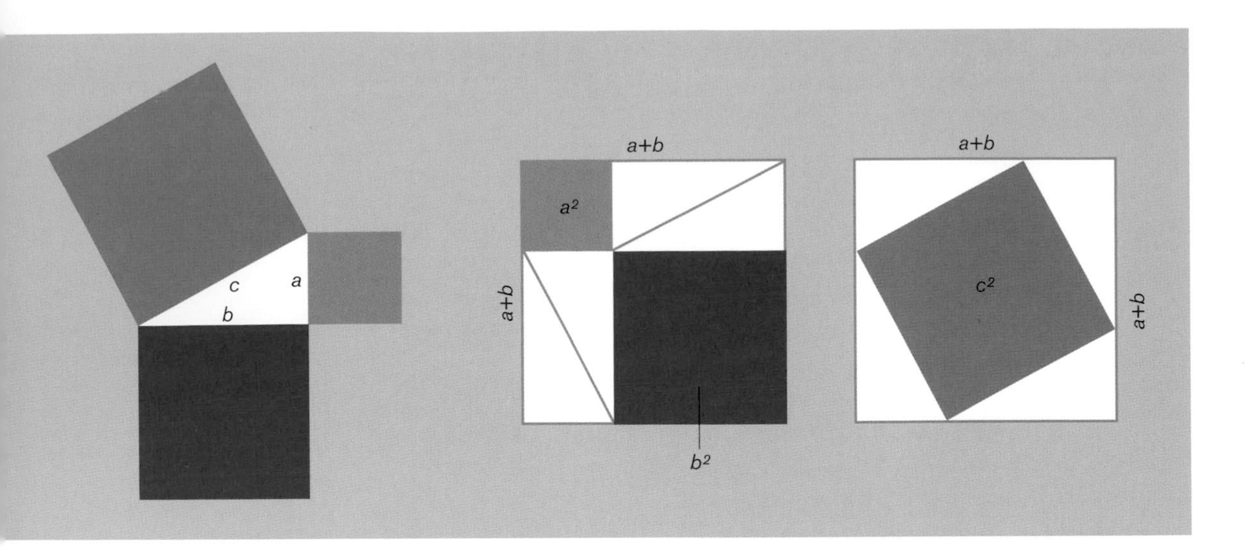

ABOVE A visual proof of Pythagoras' theorem, attributed to the Indian mathematician Bhaskara in the 12th century AD.

What we do know is that Pythagoras was a traveling mystic, philosopher and mathematician, and the leader of a religious group known as the Pythagoreans. He founded their first school in a Greek enclave in southern Italy. The Pythagoreans ate vegetarian food and believed in the transmigration of the human soul to other animals. Most importantly, however, they believed that, at the deepest level, reality is fundamentally mathematical, and especially geometrical. For them, therefore, mathematics was not only a matter of scientific curiosity, but a deeply religious duty.

Pythagorean theorem

The Pythagorean theorem concerns triangles: shapes built from three straight lines. In particular it discusses right-angled triangles, where two of the sides are perpendicular to each other. The theorem states that in a right-angled triangle, whose three sides have lengths *a*, *b* and *c*, where *c* is the longest, these numbers satisfy the law

$$a^2 + b^2 = c^2$$

which is to say $a \times a + b \times b = c \times c$. The longest side is traditionally known as the hypotenuse (roughly meaning "stretched under" in Greek), and it will always be opposite the right angle. If the shortest two sides of a right-angled triangle have lengths 3 and 4 units, then the hypotenuse will have length 5, since $3^2 + 4^2 = 9 + 16 = 25 = 5^2$.

Pythagorean proofs

In the centuries since Pythagoras was alive, many justifications of his theorem have been found. In fact, several hundred different proofs are currently known, employing a remarkable variety of techniques. One of the most elegant is due to Bhaskara, a geometer in 12th-century India. Bhaskara's proof begins with four identical copies of the right-angled triangle under discussion. These four are then laid inside a square frame, in two different ways. The width of the frame is $a + b$. In the first arrangement, the leftover space is in the form of two smaller squares, with widths a and b respectively. In the second configuration, the remaining space consists of a single square of width c. But of course, the total area in the two situations must be the same. That is to say, the area of the two smallest squares must add up to that of the larger one, or to put it another way: $a^2 + b^2 = c^2$.

Pythagoras and distance

Right-angled triangles seem rather a narrow class of shapes. So why is Pythagoras' theorem such an important landmark of mathematics? The answer is that it remains our standard way of evaluating distance. If two points on a page are 3 centimeters apart horizontally, and 4 centimeters apart vertically, then there is a hidden right-angled triangle lurking in this picture. The direct distance between the two points is given by the hypotenuse of this invisible triangle, namely 5 centimeters.

In fact, many famous pieces of geometry have Pythagoras' theorem hiding in the background. Euclid defined a circle, for example, to be the collection of points a fixed distance (r) away from the center. Of course, a circle does not look much like a right-angled triangle. But the critical notion of distance can be expressed using Pythagoras' theorem, which explains why the standard equation of a circle has more than a passing resemblance to this theorem: $x^2 + y^2 = r^2$.

Pythagoreans believed that, at the deepest level, reality is fundamentally mathematical, and especially geometrical. For them, therefore, mathematics was not only a matter of scientific curiosity, but a deeply religious duty.

Pythagoras and number theory

Pythagoras' theorem is a statement about geometry; but it has implications for another branch of mathematics: number theory. An important question is whether all three sides of a right-angled triangle can be given by whole numbers. Usually, this does not happen (see page 29). But there are some examples of Pythagorean triples: (3, 4, 5) is the first, followed by (5, 12, 13) and (7, 24, 25) and (8, 15, 17). The question of whether there are infinitely many such triples was settled in the affirmative by Euclid. Whether similar configurations are possible when the squares of the Pythagorean theorem are replaced by higher powers was the topic of Fermat's Last Theorem (see page 369).

6 Irrational numbers

BREAKTHROUGH IRRATIONAL NUMBERS ARE NUMBERS THAT CANNOT BE EXPRESSED AS FRACTIONS. THEIR DISCOVERY WAS A DRAMATIC INDICATION OF THE LIMITATIONS OF WHOLE NUMBERS.

DISCOVERER ACCORDING TO LEGEND, THE PYTHAGOREAN THINKER HIPPASUS OF METAPONTUM MADE THE DISCOVERY AROUND 500 BC.

LEGACY THIS EXPANSION OF THE NUMBER SYSTEM WAS THE FIRST EXAMPLE OF PROOF BY CONTRADICTION, A TECHNIQUE MUCH PRIZED BY MODERN MATHEMATICIANS.

Since the beginning of the subject, the positive whole numbers have been at the core of mathematics. But these numbers are not enough for many purposes, and it was not long before fractions were needed for measuring subtler quantities. Today, mathematicians know fractions as rational numbers, and for most practical purposes they are all one needs. So it was a great shock to ancient mathematicians to discover that there are also irrational numbers, meaning numbers that can never be expressed exactly as a fraction.

Pythagoras' theorem is a bedrock of geometry (see page 25), while the whole numbers are the centerpiece of number theory: 1, 2, 3, 4, and so on. But Pythagoras' theorem and whole numbers do not mesh together particularly well.

Geometry and numbers

If two sides of a right-angled triangle are given by whole numbers, it is typically the case that the third is not. The simplest example comes from slicing a square along its diagonal. This produces a triangle whose two shorter sides are each 1 unit long. It must be, therefore, that the length of the hypotenuse c satisfies $c^2 = 1^2 + 1^2$. That is to say, c is some number which when multiplied by itself produces 2. Nowadays, the square root symbol ($\sqrt{\ }$) expresses this: $c = \sqrt{2}$.

OPPOSITE The Engineering Building at Leicester University, completed in 1963. Its design reflects the appearance of irrational numbers as the lengths of diagonals across squares.

This much was obvious, at least once Pythagoras' theorem was known. But what is this number $\sqrt{2}$? It is clear that it cannot be a whole number, since neither 0 nor 1 work, and all others are too big. The shock for Greek mathematicians of the fifth century BC was that that no fraction could work either. This was a stunning discovery, because before that moment mathematicians had no inkling that numbers beyond the fractions need exist. Yet the implication was that there must be other numbers, irrational numbers, of which $\sqrt{2}$ was the first known example.

For the mystical sect who made this discovery, the Pythagoreans, the whole numbers were the foundation of all mathematics (and indeed much beyond). The revelation that there were numbers that could not be expressed as simple proportions of whole numbers caused the group considerable distress. The ultimate insult was that this fact should emerge as a consequence of their beloved theorem about right-angled triangles. Indeed, there are legends about the fate of the Pythagorean thinker who first proved the irrationality of $\sqrt{2}$, Hippasus of Metapontum. Some accounts say that he was expelled from the group for this blasphemy, or even executed for his crime. Although the truth is lost, it seems possible, at least, that this work led to a schism within the sect.

The revelation that some numbers cannot be expressed as simple proportions of whole numbers caused the Pythagoreans considerable distress. The ultimate insult was that this should emerge from their beloved theorem about right-angled triangles.

Irrational magnitudes

Later generations of Greek mathematicians came to terms with quantities such as $\sqrt{2}$; they never accepted them as true numbers, but rather magnitudes of a more abstract kind. Today we term $\sqrt{2}$ an irrational number, meaning that it is not a ratio of whole numbers: it will never be possible to write $\sqrt{2} = \frac{a}{b}$, for any pair of whole numbers a and b. What about when written as a decimal? In this case, the answer comes to approximately 1.41421356237... But irrational numbers are not easily captured as decimals either. Their decimal expansions continue forever without stopping or becoming stuck in a repetitive loop. The most famous example of this is the number π (see page 54), which was proved irrational by John Lambert in 1761, and has subsequently been calculated to trillions of decimal places. In fact, π is not merely irrational but transcendental (see page 197).

The Yale tablet

It is likely that Babylonian mathematicians were aware of this result long before the Pythagoreans. The Yale tablet (or YBC 7289) is one of the most famous examples of Babylonian mathematics, dating from around 1700 BC.

LEFT The Yale tablet, from *c.* 1700 BC, is the work of a Babylonian mathematician calculating the approximate length of a diagonal across a square. Whether they knew this was an irrational number remains a mystery.

It contains a diagram of a square whose sides are marked as being 30 units long. The diagonals are also drawn in, their lengths marked as being 42.42638 units long, which is correct to 4 decimal places.

This tells us not only that the Babylonians were familiar with Pythagoras' theorem, but also that they had a reasonable approximation to $\sqrt{2}$ (in fact, one is given on the tablet). What is not clear is if they recognized that their approximation is imperfect, or whether they believed they had an exact answer. It is conceivable that a Babylonian mathematician proved the irrationality of $\sqrt{2}$ long before Hippasus, though we cannot be sure.

Proof by contradiction

The method by which Hippasus derived his result left just as important a legacy as the theorem itself and remains one of the best-known examples of proof by contradiction. Mathematicians would exploit this technique numerous times over the coming centuries. The approach proceeds by turning the result on its head. In order to prove that $\sqrt{2}$ can never be expressed exactly as a fraction of whole numbers, Hippasus assumed the exact opposite, that it could. So he imagined that $\sqrt{2} = \frac{a}{b}$ for some whole numbers a and b. From this assumption, he was able to deduce a nonsensical conclusion—something that was self-evidently false.

On first sight, this seems like a strange argument. Nonsense is not usually on a mathematician's wish-list. Yet it worked well: if the assumption that $\sqrt{2}$ is rational leads irrevocably to an invalid conclusion, it must be that $\sqrt{2}$ is not rational after all, but irrational.

7 Zeno's paradoxes

BREAKTHROUGH ZENO'S PARADOXES PURPORTED TO SHOW THAT MOTION IS IMPOSSIBLE. BUT THEIR TRUE IMPORTANCE WAS IN OBSERVING THE TENSION BETWEEN DISCRETE AND CONTINUOUS SYSTEMS.

DISCOVERER ZENO OF ELEA (C. 490–425 BC).

LEGACY ALTHOUGH ZENO'S PHILOSOPHY NEVER GAINED WIDE ACCEPTANCE, HIS PARADOXES CONTINUE TO FASCINATE. THEY ALSO ILLUMINATE SUBTLE MATHEMATICAL ISSUES WHICH WERE NOT FULLY UNDERSTOOD UNTIL 2000 YEARS LATER.

Zeno of Elea was a philosopher famous for his list of paradoxes, which baffled thinkers for many hundreds of years. Their eventual resolution would require subtle mathematical considerations, unavailable in his own time.

Zeno's beliefs might today be branded mystical monism. He believed in the fundamental oneness of everything, and that all the apparent divisions in the world, between up and down, and between past, present and future, are illusory. Ultimately, Zeno believed that there is only one persistent, unchanging reality, which he dubbed "Being" or "*to einai*" ("το ειναι"). Zeno inherited this philosophy from his teacher (and, according to Plato, his sometime lover) Parmenides, the most famous of the pre-Socratic philosophers. In defending Parmenides' worldview, Zeno concocted a series of 40 paradoxes, of which only a handful still survive today.

Zeno's paradoxes

The most enduring of Zeno's paradoxes concern motion. Zeno held that physical movement, and indeed all forms of change, are fundamentally impossible, and set out to prove it. Unfortunately, his original texts do not survive, but come to us through the work of Aristotle. The flavor of Zeno's thinking is shown in the first paradox of dichotomy. If a boy wishes to walk from one side of a room to another, he must first cross the line halfway across the room. But before he arrives there, he must reach the point a quarter of the way across the room. Needless to say, he can't

OPPOSITE Time runs continuously: so far as we know a period of time can be sub-divided infinitely often without running into a minimum indivisible unit; yet our methods of measuring time, whether the vibrations of quartz crystals or the ticking of clockwork, are fundamentally discrete.

RIGHT According to Zeno's paradox, Achilles can never catch the tortoise. No matter how fast he runs, the creature always remains one step ahead.

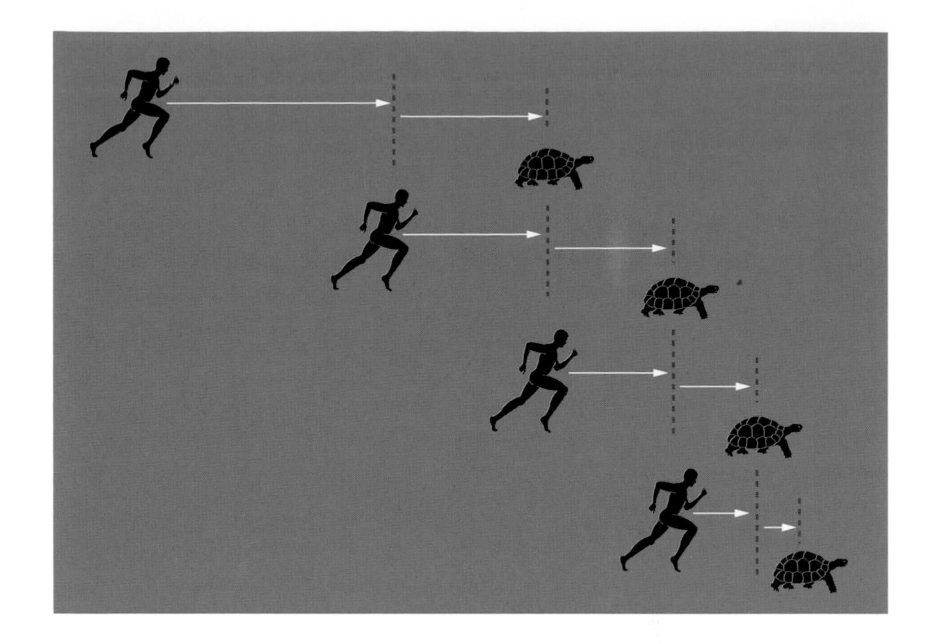

do that without first crossing the $\frac{1}{8}$ point, the $\frac{1}{16}$ point, and so on. In fact there is no obvious first step he can take.

The response of one philosopher, Diogenes the Cynic, was to stand up and walk silently across the room, eloquently making the point that the world is full of things which move with little apparent difficulty. Zeno's objections do not, in fact, seem to hold sway.

Achilles and the Tortoise

The most famous of Zeno's paradoxes is that of Achilles and the Tortoise. In this story, the legendary warrior, famed for his bloody exploits during the Trojan war, is faced with a lesser challenge: to pursue and capture a tortoise. However, in Zeno's telling of the story, Achilles finds the task more difficult than expected. Every time he reaches the point where the animal once stood, he finds that it has walked on a little. Being a slow-moving creature, it has not moved very far, yet the same problem recurs: when Achilles runs to the tortoise's new position, it has left that spot and moved forward a little more. Whatever Achilles does, the tortoise always remains one step ahead.

Discrete and continuous systems

At the core of Zeno's paradoxes is a tension that has preoccupied mathematicians for centuries: between a discrete system and a continuous one. A discrete system is one which comes in individual, separate steps. The primary example is the system of whole numbers:

beginning with 1, there is then a gap before 2 and then 3. In a continuous system, however, there are no jumps, but a smooth continuum.

The concept of a continuous number system existed before Zeno's time, but it was not until the work of Newton and Leibniz on calculus that its implications began to be revealed, and not until the formalization of the real numbers in the early 19th century (see page 173), that the laws of such a system were finally written down.

From a mathematical perspective, Zeno's great insight was that continuous and discrete systems behave radically differently. Although distance is usually measured on a continuous scale, Zeno's paradoxes are expressed in the language of discrete processes. Achilles' journey takes place in stages: first he has to reach the tortoise's initial point. Then he has to reach its second location, and so on. Of course, in these terms, the number of stages Achilles needs to complete is infinite, because the sequence of numbers 1, 2, 3, 4, 5, 6, etc. never ends.

Achilles finds it more difficult than expected to capture the tortoise: when he runs to the tortoise's new position, it has left that spot and moved forward a little more. Whatever Achilles does, the tortoise always remains one step ahead.

Yet a continuous system allows a way through. Suppose that Achilles runs ten times faster than the tortoise (this is surely an underestimate, but it makes the numbers easier). Perhaps the first leg of Achilles' chase is 9 meters. Then, while he is covering that distance, the tortoise will move on an extra 0.9 meters. While Achilles makes up that ground, the tortoise runs another 0.09 meters, and so on. So the points that Achilles needs to reach lie 9 meters, 9.9 meters, 9.99 meters, 9.999 meters, and so on away from his starting position. In the continuous world, it is easy to identify the point where he finally catches the tortoise: it is exactly 10 meters from his starting line. Critically, these distances form a convergent sequence (see page 97), meaning that they tend toward a finite limit, in this case 10.

In the dichotomy paradox, Zeno derives a contradiction by looking at shorter and shorter distances across the room: $\frac{1}{2}$, $\frac{1}{4}$, $\frac{1}{8}$, and so on. Since there is no first step, the boy is stuck. But in a continuous set-up there is never a first step. That there is no "smallest number" after 0 is a fundamental fact of the real numbers. In fact, Zeno's dichotomy paradox is highly suggestive of differential calculus as derived by Isaac Newton and Gottfried Leibniz. The decreasing sequence of distances will take the boy shorter and shorter lengths of time to cover. Centuries later, Newton and Leibniz would understand that initial speed can be calculated as the limit of average speeds over these smaller distances.

8 The Platonic solids

BREAKTHROUGH THE FIVE PLATONIC SOLIDS ARE THE MOST SYMMETRICAL 3-DIMENSIONAL SHAPES BUILT FROM STRAIGHT LINES AND FLAT FACES.

DISCOVERER THEAETETUS (C. 417–369 BC), PLATO (C. 429–347 BC).

LEGACY THIS RESULT SET THE PATTERN FOR MATHEMATICAL CLASSIFICATIONS IN THE CENTURIES AHEAD; IT REMAINS ONE OF THE MOST FAMOUS AND BEAUTIFUL TOPICS OF GEOMETRY.

Of all the achievements of the mathematicians of the ancient Greek empire, there was one they held in particularly high esteem. This celebrated theorem involves the most symmetrical and beautiful shapes of all: the Platonic solids.

The Platonic solids were contemplated by the Pythagorean school, bear the name of the philosopher Plato and form the denouement of Euclid's famous work *The Elements*. These are stellar names in the history of human thought. Yet so far as we can tell, the theorem was first proved by a geometer whose reputation is altogether less familiar today: Theaetetus. Unfortunately, Theaetetus' work has not survived, and is known to us mainly through the writings of his friend Plato. It is believed that parts of Euclid's *Elements* may be direct accounts of Theaetetus' research. His greatest achievement is known as the classification of the Platonic solids.

Geometry in two and three dimensions

Theaetetus' work illustrates how the world of three dimensions differs from the flat domain of two dimensions. A polygon is a 2-dimensional shape which can be drawn with straight lines. Examples are triangles, rectangles, pentagons, and so on. The most symmetrical of these are shapes whose sides are all the same length and whose angles are all equal. These are the regular polygons. Although there are countless different possible triangles, the only one satisfying this description is an equilateral triangle, with its three equal sides. Similarly, the only regular four-sided shape is a square. Then there are the regular pentagon,

OPPOSITE Tetrahedrite is a common ore of copper, in compound with iron, sulfur and antimony. Its name derives from its striking tendency to form crystals shaped as tetrahedra, the simplest of the Platonic solids.

hexagon, and so on. The story in two dimensions is quite simple: for each number of sides, there is exactly one regular shape.

One might naïvely expect that the same thing might apply in the 3-dimensional world. But it only took a little experimentation to work out that the situation here is subtler. In three dimensions, the most symmetrical shapes are regular polyhedra, such as the cube. It is built from flat faces, all of them identical. What is more, its faces are squares, which are themselves regular shapes. Every corner of the cube looks like every other, so if you pick the cube up and move one corner into the position of another, the shape will look exactly the same as when you started. Apart from the cube, though, what examples are there?

Plato wrote that the tetrahedron, cube, octahedron and icosahedron represent the four classical elements of fire, earth, air and water. The dodecahedron, meanwhile, was no less than God's design for the whole Universe.

Theaetetus' theorem

The mystics and geometers of the Pythagorean school were aware of two other shapes, both constructed from equilateral triangles. The tetrahedron has four faces and is essentially a pyramid with a triangular base. The octahedron, on the other hand, has eight faces. It looks like two square-based pyramids glued together at their bases.

It may be that the Pythagoreans believed these three shapes—the tetrahedron, cube and octahedron—formed the complete set. But later thinkers identified some more complex shapes which also satisfied the definition. The dodecahedron is built from 12 pentagons, while the icosahedron is again constructed from triangles, this time 20 of them, which come together in groups of five. With these discoveries, the picture was looking increasingly complicated. Might it be that there are even bigger regular polyhedra waiting to be discovered?

This was where Theaetetus proved his celebrated theorem. He showed that these five shapes really are all there are. There are no heptagonal-faced regular solids waiting to be discovered, and no regular configurations of 100 triangles.

Theaetetus' theorem is a true mathematical milestone, not only for the advance it made in our understanding of geometry, but for what it represents. It is a classification, the very highest grade of mathematical theorem. Starting with an abstract definition, in this case that of a regular polyhedron, Theaetetus was able to deduce a detailed list of all the objects that satisfy it. This was the pattern for many great theorems mathematicians would prove over the centuries. Examples are the classification of wallpaper groups (see page 221) and the classification of finite simple groups (see page 389). But Theaetetus' classification of the Platonic solids was the first, and remains the best known even today.

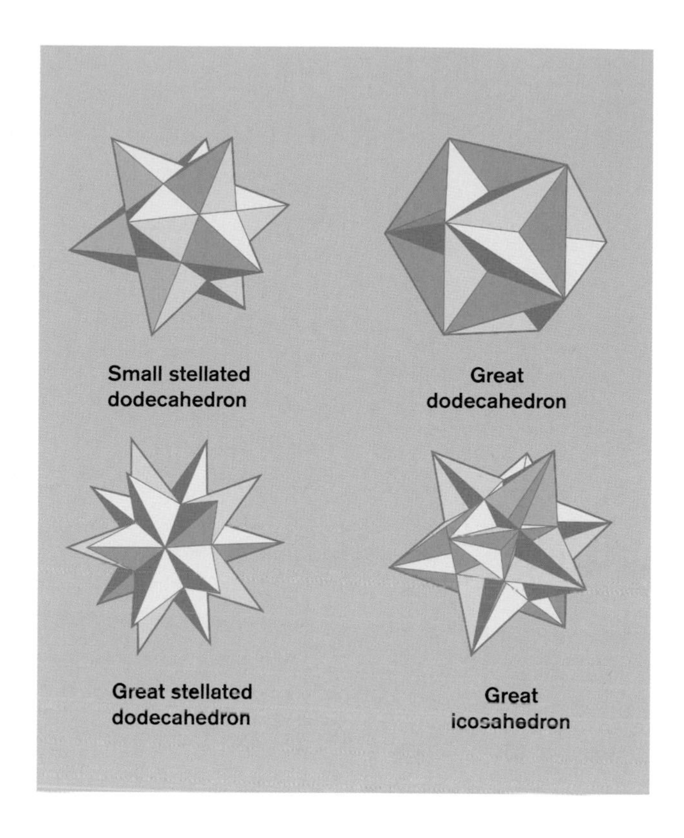

ABOVE The four Kepler–Poinsot polyhedra are the only self-intersecting 3-dimensional shapes with the same high level of symmetry as the Platonic solids.

A universe of polyhedra

Greek thinkers were somewhat fixated by these five shapes. Plato, in particular, saw a deep mystical significance to them. He wrote that the tetrahedron, cube, octahedron and icosahedron represent the four classical elements of fire, earth, air and water respectively. The dodecahedron, meanwhile, was no less than God's design for the whole Universe. This cosmological theory of the Platonic solids had an unexpected renaissance in the 16th century, when astronomer and geometer Johannes Kepler posited that our Solar System is a nested sequence of Platonic solids, in which the different planets orbit along paths dictated by the five shapes.

As we know now, neither the Solar System, nor molecules of water, air or earth have the form of Platonic solids. Once Kepler had abandoned his Platonic astronomy, he went on to make genuinely important discoveries about how the planets orbit the Sun. But Kepler also advanced the theory of polyhedra when he discovered two new shapes which seemingly satisfy the definition of a Platonic solid.

Later, Louis Poinsot would discover two more, and the four Kepler–Poinsot polyhedra posed a fresh challenge to Theaetetus' ancient theorem. How had he overlooked them? The resolution of this problem comes from the fine print. Theaetetus had assumed that polyhedra cannot self-intersect, that is, that their edges and faces cannot pass through each other. With that requirement in place, there are only five possible shapes. But dropping it, there are nine: the five Platonic solids, plus the four Kepler–Poinsot polyhedra.

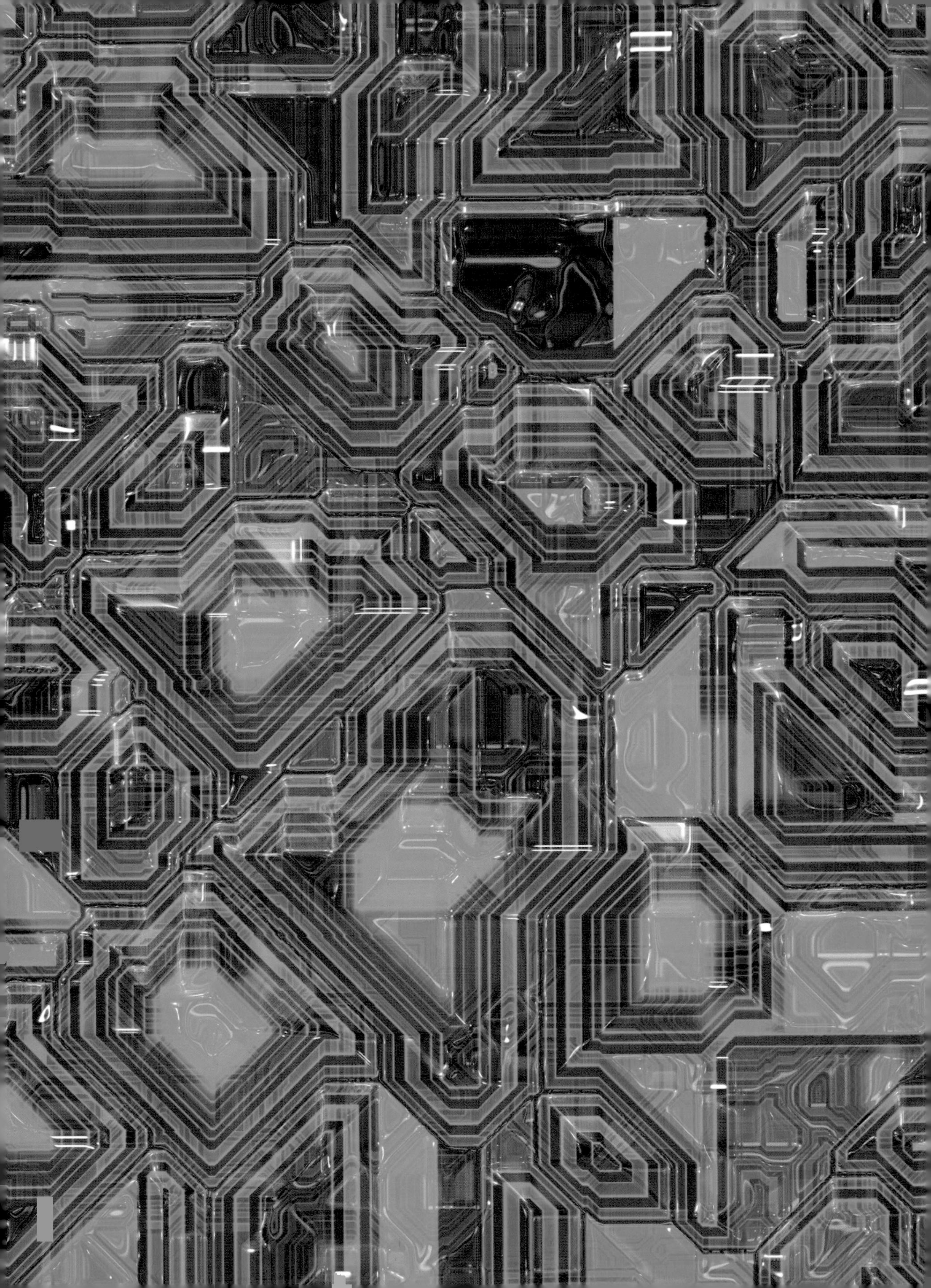

9 Logic

BREAKTHROUGH ARISTOTLE'S ANALYSIS OF VALID ARGUMENTS BEGAN THE STUDY OF LOGIC AS A SUBJECT IN ITS OWN RIGHT.

DISCOVERER ARISTOTLE (384–322 BC).

LEGACY ARISTOTLE'S ANALYSIS REMAINED THE STATE OF THE ART FOR 2000 YEARS. DURING THE 19TH CENTURY, IT BEGAN TO BE SUBSUMED INTO A LARGER BODY OF WORK, ULTIMATELY LEADING TO THE COMPUTER AGE.

Logic is what holds mathematics together. What distinguishes a theorem from a conjecture (or a mere guess) is that it has been provided with a logical justification. Of course, the use of logic is by no means confined to mathematics, and this has never been more true than in the current age of the internet, which is built on logical principles. The first serious analysis of logic was by the philosopher Aristotle in around 350 BC, and it remained the state of the art for thousands of years.

Aristotle hailed from Stagira, in the kingdom of Macedon on the Greek mainland. As a teenager, he went to study at Plato's school in Athens, where he developed a reputation as a brilliant thinker and forged a strong relationship with his teacher and colleagues. Following Plato's death, Aristotle traveled to Assos in modern Turkey, where he founded his own school. He married Pythias, the adopted daughter of the king of Assos, but Aristotle and his wife were forced to flee their new home when the Persian army invaded.

Aristotle returned to Macedon to work as a tutor to the son of King Philip, a thirteen year old boy named Alexander. This young prince would grow up to be the greatest military leader of the ancient world, Alexander the Great, who would construct one of the largest empires the world had seen. It is impossible not to wonder about the relationship between these two giant figures in human history; regrettably, there is little solid evidence to go on. Aristotle was highly prolific, writing as

OPPOSITE The microchips which underlie modern technology are logic rendered in physical form. Built from circuits of thousands or millions of logic gates which apply simple principles such as "AND," "OR," and "NOT" to their electrical inputs, the result is devices capable of highly sophisticated computation.

many as 200 works on various topics in science, politics and philosophy, most of which have sadly been lost. But his greatest achievement was in the group of six works which made up *The Organon* (meaning organ or instrument). Here he performed a rigorous analysis of arguments of various forms. This was the moment that logic was born as a formal subject in its own right, rather than an intellectual tool for studying other matters.

Aristotelian syllogisms

Aristotle analyzed a certain type of argument known as a syllogism. The most famous example (although not one Aristotle actually used) runs as follows:

Socrates is a man
All men are mortal
Therefore Socrates is mortal

Here, the conclusion in the third line is logically implied by the premises contained in the first two. What is more, the validity of the argument actually has nothing whatsoever to do with Socrates. Its truth is not reliant on Socrates' actual species or sex, or indeed on the nature of mortality. The argument is determined entirely by its form:

X is *Y*
All *Y*s are *Z*
Therefore *X* is *Z*

These three-line arguments are known as syllogisms. But not all are equally valid. For example, this is an example of a false syllogism:

X is *Y*
All *Z*s are *Y*
Therefore *X* is *Z*

An instance of this might be:

Socrates is mortal
All chimpanzees are mortal
Therefore Socrates is a chimpanzee

Aristotle's analysis took him through an exhaustive list of possible arguments of this form. To begin with, he divided the types of statements which might appear in a syllogism into four forms. The first pair are: "All *X*s are *Y*," which is contradicted by "Some *X* is not *Y*." The second pair are "Some *X* is *Y*," whose opposite is "No *X*s are *Y*." Each of these four can appear at any stage in a syllogism, either as one of the premises, or as the conclusion. What is more, each syllogism refers to three different entities, in general referred to as *X*, *Y*, *Z* (such as Socrates, mortality and chimpanzees).

Aristotle set about analyzing the possible arrangements of these three objects within the structure of each of the four sentence types. Then each of these could be positioned as either the premise or conclusion of a syllogism. This analysis gave rise to exactly 256 possible syllogisms. Of these, Aristotle concluded that only 24 are truly valid arguments.

Leibniz, Boole and De Morgan

Aristotelian logic remained the standard until the 19th century. Despite Aristotle's achievement in completely classifying the valid syllogisms, there were still challenges. After all, not many complex questions can be answered by a three-line argument of the "Socrates is a man" type. But by combining several syllogisms, it is possible to construct answers to questions that are far from obvious.

ABOVE Because syllogisms have only finitely many possible forms, they are amenable to mechanical analysis. Charles Stanhope's "Demonstrator" of around 1777 was capable of deducing the correct conclusion from two premises of a syllogism. A second version built almost thirty years later was also able to solve mathematical and probability problems.

In the years between Aristotle and 19th-century thinkers such as George Boole and Augustus de Morgan, only one person made serious advances in the study of logic: Gottfried Leibniz. What Boole and de Morgan did was to deconstruct the ingredients of logical argument still further, beginning with logical particles such as "AND," "OR" and "NOT." For instance, the statement "*X* AND *Y*" is true precisely when *X* and *Y* are each individually true. Whereas "*X* OR Y" is true so long as one of them is true. But when these connectives are combined, there can be surprises. De Morgan gave his name to the famous law: "NOT [*X* AND *Y*]" is logically equivalent to "[NOT *X*] OR [NOT *Y*]."

Just as Aristotle had done, these logicians sought to replace the uncertainties of human intuition with a rigid set of rules about logical validity. This essentially entailed the creation of new logical languages, with a strong algebraic flavor to them. It was on these foundations that the 20th century's leaps in understanding logic took place.

10 Euclidean geometry

BREAKTHROUGH EUCLID PIONEERED THE AXIOMATIC APPROACH TO MATHEMATICS, EXTENSIVELY INVESTIGATING THE GEOMETRY OF A FLAT PLANE.

DISCOVERER EUCLID OF ALEXANDRIA (C. 300 BC), ALTHOUGH MUCH OF HIS WRITING WAS A COLLATION OF THE WORKS OF EARLIER SCHOLARS.

LEGACY EUCLID'S *ELEMENTS* WAS THE STANDARD TEXTBOOK ON GEOMETRY FOR 2000 YEARS. IT CONTAINS ALMOST EVERYTHING TAUGHT IN SCHOOL GEOMETRY CLASSES EVEN TODAY.

From Pythagoras to Plato, geometry was extremely important to the intellectual life of the ancient Greek empire. Many studied it, but the defining moment came through the work of Euclid of Alexandria. He assembled the knowledge of the Greek geometers into a single coherent whole in his work *The Elements*.

Euclid was based in Alexandria, the capital of Egypt, and part of the Hellenistic empire which encompassed much of the modern Middle East. Alexandria was a wealthy metropolis and a renowned center of culture and learning.

The Library of Alexandria

The city was especially famed for its library, the largest in the world, which was founded around the year 350 BC and adjoined by a museum and a temple. From around the empire intellectuals of the era flocked to the library to match wits with their contemporaries and study the writings of their forebears. In 48 BC, the library was eventually burned down by Julius Caesar, as the Hellenistic era gave way to the Roman empire. It was in the early years of the library that Alexandria's most famous scholar, Euclid, lived and worked.

We know very little about Euclid's life. It is possible that he was born and raised in Alexandria, or he may have traveled there to study. It seems likely, however, that he spent some time in Athens. This was where Plato's students were working—the foremost experts in geometry of the age—

OPPOSITE In physics, beams of light are often modeled as parallel lines, meaning that they may be extended indefinitely without ever meeting. The pervasiveness of this idea to modern science can be traced back to Euclid's work.

ABOVE A page from Euclid's *Elements* illustrating the inscribed angle theorem. In the right-hand circle, the angle formed at the top of the circle is exactly half that in center.

and Euclid was certainly familiar with their work.

Euclid founded his own school of geometry in Alexandria and became respected as a teacher of mathematics. It seems reasonable to imagine that he spent much of his time in and around the great library. His students would continue his legacy and were a major influence on Apollonius, among others, as he developed the theory of conic sections (see page 57).

Euclid's *Elements*

Euclid's great achievement was a work in 13 volumes called *The Elements* (or *Stoicheia*). Although it addressed various mathematical topics, notably the theory of prime numbers (see page 49), its principal concern was geometry. *The Elements* is arguably the most important single work of mathematics ever published, remaining the standard textbook on geometry until the 19th century. It passed through numerous publications, translations and commentaries, only outnumbered by editions of the Bible.

It is hard to tell how many of the theorems contained within the work are Euclid's own discoveries, since his intention was to collate all the geometry which was known at the time. In any case, the importance of *The Elements* lies not only in the theorems it contains, but also in the approach that Euclid adopted. For the first time, Euclid built up a coherent body of knowledge from the bottom, starting from five fundamental principles known as Euclid's postulates. He then proceeded step by step, carefully deducing each theorem from those which went before. This is exactly how modern mathematicians approach the subject: beginning with the fundamental axioms, then building up from there.

Euclidean geometry

Euclidean geometry takes place on a flat 2-dimensional plane. In this setting, he investigates the properties of straight lines and circles, and the shapes which can be constructed from them. He explores fundamental notions of distance, angle and area. Most of the theorems and topics taught in geometry classes around the world are taken directly from Euclid.

Euclid presented proofs of several important facts about triangles, including Pythagoras' theorem itself, and the critical fact that the three angles inside any triangle will always add up to 180 degrees

Parallel lines, meaning pairs of lines that can extend indefinitely without ever meeting, had a foundational status in Euclid's work. Indeed, their exact role would be debated for many centuries, culminating in the discovery of new, non-Euclidean geometries (see page 189). Euclid's analysis of parallel lines includes all the most familiar facts about them. For instance, the alternate angle theorem describes what happens when a pair of parallel lines are crossed by a third line: the angles formed on the two lines must be equal.

In constructing shapes from straight lines, the first to appear are triangles. The Pythagorean tradition meant that triangles were close to the hearts of all Greek geometers. Euclid presented proofs of several important facts about triangles, including Pythagoras' theorem itself, and the critical fact that the three angles inside any triangle will always add up to 180°. He proved more advanced theorems, too, including results on the threshold of trigonometry (see page 61).

Circles were also among Euclid's considerations. He gave their modern definition: the shapes that appear when you pick a point on the plane and mark all the positions a fixed distance away. The inscribed angle theorem is one of Euclid's famous results in the geometry of circles. It says that if you pick two points on a circle and connect them with straight lines to a third point on the circle, then the angle created will be exactly that formed by connecting the two points to the center.

Elsewhere in *The Elements*, Euclid covers various topics that would flourish over the centuries, including the golden section (see page 93). He devotes a considerable amount of time to the analysis of ruler-and-compass constructions, yet another topic which withstood centuries of enquiry, not being resolved until the 19th century (see page 193). In the final volume of *The Elements*, Euclid moves away from the geometry of a flat plane into three dimensions, presenting a proof of that centerpiece of classical geometry, the classification of the Platonic solids (see page 37).

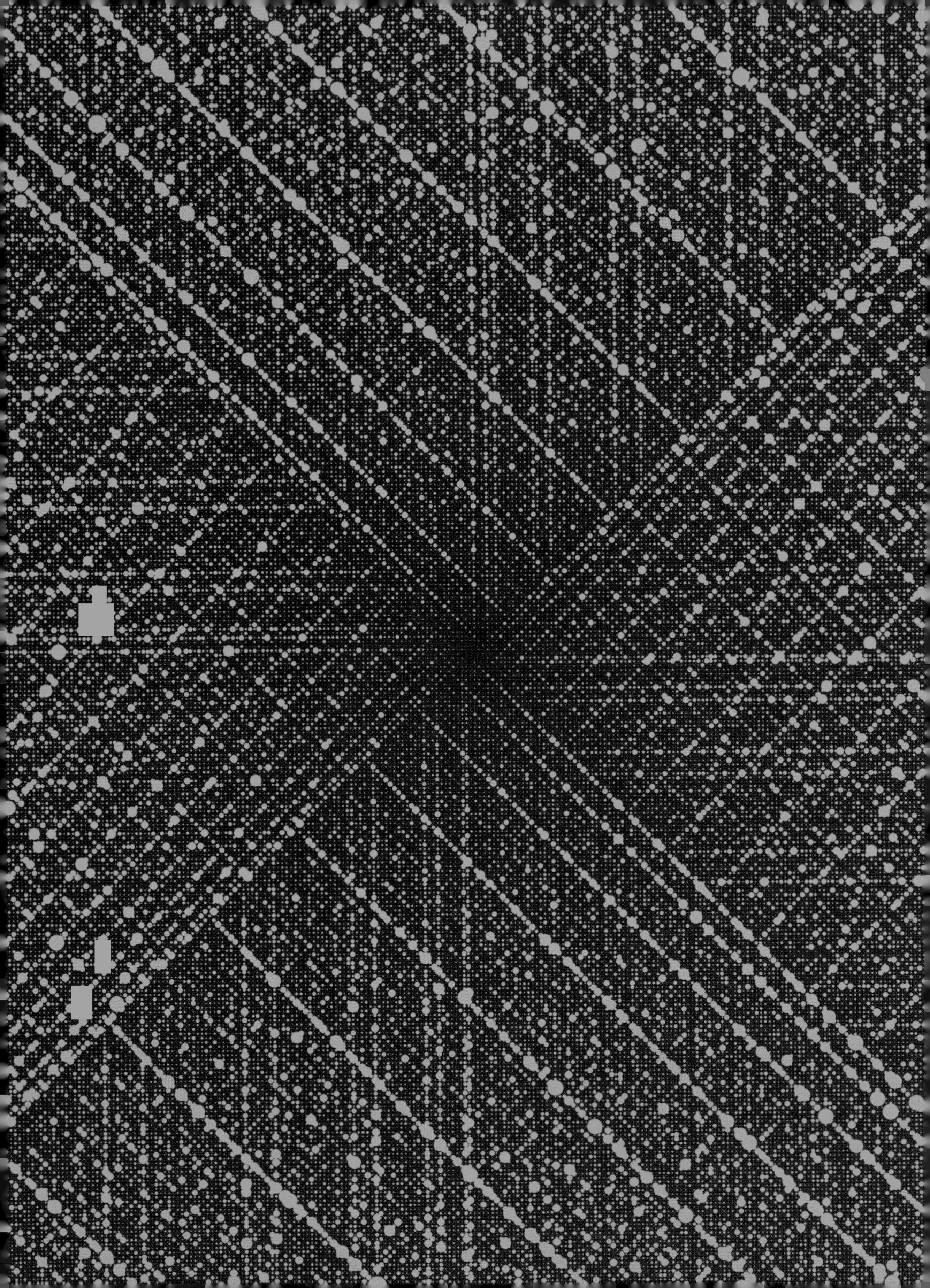

11 Prime numbers

BREAKTHROUGH AROUND 300 BC, EUCLID PROVED TWO OF THE MOST IMPORTANT FACTS IN MATHEMATICS: THAT THE LIST OF PRIME NUMBERS IS INFINITE, AND THAT EVERY NUMBER CAN BE SPLIT INTO PRIMES.

DISCOVERER EUCLID OF ALEXANDRIA (300 BC).

LEGACY EUCLID'S THEOREM MADE PRIME NUMBERS A FOCUS FOR MATHEMATICIANS, WHICH THEY HAVE REMAINED EVER SINCE. MANY THEOREMS ON THE SUBJECT OF PRIMES HAVE SUBSEQUENTLY BEEN PROVED, THOUGH A GREAT DEAL REMAINS MYSTERIOUS.

Euclid's most famous work, *The Elements,* was a book of such thoroughness and detail that it remained the standard text on geometry for over 2000 years. But *The Elements* was not restricted to geometry. In book nine, Euclid also included a section on number theory, and his findings on that topic were among some of the most important in the history of mathematics. At their heart was a study of prime numbers.

Today, prime numbers play a similar foundational role in mathematics that atoms do in chemistry. Chemical elements, such as oxygen or hydrogen, consist of atoms of just one type. More complex chemicals, such as water, are compounds of these (one water molecule comprising two hydrogen atoms and one of oxygen). Just as chemists try to understand the elements and the ways they combine, so mathematicians study prime numbers. For such fundamental objects, prime numbers larger than 3 occur in nature surprisingly infrequently. But they do appear. For instance in 2001, the mathematical biologist Glenn Webb realized that the 13 and 17 year life-cycles of certain species of cicadas helped them avoid resonances with the life-cycles of their predators.

OPPOSITE The Ulam Spiral, discovered in 1963, reveals unexpected patterns among the primes. Here, whole numbers are represented as blue dots spiraling out from the center, starting with 1. The size of each dot represents the number of proper divisors the number has. So prime numbers, which have just one, appear as dark patches in the pattern.

The study of primes

A prime number is a whole number that cannot be broken down into other whole numbers multiplied together. So 6 is not prime, because it can be written as 2×3. On the other hand, 5 is prime. The only way of writing 5 as two whole numbers multiplied together is as 5×1 (or 1×5).

Euclid was not the first to study primes; the Pythagorean school had also taken an interest in them, seeing some mystical significance in their indecomposable, atomic nature. But it was in Euclid's work that their centrality to mathematics was established.

The first few primes are: 2, 3, 5, 7, 11, 13, 17, 19, 23 and 29. In his investigations into the prime numbers, Euclid proved two of the most important facts about them. The first addressed the question of whether there is a limited supply of primes, or whether the list goes on forever. Euclid proved that the list of primes continues indefinitely: there are infinitely many primes.

The infinity of the primes was the greatest insight into the prime numbers that there had been. But Euclid's investigation did not stop there. He went on to prove a second result, which was arguably of even greater importance. In more recent years, it has become known as the fundamental theorem of arithmetic. It states that if you begin with any number (such as 100), you can always break it down into primes: $2 \times 2 \times 5 \times 5$. What is more, this representation is unique: if you want to write 100 as a product of primes, there is only one way of doing so—aside from trivial reorderings such as $2 \times 5 \times 2 \times 5$.

> **Prime numbers are the basic units from which all other numbers are built. Yet these important numbers continue to harbor many mysteries; even seemingly simple observations about them can turn into deep conundrums.**

Goldbach's conjecture

Euclid's work was the first serious investigation of the prime numbers, and his fundamental theorem explains why they have been such a preoccupation for the mathematicians who have followed him. It is because they are the basic units from which all other numbers are built. Yet these important numbers continue to harbor many mysteries; even seemingly simple observations about them can turn into deep conundrums. There is no better example of this than the observation discussed in 1742 in correspondence between Christian Goldbach and Leonhard Euler. Goldbach noticed that every even number from 4 onward seemed to be expressible as two primes added together: $4 = 2 + 2$, $6 = 3 + 3$, $8 = 5 + 3$, and so on. At higher levels, $100 = 29 + 71$, and $1000 = 491 + 509$. In reply to his friend's observation, Euler wrote "every even integer is a sum of two primes. I regard this as a completely certain theorem, although I cannot prove it."

Euler's confidence is backed up by a considerable body of evidence. In recent years, Tomás Oliveira e Silva has been running a distributed computing project, which has verified Goldbach's conjecture for every even number

LEFT Certain species of cicada have life-cycles given by prime numbers. Some emerge en masse every 13 years, while others adopt a 17-year cycle. This strategy helps them avoid predators. If they instead adopted a 10-year cycle, this would risk syncing with a predator's 2- or 5-year cycle.

as far as 1,609,000,000,000,000,000. Yet for all the attention this apparently simple problem has attracted over two and half centuries, a full proof, valid for every possible even number, still remains a distant prospect.

Bertrand's postulate

Goldbach's conjecture might suggest that the study of the prime numbers is a hopeless cause. But there have also been several success stories since Euclid. One such followed from Joseph Bertrand's observation about the spread of the primes. Exactly how frequently primes appear among the whole numbers is one of the deepest questions about primes, and is addressed by the Riemann hypothesis (see page 206). But Bertrand noticed that the gaps between successive primes were never larger than the numbers themselves. That is to say, starting at the number 100, there cannot be a gap between primes of more than 100; in other words, there is bound to be a prime somewhere between 100 and 200. More generally, if you start at any number (say n) and start counting upward, you are bound to hit a prime at some point before you reach $2 \times n$. Unlike Goldbach's conjecture, Bertrand's postulate quickly succumbed to effort and ingenuity: in 1850, the great Russian mathematician Pafnuty Chebyshev provided a proof.

Of course, Bertrand's postulate is something of an underestimate; in general, there is likely to be far more than one prime between n and $2n$. (In fact, between 100 and 200, there are 21.) Later, two of the 20th century's greatest mathematicians, Srinivasa Ramanujan and Paul Erdős, would each improve on Bertrand's result, showing that you are guaranteed to find as many primes between n and $2n$ as you like, so long as you look at large enough values of n.

12 The area of a circle

BREAKTHROUGH ARCHIMEDES DISCOVERED FORMULAE FOR CALCULATING THE AREA OF A CIRCLE AND THE VOLUME OF A SPHERE, AMONG OTHER SHAPES.

DISCOVERER ARCHIMEDES OF SYRACUSE (287–212 BC).

LEGACY THESE FORMULAE HAVE FORMED A CORNERSTONE OF GEOMETRY FOR THOUSANDS OF YEARS, PROVIDING ENGINEERS WITH TOOLS AND ANTICIPATING THE MUCH LATER DISCOVERY OF CALCULUS.

Archimedes of Syracuse was one of the greatest minds of the ancient world, inspiring devotion among his contemporaries and wonder among the scientists who followed him. His many brilliant insights have had a lasting impact on the history of mathematics, in both its purest and most practical branches. But it was the very simplest of shapes, the circle, that inspired him to his greatest discoveries.

Even a bathtub could serve as a source of inspiration to Archimedes. While bathing one day, Archimedes noticed that objects held underwater feel lighter than they do in the air. Pondering this, he came to the conclusion that this difference in weight is exactly equal to the weight of the displaced water, a phenomenon now known as the first law of hydrostatics. It was this that prompted him to shout "Eureka!," meaning "I have found it!"

Throughout his life, Archimedes would have many other opportunities to utter his famous cry. He performed the first studies of the principle behind pulleys and levers, the latter being seized upon by the Syracusian authorities, who demanded that Archimedes design catapults for their army. But his discoveries in the field of geometry surpassed even his achievements in mechanics. Chief among them was a pioneering technique for finding the area enclosed by a curve. This method anticipated the work of Isaac Newton and Gottfried Leibniz by almost 2000 years (see page 134). The idea was to subdivide the region into strips of very small width, calculate the area of each strip and then add

OPPOSITE Circles are such fundamental shapes that they can appear, in approximate form, in countless places in the natural world, from the rings of Saturn to the whorl of a shoal of fish.

up the total. Centuries later, this technique would become known as integral calculus.

Archimedes applied his technique to several curves with great success: ellipses, spirals, parabolas, hyperbolas and so on. But when he turned to the most famous curve of all, the circle, it led him to perhaps the best-known formula in the whole of mathematics: $A = \pi r^2$

Circles and squares

Archimedes' equation describes a method for calculating the area of a circle (denoted A in the formula), so long as we know its radius (r). The radius is the length of a straight line between the center of the circle and its outside edge.

In particular, Archimedes' formula relates the area of the whole circle to that of a small square whose sides are the same length as the circle's radius. If the radius is 3 centimeters, for example, then the resulting square has an area of $3 \times 3 = 9$ cm^2. In general, if a circle has radius of length r, then the area of the corresponding square will be r squared or r^2 (shorthand for $r \times r$).

In Archimedes' hands, π began its meteoric rise to fame. His new formula demonstrated that π is the key to determining not only the length of the circle, but its area too.

Archimedes' formula then tells us how many times that small square fits inside the corresponding circle. The answer is the same for every circle, regardless of size, and is provided by that superstar of the geometric world, the number π. So, to find the area of any circle, we must multiply the radius by itself, and then by π. If the radius is 3 centimeters, for example, then the area of the circle is $\pi \times 3 \times 3$ which comes out around 28.3 cm^2.

Approximating π

Today, the number π is every mathematician's best friend. The concept is thoroughly understood and has countless applications throughout science. In Archimedes' time, however, π was a far more mysterious customer. For centuries, it had been known that there was one single number which would tell us how many times the radius of any circle fits into its circumference, that is to say the distance around the circle. Much later, that special number was christened "π" (the Greek letter P, standing for periphery).

But what was this number π? Nowadays, we know its value to trillions of decimal places, but in the ancient world its true value was a matter of deep uncertainty. With this question, too, Archimedes made great progress. By carefully comparing the lengths of polygons inside and outside a circle, he was able to pin π down with remarkable accuracy, to between 3.141 and 3.143. Nowadays, we know that the true value is near 3.142 (although π can never be expressed exactly as a decimal, being an irrational number; see page 29).

ABOVE *The Death of Archimedes*, an 18th-century copy of a Roman mosaic dating from 212 BC showing the great geometer's murder when his circles are disturbed.

Archimedes' obsession with circles eventually brought about his downfall. Aged 75, he was sitting poring over yet another geometrical diagram when a soldier of the invading Roman army walked by and carelessly knocked it. Archimedes was annoyed by the intrusion and snapped, "Don't disturb my circles!"—at which the soldier drew his sword and unceremoniously killed him.

Spheres and cylinders

Archimedes' geometrical investigations were not limited to curves and circles. He also tackled the circle's big brother: the sphere. By adapting his techniques to three dimensions, Archimedes was able to concoct formulae for both the surface area of a sphere $A = 4\pi r^2$ and its volume $V = \frac{4}{3}\pi r^3$. From these formulae, he was able to show that when a sphere sits inside a cylinder of the same height and width, it occupies precisely $\frac{2}{3}$ of the space inside, as well as having exactly $\frac{2}{3}$ of the cylinder's surface area.

From his long list of tremendous achievements, it was this elegant relationship between the sphere and the cylinder of which Archimedes was most proud. Indeed, in the centuries after his death, Archimedes' tomb was recognizable by its adornments of a sphere and a cylinder, together with an inscription of the formulae linking their dimensions.

13 Conic sections

BREAKTHROUGH THE CONIC SECTIONS ARE THE BEAUTIFUL CURVES THAT EMERGE WHEN A CONE IS CUT WITH A PLANE. APOLLONIUS WAS ABLE TO GIVE A COMPLETE ACCOUNT OF THEM.

DISCOVERER MENAECHMUS (C. 380–320 BC), APOLLONIUS OF PERGA (C. 262–190 BC).

LEGACY CONIC SECTIONS APPEAR THROUGHOUT GEOMETRY AND PHYSICS, NOTABLY AS THE ORBITS OF PLANETS AND COMETS IN ASTRONOMY.

Geometry began with the very simplest of shapes: straight lines and circles. There is a great deal that can be said about these, and geometers of the ancient world such as Euclid and Archimedes devoted an enormous amount of effort to their study. Nevertheless, the moment soon came for geometry to move beyond such elementary shapes and tackle figures of greater sophistication. When geometers first stepped away from the safe territory of straight lines and circles, the first shapes they encountered were a family of beautiful curves, collectively known as the conic sections.

Discovered by Menaechmus, around 350 BC, the conic sections were contemplated by several Greek mathematicians, including Euclid. But it was Apollonius of Perga, known as "the great geometer" to his contemporaries, who in around 250 BC performed the first definitive analysis, truly taming these elegant figures.

OPPOSITE The solar furnace at Odeillo in France. Its parabolic surface is covered with 9500 mirrors, which reflect the sunlight to a single spot at its focus, producing temperatures of up to 3500°C.

Apollonius, the great geometer

Appolonius's eight books on conic sections are a geometrical tour de force, systematically investigating the curves from several different angles. He also provided the three principal types of conic section with the names that they still carry today: the ellipse, the parabola and the hyperbola.

RIGHT The conic sections can be defined as the curves which arise when an infinite cone is sliced by a flat plane. The plane's angle will determine the type of curve.

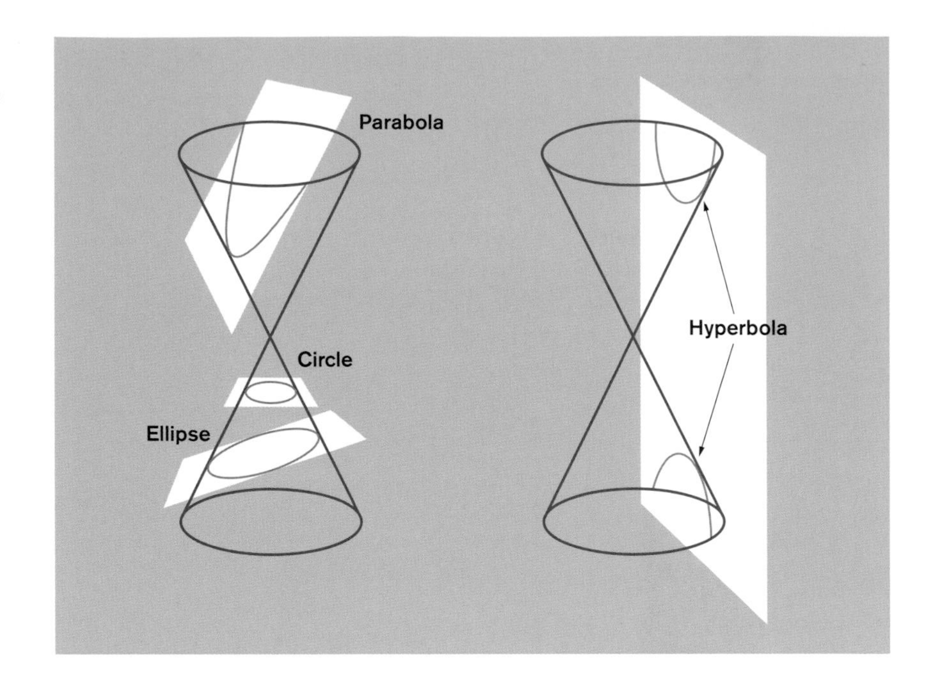

This new, advanced form of geometry actually arose directly from considerations of the old shapes. Apollonius's construction begins with a circle and a single point floating directly above its center. Now imagine that every point on the edge of the circle is connected with a straight line to that floating point. The resulting surface is a cone. But if the lines are extended indefinitely, it is not just a single cone, but two infinitely long cones joined at their tips.

This conic surface is one of the simplest 3-dimensional shapes. But contained within it are some sophisticated 2-dimensional curves. Appolonius's question was this: suppose you make a flat cut through that conic surface. What might the cross-section look like? One possibility is that you will get back the original circle, if you cut at exactly the right place. A little below or above will produce larger or smaller circles. But the story becomes more interesting when the cut is not made horizontally, but at an angle. The first shape to emerge looks like a circle which has been stretched or foreshortened. This is the commonest of the conic sections: the ellipse.

Conic sections in nature

Almost two millennia after Apollonius, the role of ellipses in our universe was finally revealed. For centuries, the relative motion of the Earth, Sun and other planets was debated and studied. But as the telescopes and scientific equipment became even more sophisticated, the measurements that were taken seemed ever more strange and contradictory. This continued even after Nicolaus Copernicus' revelation that the Earth orbits the Sun, and not vice versa.

It was the astronomer and mathematician Johannes Kepler who finally corrected the mistake in the early 17th century. Copernicus, and indeed all other astronomers, had erroneously assumed that planetary orbits must be circles. In fact, Kepler realized, they are ellipses. What is more, the Sun is not at the center of the ellipse, but at a critical point inside it known as a "focus." As Apollonius had already realized, an ellipse has two foci, positioned symmetrically on either side of its center. These lead to a nice way to draw it: if you take a piece of string, and pin both ends of it to a table, the places where the string is pulled taut trace out an ellipse, with foci at the two pins.

As the telescopes and scientific equipment became even more sophisticated, the measurements that were taken seemed ever more strange and contradictory.

Gravity also gives rise to the second conic section: the parabola. Apollonius originally derived this shape by slicing through a conic surface with a cut exactly parallel to the edge of the cone. This produces a line which, unlike an ellipse, is not closed, but an infinitely long U-shape.

Parabolas, too, are all around us. If you throw a stone through the air, the path it traces out will be parabolic (at least if you ignore the effects of air resistance). This was demonstrated by Galileo Galilei in 1638, confounding the old physics of Aristotle, which held that that the stone would travel in a straight line, gradually slowing down until it dropped to the ground. Parabolas also appear on the astronomical scale. Some comets, like Halley's, appear in the night sky with predictable regularity. This is because they are on large elliptical orbits around the Sun. Halley's comet reappears around every 75 years, while the Hale-Bopp comet is on an orbit over 2000 years long. But there are other comets that never reappear, simply flying off into space. One such was the Great Comet of 1577, a major astronomical event of Johannes Kepler's childhood, which would go on to shape his work on celestial orbits. Single-apparition comets such as these usually have parabolic paths around the Sun.

The last of the three conic sections is the hyperbola. Apollonius created this shape by cutting through his double cone at a steep angle, so that the plane hit both halves of the surface. The resulting curve has two separate branches which never meet but mirror each other perfectly. The hyperbola is the rarest of the conic sections in nature, but it too is useful, for example in the shapes of the mirrors in the Hubble space-telescope, which allow astronomers to scan the Universe.

14 Trigonometry

BREAKTHROUGH THE ANALYSIS OF DISTANCES AND ANGLES, ESPECIALLY WITHIN TRIANGLES, WAS FIRST STUDIED BY THE ASTRONOMER HIPPARCHUS IN AROUND 150 BC.

DISCOVERER HIPPARCHUS (190 BC–120 BC).

LEGACY TRIGONOMETRY IS USED THROUGHOUT SCIENCE AND ENGINEERING TODAY, ESPECIALLY SINCE THE ADVENT OF THE POCKET CALCULATOR. YET THIS TOPIC EXTENDS FAR BEYOND THE ELEMENTARY GEOMETRY OF TRIANGLES TO SOME OF THE MOST SOPHISTICATED TOPICS IN MODERN MATHEMATICS.

In elementary geometry, two of the most fundamental topics are distances and angles. Yet the relationship between them is far from obvious. Although the connections were investigated early within Euclid's *Elements*, it was not until the astronomer Hipparchus turned his attention to the subject that trigonometry became a practical technique.

In his *Elements* (see page 46), Euclid spends a great deal of time contemplating triangles. There are two major topics he studies: the relationship between the lengths of the sides of the triangle, and the connection between its angles. In the first context, the great result is Pythagoras' theorem, which gives an exact relationship between the sides of a triangle. Once you know the lengths of two of the sides, you can easily calculate the third. But famously, this result does not apply to all triangles—only to those containing a right angle. In his study of angles, Euclid's major result is that the three angles in any triangle must always add up to 180°. The fact that Pythagoras' theorem, which is about length, is only valid for triangles containing a certain angle (namely 90°) is a strong indication of the subtle dependency between angles and lengths. This would be fleshed out in the study of trigonometry.

Similarity and proportion

Although Euclid did not study what we would now call trigonometry, he did realize the importance of similar triangles, meaning triangles that are the same shape, without necessarily being the same size. Crucially, two similar triangles must share all the same angles. The very fact that

OPPOSITE Trigonometry is an essential tool for today's engineers, whether triangles are explicitly visible or concealed within the structure of a building.

similar triangles can exist illustrates that the relationship between length and angle is not entirely straightforward: length does not determine angle, nor vice versa.

Euclid's key insight was that similar triangles must have the same proportions as each other. So, if you know the angles within a triangle, and you know the length of one side, then it should be possible, in principle, to determine the remaining two sides.

Hipparchus' chord table

It is one thing to observe that fixing the angles of a triangle and fixing one of its sides must theoretically determine the lengths of the remaining two sides. But it was something else altogether actually to calculate them. How could this be done? The man who answered this question was Hipparchus, born in Nicaea in what is now Turkey. He achieved his fame working as an astronomer in Alexandria and on the island of Rhodes. Although little of his work has survived, we know that his astronomical observations included several great achievements. He calculated the length of a year to within seven minutes' accuracy, and produced the most accurate map of the skies the world had seen.

For over a thousand years, thinkers from Greece to China gradually improved Hipparchus' trigonometric tables.

Hipparchus took a mathematical approach to his astronomy, striving to understand the relationships between the Earth, Sun and other heavenly bodies. But he found the geometry of his day inadequate. In particular, he needed to translate information about the angles between astronomical bodies into data about the distances between them.

Hipparchus adopted a highly practical approach. By drawing and measuring various triangles, he was able to assemble a table of the data he needed. This was the first true exploration of trigonometry. The tool Hipparchus used was the chord function, which has subsequently been surpassed by the sine and cosine functions found on modern calculators (known as sin and cos for short). The essential idea is the same in all cases, however.

In a right-angled triangle, suppose we know one of the other angles (perhaps 30°), and the length of the longest side, say 10cm. In principle, this data completely determines the triangle; there is only one triangle that can be drawn fitting this description. If you want to know the length of the side opposite the angle, it turns out that we can simply multiply 10cm by $\frac{1}{2}$, to get an answer of 5cm. This magic number $\frac{1}{2}$ depends on the angle 30°; for other angles, different values are needed. Today we write this as $\sin 30° = \frac{1}{2}$.

Hipparchus used chord instead of sine, and tabulated its values for different angles x. With these known, it was a simple matter to derive the dimensions of a triangle containing the given angle.

ABOVE Modern communication relies on a network of satellites, through which huge quantities of information are constantly flowing. This system relies on trigonometry to analyze the angles and distances between the satellites and points on the Earth.

Madhava and transcendental functions

Since its inception, trigonometry has been intertwined with astronomy and technology. This remains true today. For example, satellite communication relies on exact calculations of the angles and distances between the satellites and fixed points on the Earth. So thinkers from Greece to China were drawn to improve Hipparchus' tables. Yet it is striking that they should need to rely on tables at all; it would be far better to find a single formula which could produce the value of $\sin x$ for every possible angle x. Yet with no one able to find such a method, there was no choice but to rely on the somewhat primitive approach of constructing detailed tables.

It was in Kerala in southern India, around the year 1400, that such a formula was finally found. Again it was an astronomer who made the breakthrough. Madhava realized that to calculate the sine function required an infinite process. Specifically, with x measured in radians,

$$\sin x = x - \frac{x^3}{3 \times 2} + \frac{x^5}{5 \times 4 \times 3 \times 2} - \frac{x^7}{7 \times 6 \times 5 \times 4 \times 3 \times 2} + \dots$$

Or more concisely: $\sin x = \frac{x}{1!} - \frac{x^3}{3!} + \frac{x^5}{5!} - \frac{x^7}{7!} + \dots$

This discovery would not be made in Europe until the discovery of calculus in the 17th century (see page 132). Yet, in time, its consequences would prove dramatic: sine (and the other trigonometric functions) are what is now termed transcendental functions, meaning that they necessarily involve infinite processes to calculate exactly. For this reason, trigonometric functions would eventually outgrow their humble roots in triangular geometry, becoming critical ingredients in complex analysis (see page 154) and the study of abstract waveforms (see page 169).

15 Perfect numbers

BREAKTHROUGH A PERFECT NUMBER IS ONE WHOSE FACTORS TOTAL TO THE ORIGINAL NUMBER. PERFECT NUMBERS WERE A SOURCE OF WONDER IN THE ANCIENT WORLD.

DISCOVERER PYTHAGORAS (C. 570–475 BC), EUCLID (C. 300 BC), NICOMACHUS (C. AD 60–120).

LEGACY LEONHARD EULER ESTABLISHED THE RELATIONSHIP BETWEEN PERFECT NUMBERS AND MERSENNE PRIMES, BUT PERFECT NUMBERS ARE STILL FAR FROM BEING FULLY UNDERSTOOD TODAY.

The two most fundamental operations on numbers are addition and multiplication. The Pythagoreans were intrigued by special numbers where these two operations seem to balance each other out exactly. They endowed these unusual numbers with great mystical importance, describing them as "perfect."

The first perfect number is 6. Whether or not a number is perfect is determined by its factors, those smaller numbers which divide it exactly. In the case of 6, its factors are 1, 2 and 3. (Six itself is also a factor, but is omitted for these purposes.) The Pythagoreans spotted that adding these factors together reveals something special: $1 + 2 + 3 = 6$.

Perfect numbers are rare: this trick does not work with most numbers. For instance, the factors of 8 are 1, 2 and 4, which add up to 7. After 6, the next perfect number is 28, which has factors 1, 2, 4, 7 and 14. Following 28, the next perfect number is 496. The fourth in the series, 8128, was discovered by Nicomachus in around AD 100. (As we shall see later, Nicomachus had a somewhat unusual interpretation of his mathematics, seeing this numerical perfection as having moral overtones.) However, it was not until the 15th century that the fifth perfect number, 33,550,336, was identified.

OPPOSITE The number 6 was the first to be deemed "perfect," having both interesting arithmetical properties, and describing many of the symmetries of patterns and the natural world. Notable examples are snowflakes, whose beauty comes from their hexagonal symmetry.

Because perfect numbers are elusive, they have been the subject of a lot of speculation and guesswork. Nicomachus, for example, wrote that perfect numbers alternate between ending in 6 and ending in 8. But this

1040793219466439908192524032736408 5
5386152622472667048053191123504036 0
8059673360298012239441732324184842 4
2161395428100779138356624832346490 8
1399066056773207629241295093892203 4
5773183349661583550472959420547689 8
1121169367714754847886696250138443 8
2602917323488853111608285384165850 2
8255604666224831890918801847068222 2
0314052102669843548873295802887805 0
8697361869007147207105557031687290 87

conjecture was refuted in 1588 with the discovery of the sixth perfect number: 8,589,869,056—the fifth also ended in a 6.) Iamblichus, meanwhile, wrote that there is exactly one perfect number between 1 and 10, between 10 and 100, between 100 and 1000, and so on. In time this, too, was seen to be false.

The biggest mystery in the study of perfect numbers is whether or not the list continues for ever, or whether there are only a finite collection of perfect numbers. This question remains a major challenge to today's number theorists.

ABOVE Perfect numbers are intimately tied to the search for large primes. This is the 15th Mersenne prime, discovered in 1952 by Raphael Robinson, and is equal to $2^{1279} - 1$. The largest prime currently known is the 47th Mersenne prime, $2^{43,112,609} - 1$, which is 12,837,064 digits long.

Mersenne primes

The true importance of perfect numbers was cemented in around 300 BC, when Euclid discovered that they are intimately related to those other celebrities of the mathematical world, the prime numbers (see page 49). It is sometimes true that if p is a prime number, then $2^p - 1$ is also prime (2^p means $\underbrace{2 \times 2 \times \ldots \times 2}_{p \text{ many } 2\text{s}}$). For example, 3 is prime, as is $2^3 - 1 = 7$. This does not always work, however: $2^{11} - 1 = 2047$, which is non-prime (since $2047 = 23 \times 89$). Primes of the form $2^p - 1$ are known as Mersenne primes, after Friar Marin Mersenne, who began to list them in the 17th century. Most of the largest known primes are Mersenne, since these numbers are much easier to test for primality.

Euclid was aware of these special primes thousands of years before Mersenne. In particular, he spotted the connection between Mersenne primes and perfect numbers. He proved that if M is a Mersenne prime, then $\frac{M \times (M + 1)}{2}$ will always be a perfect number, e.g. 3 is a Mersenne prime, and $\frac{3 \times 4}{2} = 6$, which is perfect. Similarly, 7 is a Mersenne prime, and $\frac{7 \times 8}{2} = 28$ is perfect. It was not until the 18th century that Leonhard Euler proved that this correspondence is exact: every even perfect number must be of the form $\frac{M \times (M + 1)}{2}$, where M is a Mersenne prime. This theorem tied together the search for Mersenne primes with the search for even perfect numbers. In both cases, we still do not know whether the list is finite or infinite.

The Euclid–Euler theorem concerns only even perfect numbers. What, then, of odd perfect numbers? Like yetis, no one has ever seen one, and most people doubt their existence. At the same time, no one has managed to rule out the possibility altogether.

Deficient and abundant numbers

Whether there are finitely or infinitely many of them, the perfect numbers are certainly in a tiny minority among the whole numbers. When you add up their factors, many numbers will fall short (as happens with 8). Nicomachus identified these deficient numbers with "wanting, defaults, privations and insufficiencies." Other numbers may overshoot, for instance the factors of 12 are 1, 2, 3, 4 and 6, which add up to 16. Nicomachus considered these abundant numbers indicative of "excess, superfluity, exaggerations and abuse."

The biggest mystery in the study of perfect numbers is whether or not the list continues for ever, or whether there are only a finite collection of perfect numbers. This question remains a major challenge to today's number theorists.

Occasionally, though, deficient and abundant numbers can cancel each other out. For instance, the factors of 220 are 1, 2, 4, 5, 10, 11, 20, 22, 44, 55 and 110, which add up to 284 (making 220 abundant). However, when you add up the factors of 284 (namely 1, 2, 4, 71 and 142) you arrive back at 220. "Amicable pairs" such as 220 and 284 were studied in depth by Arabic mathematicians including Thabit ibn Qurra, who devised a method for finding new pairs.

Aliquot sequences

In 1888, Eugène Catalan asked what would happen if you started with some number, added up its factors, then did the same with the new number, repeating the process over and over again. The result of this process is known as the original number's "aliquot sequence." One possible outcome is that the sequence will hit a perfect number and stay there forever. Alternatively, if it lands on a member of an amicable pair, the sequence will simply flit back and forth between those two. Longer repeating cycles are possible, too, known as "sociable numbers." On the other hand, the cycle might hit a prime (such as 7), in which case it will fall to 1 (since the prime has no other factors), and then to 0. Catalan asked whether every sequence must end in one of these three ways. The answer is still unknown. The eventual outcome of the number 276, for instance, is not yet known. So far, the sequence just seems to be growing ever larger; making sense of this seems an extremely hard problem.

16 Diophantine equations

BREAKTHROUGH DIOPHANTUS MADE EARLY BREAKTHROUGHS IN ALGEBRA, BUT HE IS BEST KNOWN FOR THE ANALYSIS OF WHOLE NUMBERS THROUGH WHAT ARE NOW CALLED DIOPHANTINE EQUATIONS.

DISCOVERER DIOPHANTUS OF ALEXANDRIA (C. AD 200–84).

LEGACY DIOPHANTUS' *ARITHMETICA* WAS A MAJOR INSPIRATION TO LATER CLASSICAL MATHEMATICIANS, THINKERS OF THE ISLAMIC GOLDEN AGE, AS WELL AS MODERN EUROPEAN MATHEMATICIANS SUCH AS PIERRE DE FERMAT.

To understand the system of whole numbers, we need to know what relationships are possible between them. These are expressed in terms of Diophantine equations, named after Diophantus of Alexandria. The work in which he introduced them, *Arithmetica*, has been a source of inspiration to thinkers down the centuries.

We know very little about Diophantus. Like Euclid before him, he lived in that great intellectual center of the classical world, Alexandria in Egypt. Although this was now a part of the Roman Empire, Diophantus would have spoken and written in Greek. His exact dates are a matter of historical conjecture, though later accounts suggest that he died at the age of 84.

Diophantus' great work was entitled *Arithmetica*. While Euclid's *Elements* was a detailed treatise on a particular topic, *Arithmetica* was a far less coherent work, consisting of 130 assorted questions divided between 13 volumes. Regrettably, only six of the books survived, and the missing books appear to have been lost very early. In 1968, however, some ancient Arabic texts were discovered, which may include translations of some of the lost parts of *Arithmetica*, though this is not universally accepted.

OPPOSITE Diophantine equations are the primary way of expressing relationships between whole numbers. Solving these equations is a difficult task, but tells us the relationships that whole numbers may, or may not, satisfy.

Diophantine equations

Diophantus is sometimes described as the Father of Algebra, since *Arithmetica* was the first book devoted to the art of solving equations. Diophantus investigated topics such as linear and quadratic equations in *Arithmetica*; they would later be given a more thorough treatment by Al-Khwarizmi (see page 85).

However, a point of ideology led Diophantus to adopt a particular focus for his equations that would make his questions of particular fascination in the coming years. Like many mathematicians of the classical era, Diophantus was highly skeptical of irrational numbers (see page 29). So, in solving his equations, Diophantus was eager always to stay within the bounds of the whole numbers, or at least the rational numbers (which is to say fractions). This made the problems considerably more difficult to solve, and led a frustrated later scholar to scrawl in his copy of the book: "Thy soul, Diophantus, be with Satan because of the difficulty of your theorems."

It was while Pierre de Fermat was reading through *Arithmetica* that he had the idea of extending Diophantus' analysis of Pythagorean triples to higher powers. The consequences would form one of the great tales in mathematics, Fermat's Last Theorem.

Diophantus was particularly interested in the arithmetic of powers, especially squares. Squares are numbers like 9 or 16, which consist of some number multiplied by itself ($9 = 3 \times 3$, $16 = 4 \times 4$). Higher powers involve longer strings of repeated multiplication. For example, in book 3 of *Arithmetica*, he wished to find a number x so that both $10 \times x + 9$ and $5 \times x + 4$ are square numbers. He produced an answer of 28. This works because $10 \times 28 + 9 = 289 = 17 \times 17$ and $5 \times 28 + 4 = 144 = 12 \times 12$.

Questions like this, which seek whole or rational numbers satisfying some specific criteria, would later come to be called Diophantine equations, and would come to be seen as the primary way of understanding the possible and impossible relationships between whole numbers.

Hypatia's commentary

Around 100 years after Diophantus wrote it, *Arithmetica* became a focus of interest for the first female mathematician we know by name, Hypatia. Also a resident of Alexandria, Hypatia studied and taught astronomy and physics, and gave public lectures on the philosophy of Plato and Aristotle. She wrote a commentary to the six surviving books of *Arithmetica*, suggesting that seven were already lost by this time. (Writing commentaries on the works of earlier scholars was common practice. Hypatia also produced a commentary on the *Conics* of Apollonius; see page 57.)

As well as being a famous female intellectual, Hypatia was an outspoken critic of the state of Christianity of her era, preferring to identify with a

Platonist philosophy. The religious politics between the Christians, Jews and pagans of Alexandria was a tense and intermittently bloody affair, and for her public stance against the Church, Hypatia suffered a cruel death. In AD 415, she was seized from her carriage by a mob of Christian fanatics, who dragged her into a church and murdered her.

ABOVE *Saint Catherine Hypatia* by Onorio Marinari (1627–1715). Hypatia was a distinguished philosopher, teacher and scientist, and the first female mathematician we know by name. Her murder was said to have marked the end of the dominance of Alexandria as an intellectual center.

Rediscovering Diophantus

As the golden era of classical mathematics faded, Indian and Arabic mathematics came to the fore. *Arithmetica* was translated and heavily studied in the Islamic world, the most enduring translation being in the tenth century by Mohammad Abu'l-Wafa Al Buzjani, who also wrote his own commentary to the work. It was not until the 16th century that the work was translated into Latin and thus became available to European researchers.

Several important mathematical stories have their origins in the pages of *Arithmetica*, though they would not emerge until thousands of years later, when the importance of Diophantine equations was finally realized. It was while Pierre de Fermat was reading through Claude Bachet's recent translation of the work that he had the idea of extending Diophantus' analysis of Pythagorean triples to higher powers. The consequences would form one of the great tales in mathematics, Fermat's Last Theorem (see page 369). The roots of another major topic in number theory also lie with Diophantus: the question of how whole numbers can be split up into powers. This would later become immortalized as Waring's problem (see page 241).

Indirectly, Diophantus' work would also become highly important to 20th-century logic and computer science. In his tenth problem (see page 329), David Hilbert asked whether Diophantine problems might all be solvable through some automatic procedure. The unexpected answer was a milestone in modern mathematics.

17 Hindu–Arabic numerals

BREAKTHROUGH THE SYSTEM WE KNOW AS THE HINDU–ARABIC NUMERALS DEVELOPED IN ANCIENT INDIA BETWEEN 150 BC AND AD 600.

DISCOVERER PERHAPS THE EARLIEST DOCUMENTED CASE OF THESE NUMERALS AT WORK IS THE BAKHSHALI MANUSCRIPT, POSSIBLY DATING FROM AROUND AD 250.

LEGACY TODAY, HINDU–ARABIC NUMERALS ARE THE GLOBAL STANDARD FOR REPRESENTING NUMBERS.

The symbols that we use to write numbers have their origins in ancient India. In particular, the characters 1, 2, 3, 4, 5, 6, 7, 8, 9 and 0, which are so familiar to us today, first developed there before being adopted in the Islamic world and then brought to Europe and the New World.

As in many other parts of the world, the mathematicians of ancient India had different ways of representing numbers. Most were, by modern standards, inefficient, sharing many of the disadvantages of Roman numerals, which were in use across Europe. Yet some time before the third century AD, a system began to evolve in India that would develop into the system of numerals which has become the worldwide standard.

While the ancient Babylonians wrote their numbers in base 60 (see page 17), in India and elsewhere decimal systems were typical, that is to say numbers written in base 10. Although there is nothing mathematically significant about this number, it is easy to see why this should be. Humans have evolved to have ten fingers; long before the first tally stick was carved, our hands were our first tools for counting.

OPPOSITE Devanagari numerals at the 18th observatory in Jantar Mantar in Jaipur, India. These numerals remain in use in modern Hindi, and share common ancestry with Hindu–Arabic numerals 0–9. The illustrated numeral represents the number 6.

From around 150 BC, the residents of central India developed a system known as Brahmi numerals, which are the direct forebears of the modern system. It contained symbols for the numbers 1 to 9, and new symbols for the numbers 10, 20, 30, … and 100, 200, 300, … Putting these together allowed a range of numbers to be expressed. In time, however, the quantity of symbols needed would reduce.

Vedic and Jain mathematics

One of the driving forces toward a more economical system of numerals was a religious fascination with very large numbers. The Vedic period of Indian history began around 1000 BC and is named after the Vedas, the first texts of the Hindu religion, written in the earliest variety of Sanskrit. Some of the scriptures of this period discuss gigantic numbers. In the epic poem *Ramayana*, which is thought to date from around 250 BC, the hero Rama was said to have commanded an army of 1000010000100000 0010000100000100010000010001000001000010000000005 fighters!

Vedic-era mathematicians had names for many powers of 10, all the way up to the one needed for Rama's army—10^{62}. (Meanwhile, the largest number the ancient Greeks had an individual name for was the myriad, or 10,000.) It is very difficult to contemplate numbers like this without being forced into some sort of place-value notation. Indeed, scholars believe that this is the era when decimal place-value notation first developed.

BELOW This table charts the development of the symbols 0–9, now the internationally standard method for writing numbers.

European.		Gobar.	Indian.		
14th cent.	12th c.	(Arab.)	10th c.	5th c.	1st c.
1	1				
2	2				
3	3				
4	4				
5	5				
6	6				
7	7				
8	8				
9	9				
0	0				

Another religious tradition, Jainism, founded around the sixth century BC, pushed this idea even further. Jains contemplated extremely long time scales, based on units such as the *shirsha prahelika*, defined as $756 \times 10^{11} \times 8{,}400{,}000^{28}$ days. (As well as huge finite numbers, Jains also developed a theory of different denominations of infinity, strikingly anticipating Georg Cantor's theory of cardinal numbers; see page 217.)

The Bakhshali manuscript

Arguably the first documented use of what we now call Hindu–Arabic numerals is the Bakhshali manuscript. Discovered in 1881 in the village of Bakhshali in what is now Pakistan, the surviving portion consists of 70 leaves of birch bark on which various mathematical rules are laid out, with examples of each provided. Several of the leaves are in tatters, so it has been difficult for scholars to study, and especially hard to date accurately. In the past, the manuscript's date has been variously given as 200 BC and

AD 1200. But recent scholarship suggests that it is likely to date from around AD 400 (although it may be a later copy of an original document from this period).

The Bakhshali manuscript is principally interesting for the notation it uses. Although the symbols used are not recognizable to modern eyes, it is possibly the first surviving example of a fully fledged decimal place-value notation in action, complete with a symbol for zero (namely a dot). It also contains examples of fractions, expressed much as they are today, though with the dividing line omitted. It contains discussion of profit and debt, using negative numbers (which confusingly to modern eyes, are denoted using the symbol "+").

Whatever the truth of the Bakhshali manuscript, it is certainly the case that during the classical period of Indian mathematics, of which the work of Brahmagupta was perhaps the greatest highlight (see page 81), a system of decimal place-value notation was flourishing.

Uptake by Arabs and Europeans

The new system was taken up by Persian mathematicians in northern India, via whom it spread around the Islamic world. Around AD 825, Al-Khwarizmi (see page 85) wrote an influential work entitled *On Calculation with Hindu Numerals* (which only survives in Latin translation), which was principally responsible for its uptake among Islamic scholars, and perhaps cemented the system in roughly its modern form.

One of the driving forces toward a more economical system of numerals was a religious fascination with very large numbers.

The book which introduced these numerals to Europe was the 1202 work *Liber Abaci* (Book of Calculation) by Leonardo da Pisa, better known as Fibonacci (see page 93). His father was a wealthy merchant and diplomat who spent much of his time in Algeria in northern Africa. The young Fibonacci traveled widely with his father and came to appreciate the efficiency of the mathematical notation used by the Islamic scholars around the Mediterranean.

For a few centuries, "Arabic numerals," as they were known, and Roman numerals coexisted throughout Europe, with the modern system gradually gaining prominence. It was not until the printing press revolution of the 15th and 16th centuries, and the standardization and increase in literacy that it brought, that these ten characters settled into the familiar forms that we recognize today, growing into a global system for writing mathematics.

11
12
55
5
50
10
45
15
40
20
25

18 Modular arithmetic

BREAKTHROUGH MODULAR ARITHMETIC IS A CYCLICAL SYSTEM OF NUMBERS, LIKE THAT ON A CLOCK FACE.

DISCOVERER SUN-ZI (C. AD 400–60), PIERRE DE FERMAT(1601–65).

LEGACY MODULAR ARITHMETIC IS IMPORTANT TO TODAY'S NUMBER THEORISTS; IT IS ALSO A SIGNIFICANT PART OF COMPUTER SCIENCE, WHERE IT FORMS THE BASIS OF MODERN CRYPTOGRAPHY.

What is 11 + 7? In some contexts, the obvious answer may not be the right one. For instance, if it is now 11 o'clock, then in seven hours' time it will be 6 o'clock. Here, then, it is perfectly correct to write 11 + 7 = 6. This clock-face mathematics is known as modular arithmetic, and it was first seriously investigated by Sun-Zi in around AD 450.

In the case of a clock face, arithmetic can be said to take place modulo 12, but of course any other whole number is equally possible. If today is Tuesday, what day of the week will it be in fifteen days' time? This can be interpreted as a question about arithmetic modulo 7, while consideration of the minutes past an hour leads to arithmetic modulo 60.

Minutes, hours and days

Modular arithmetic has been contemplated since humans first began to divide periods of time into intervals such as minutes, hours and days. If the notches of the Lebombo bone (see page 12) represent a lunar calendar, then it could be argued that arithmetic modulo 29 is the very first human mathematics we have evidence for.

OPPOSITE Modular arithmetic is well-known as the mathematics of time-keeping. The hours on a clock face illustrate arithmetic modulo 12, while the seconds in a minute work by arithmetic modulo 60.

Yet it took many years for this topic to receive serious mathematical attention. Its first appearance was as a puzzle in the work of Sun-Zi in around AD 450. In the centuries since, modular arithmetic has become hugely important in mathematics. It revolves around the idea of a remainder. If it is 8 o'clock now, what time will it be 20 hours from now? This amounts to working out 8 + 20 modulo 12. This is answered by the observation that 12 fits into 28 exactly 2 times (as $12 \times 2 = 24$), leaving a remainder of 4. It is this remainder which provides the answer: $8 + 20 = 4$ modulo 12.

All the basic arithmetical operations—addition, subtraction, multiplication and division—carry over into the modular setting, so we can write that $3 \times 4 = 1 \bmod 11$, while $7 - 9 = 15 \bmod 17$, and even $5 \div 4 = 3 \bmod 7$ (since $5 = 12 \bmod 7$). Because ordinary arithmetic works just as well in the modular setting as it does for ordinary whole numbers, modular arithmetic is much studied by number theorists. It is convenient that modular arithmetic is a finite system. Modulo 5, every number must be equal to one of 0, 1, 2, 3, 4. So, modulo 5, there really are only five numbers, just as a clock face only needs 12.

Sun-Zi's Chinese remainder theorem

In *Sunzi Suanjing* (Sun-Zi's Manual of Mathematics), the author posed the following puzzle: "Suppose we have an unknown number of objects. When counted in threes, 2 are left over, when counted in fives, 3 are left over, and when counted in sevens, 2 are left over. How many objects are there?"

Nothing is known about Sun-Zi apart from this one riddle. Yet it would go on to be of incalculable influence, since it opened the door to the mathematics of modular arithmetic.

Nothing is known about Sun-Zi apart from this one work. Yet this riddle would go on to be of incalculable influence, since it opened the door to the mathematics of modular arithmetic. What Sun-Zi asks for is a number n, such that $n = 2 \bmod 3$, while simultaneously $n = 3 \bmod 5$, and $n = 2 \bmod 7$. Sun-Zi's breakthrough, in fact, is the basic insight that such a number must exist. It will always be possible to find a suitable number that simultaneously satisfies multiple modular conditions. In this case, an answer is 23. For no better reason than the nationality of the mathematician involved, this celebrated fact became known as the Chinese remainder theorem.

There is a technical proviso. The bases, in this case the numbers 3, 5 and 7, must be coprime. This means there is no number which divides all of them. Otherwise, the conditions may contradict each other. It is impossible, for instance, to find any value of n so that $n = 1 \bmod 3$, but $n = 2 \bmod 6$.

Fermat's Little Theorem

In later years, modular arithmetic became bound up with the analysis of prime numbers. A famous example of this was Pierre de Fermat's so-called Little Theorem (not to be confused with his Last Theorem; see page 369).

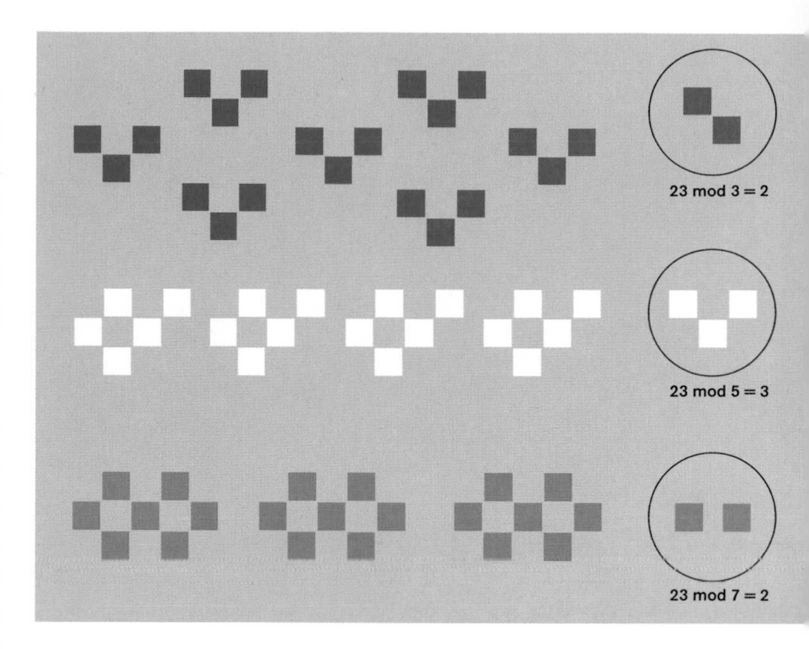

ABOVE The solution to Sun-Zi's original problem: 23 squares grouped in 3s, 5s and 7s, leaving remainders of 2, 3 and 2 respectively.

Fermat's Little Theorem considers a prime number, say p, and another whole number, n, which need not be prime. Fermat contemplated n^p, which is $\underbrace{n \times n \times \ldots \times n}_{p\text{ times}}$. Fermat's theorem states that n and n^p will always be equivalent modulo p, that is to say $n^p = n \bmod p$.

This fact is of great use for testing numbers to see whether they are prime. If you want to test a number q to see whether or not it is prime, and you can find any number at all such that $n^q \neq n \bmod q$, then it must have been that q was not prime after all. This is the basis of modern computerized primality tests, which are the means by which new large prime numbers are checked. It is also the foundation for modern public key cryptography (see page 353).

Gauss's golden theorem

The modular arithmetic of prime numbers is also the setting for what Carl Friedrich Gauss called his golden theorem, which is now mostly known as quadratic reciprocity. Given two prime numbers, called p and q (neither equal to 2), Gauss looked at two quantities: $p \bmod q$ and $q \bmod p$. In 1801, he proved a beautiful symmetry between the two. He showed that one of these is precisely a square if the other is. (There is an exceptional case, however. If both p and q are three more than a multiple of 4, then one is precisely a square if the other isn't.) This theorem has been extraordinarily influential. Attempts to generalize it led directly to the deep waters of the Langlands program (see page 393).

19 Negative numbers

BREAKTHROUGH BRAHMAGUPTA EXPANDED THE NUMBER SYSTEM TO INCLUDE ZERO AND NEGATIVE NUMBERS. FOR THE FIRST TIME, ZERO WAS RECOGNIZED AS A NUMBER IN ITS OWN RIGHT.

DISCOVERER BRAHMAGUPTA (AD 598–C. 665).

LEGACY BRAHMAGUPTA'S SYSTEM OF ARITHMETIC WAS A KEY MOMENT IN THE DEVELOPMENT OF THE NUMERICAL FRAMEWORK THAT IS UNIVERSAL TODAY.

Brahmagupta's *Brahmasphutasiddhanta* was a remarkable document. It described a world of new numbers, together with the laws that govern them. Particularly significant was that, for the first time, Brahmagupta elevated zero to the status of a true number. He then expanded the number system even further to include what today we call negative numbers.

In India, from at least the time of the Bakhshali manuscript (see page 74), mathematicians wrote numbers using a place-value system similar to the one used worldwide today. In such a system, the meaning of a numeral, such as 3, depends on its position. The 3 in the number 37 means something different from the 3 in 73. This development echoed ancient Babylonian mathematics (see page 18), and was a markedly better system than the numerical representations used by the Greeks and Romans. The arrangement led both Babylonian and Indian mathematicians to invent symbols for zero. They needed something to represent an empty column, so as to distinguish 307 from 37. Nevertheless, during this period, zero remained closer to a punctuation mark than a true number in its own right.

OPPOSITE In the Celsius scale, 0° is the point at which water freezes into ice at ordinary atmospheric pressure. Negative numbers represent temperatures colder than this. The lowest temperature recorded on Earth is –89.2°C, at the Vostok research station in northern Russia.

Brahmagupta's *Brahmasphutasiddhanta*

The first real attempt to expand the number system came in AD 628, in the work of the scholar Brahmagupta. He was a renowned astronomer in the sacred Hindu city of Ujjain, in what is now Madhyar Pradesh in central India. His great mathematical breakthrough formed part of his work *Brahmasphutasiddhanta* or "The Correctly Established Doctrine of Brahma" (Brahma being the Hindu god of creation). The volume was written in verse and contained several important insights into algebra, number theory and geometry. But in the most significant section, Brahmagupta set out a new set of rules of arithmetic. These were laws of adding, subtracting, multiplying and dividing. Critically, these laws were drafted to encompass the number zero.

In Brahmagupta's hands, zero became far more than a mere punctuation symbol. In fact, he gave a formal definition of it as a number. He said that it is the result of subtracting any number from itself. So $7 - 7 = 0$, for example. This seems completely obvious to us now, but it required a leap of imagination to see this as a genuinely meaningful statement. In the past, numbers were adjectives for describing collections of objects. When all objects have been removed, why should a number remain?

This objection was reinforced by zero's refusal to conform to the usual laws of arithmetic. When you multiply any number by 2, for example, it doubles in magnitude. Yet zero does not. Brahmagupta was not fazed by this; he carefully noted the laws that describe zero: firstly, if zero is added or subtracted from any number, the number is left unchanged. Secondly, any number multiplied by zero is equal to zero.

BELOW Split tallies were used for registering debts in Europe until the early 19th century. The two participants would carve out the sum of money owed in notches, and then split the stick and take one half each. One half represented a positive number, and the other the corresponding negative number.

Negative numbers

Brahmagupta did not only advance the cause of zero. He went further, incorporating into his treatise negative numbers—entities which are even less intuitive than zero. While "zero mangoes" could be understood as the absence of any mangoes, what might "−4 mangoes" mean?

Although Brahmagupta wrote his work as a piece of abstract mathematics, the language he used was suggestive of finance. In this setting, negative numbers represent debt. To have −4 mangoes means to be in debt by 4 mangoes. Under these terms, 0 is the break-even point, where one neither has nor owes anything. Negative numbers were not a completely new concept: Chinese mathematicians had also considered them in the past, for the purposes of trade. But Brahmagupta's genius was to unite all numbers—positive, negative and zero—into a single new number system.

In the past, numbers were adjectives for describing collections of objects. When all objects have been removed, why should a number remain?

He was the first to write down many laws of arithmetic that we now consider standard. For instance, when you add together two negative answers, the answer is negative, and when you multiply a positive number by a negative number, you get a negative result, $4 \times (-3) = -12$. But Brahmagupta also grasped one of the most misunderstood aspects of arithmetic: when you multiply two negative numbers together, you get a positive result: $(-4) \times (-3) = 12$.

Division by 0

Brahmagupta's theory of addition, subtraction and multiplication has come down to us almost unchanged. But when it came to division, his thinking was not quite in tune with the modern approach. If you start with eight books, and divide them into four piles, each pile contains two books. So we write "$8 \div 4 = 2$." But what happens when zero arises in this context? If you start with zero books, and divide them into four piles, how many are in each pile? The answer, it seems clear, is zero. Brahmagupta understood this much, but he encountered a problem when the problem was reversed.

What if eight books are divided into zero piles: how many books must there be in each pile? If this question seems nonsensical, that's because it is. Nor does it make any more sense when expressed purely arithmetically. We say that $8 \div 4 = 2$, because 2 is the unique number which, when multiplied by 4, gives 8. So, to calculate $8 \div 0$, we need a number which gives an answer of 8 when multiplied by 0. But there is no such number!

Faced with this problem, Brahmagupta made the bold decision to invent a whole raft of new numbers to fill this need. These were represented by $\frac{1}{0}$, $\frac{2}{0}$, and so on. Then, by definition, $0 \times \frac{8}{0} = 8$. This is not the approach taken today. Today's mathematicians stick to the initial observation that the question is nonsensical, and expressions such as $\frac{8}{0}$ are meaningless. So a fundamental law of modern mathematics is that you cannot divide any number by zero.

20 Algebra

BREAKTHROUGH AL-KHWARIZMI ESSENTIALLY BEGAN THE ABSTRACT STUDY OF EQUATIONS. HIS INTENTION WAS TO WRITE A PRACTICAL GUIDE TO ASSIST WITH CALCULATION; THE EFFECT WAS THE FOUNDING OF ALGEBRA.

DISCOVERER ABU JA'FAR MUHAMMAD IBN MUSA AL-KHWARIZMI (AD 780–850).

LEGACY AL-KHWARIZMI'S WORK ON QUADRATIC EQUATIONS CONTINUES TO BE TAUGHT TO SCHOOL STUDENTS AROUND THE WORLD. HIS WORK LAID THE FOUNDATIONS FOR LATER DEVELOPMENTS IN ARS MAGNA (SEE PAGE 101) AND BEYOND.

Algebra can be thought of as the science of solving equations. Al-Khwarizmi's book *Hisab al-jabr w'al-muqabala* (The Compendious Book on Calculation with Completion and Balancing) not only gave us the word "algebra," it also founded the mathematical subject with that name.

Al-Khwarizmi was a scholar at the House of Wisdom in ninth-century Baghdad, then a wealthy, bustling city and the intellectual capital of the Islamic world. The House of Wisdom was a library and center of research sponsored by the Caliph, the region's religious leader.

The House of Wisdom was especially concerned with translation. Scholarly works from around the world passed through its doors, to be translated from Persian, Sanskrit, Chinese and other languages into Arabic. Al-Khwarizmi had a particular responsibility for ancient Greek scientific and mathematical volumes. The works of Euclid (see page 45), among many others, are likely to have passed through his hands, forming the basis for his own mathematical investigations. Like the Library of Alexandria before it, the House of Wisdom was eventually destroyed by an invading force, in this case the catastrophic Siege of Baghdad by the Mongol army in 1258.

OPPOSITE When faced with an equation such as $4x^2 + 3x - 7 = 0$, a mathematician's immediate response is to search for solutions–values of x which make it true. Bahman Kalantari of Rutgers University realized that methods for finding such solutions also generate beautiful pictures, a technique he calls polynomiography.

The beginnings of algebra

The two major preoccupations of Hellenistic mathematicians were number theory and geometry. These two came together in the discovery of irrational numbers, or, as they were termed, "magnitudes." This threw

Greek mathematics into turbulence. Although mathematicians such as Euclid and Diophantus would go on to make wonderful progress, they were never quite at ease with the fundamental nature of these numbers.

In fact, Arabic mathematicians such as Al-Khwarizmi were somewhat wary of negative numbers. Yet the approach he took was simply not to worry unduly. The philosophical interpretation of different kinds of numbers, whether positive, negative, rational or irrational, was not actually that important. What really mattered was what you could do with them: addition, subtraction, multiplication and division. These four operations would allow all manner of practical problems to be solved, irrespective of any further significance the numbers may have.

To Al-Khwarizmi, the philosophical interpretation of different kinds of numbers was not that important. What really mattered was what you could do with them.

In *Hisab al-jabr w'al-muqabala*, Al-Khwarizmi provided the first serious analysis of equations and gave detailed instructions on how to solve them. This initiated algebra as a subject in its own right, as well as providing a valuable guide to the scientists and bureaucrats of his age, who used mathematical techniques to solve practical and financial problems.

Equations and unknown numbers

If some number has four added to it and the result is nine, what was the number? Al-Khwarizmi contemplated questions like this, always writing them out in prose. Nowadays, we prefer a symbolic approach, typically denoting the mystery number by the letter x. Then the problem becomes to find x, given that $x + 4 = 9$. This is an equation. Solving it entails finding the value of the number x.

Of course, in this case the answer is obvious. Yet Al-Khwarizmi's insight was that there is a definite procedure to follow which will yield the solutions to any equation of this type. Al-Khwarizmi described the procedure as completing (*al-jabr*) or balancing (*al-muqabala*) the equation. This is still the method taught to schoolchildren around the world. Essentially, it means that the same thing must be done to both sides of the equation in order to preserve its truth. Starting with $x + 4 = 9$, the irritant is the "+4." If that could be eliminated, then the result would be a clear assertion of the value of x. Al-Khwarizmi's balancing act was to subtract 4 from both sides of the equation. Subtracting 4 from the left would eliminate the rogue "+4." To keep the equation true, the same thing should be done to the right: $x = 9 - 4$, and so $x = 5$. Similarly, if an equation reads $2x = 12$ (this is short for $2 \times x = 12$), then Al-Khwarizmi's instructions are to divide both sides by 2, giving us $x = 12 \div 2 = 6$.

Often, these techniques need to be combined. Starting with $3x + 5 = 11$, first 5 should be subtracted from both sides, giving $3x = 6$, and then both sides must be divided by 3, to get $x = 2$. The value of Al-Khwarizmi's writing in this area was in carefully laying out techniques that mathematicians had quietly relied upon for centuries. But he did not stop there.

ABOVE A page from Al Khwarizmi's *Hisab al-jabr w'al-muqabala*, the book that gave us the word "algebra" and laid down rules for solving quadratic equations.

Quadratic equations

Equations become dramatically more complicated when the unknown number x may be multiplied by itself. It was with quadratic equations such as this that Al-Khwarizmi made his greatest breakthrough. A simple example is $x^2 = 16$ (where x^2 stands for $x \times x$. How can we find x? In this case, we need a number that squares (that is to say, multiplies by itself) to give 16. Nowadays, we call this the square root of 16, written as $\sqrt{16}$. A little thought is enough to come up with its value: 4. (In fact, it is important to notice that according to the rules of negative numbers (see page 18), -4 is also a valid answer.)

Yet some quadratic equations are trickier, such as $x^2 - 5x + 6 = 0$. There is no obvious procedure to derive the answer here, even with the aid of a modern calculator. Though written out in words, Al-Khwarizmi's method amounts to the formula used in modern classrooms around the world. The two solutions to any equation of the form $ax^2 + bx + c = 0$ (where a, b and c are some fixed numbers) are given by the formula

$$x = \frac{-b \pm \sqrt{b^2 - 4ac}}{2a}.$$

In the above example, this gives the answers $x = 2$ or $x = 3$.

The effort to extend Al-Khwarizmi's work to even more complex equations would be a major driving force in algebra in both the Islamic world and Europe, culminating in Girolamo Cardano's *Ars Magna* of 1545 (see page 101).

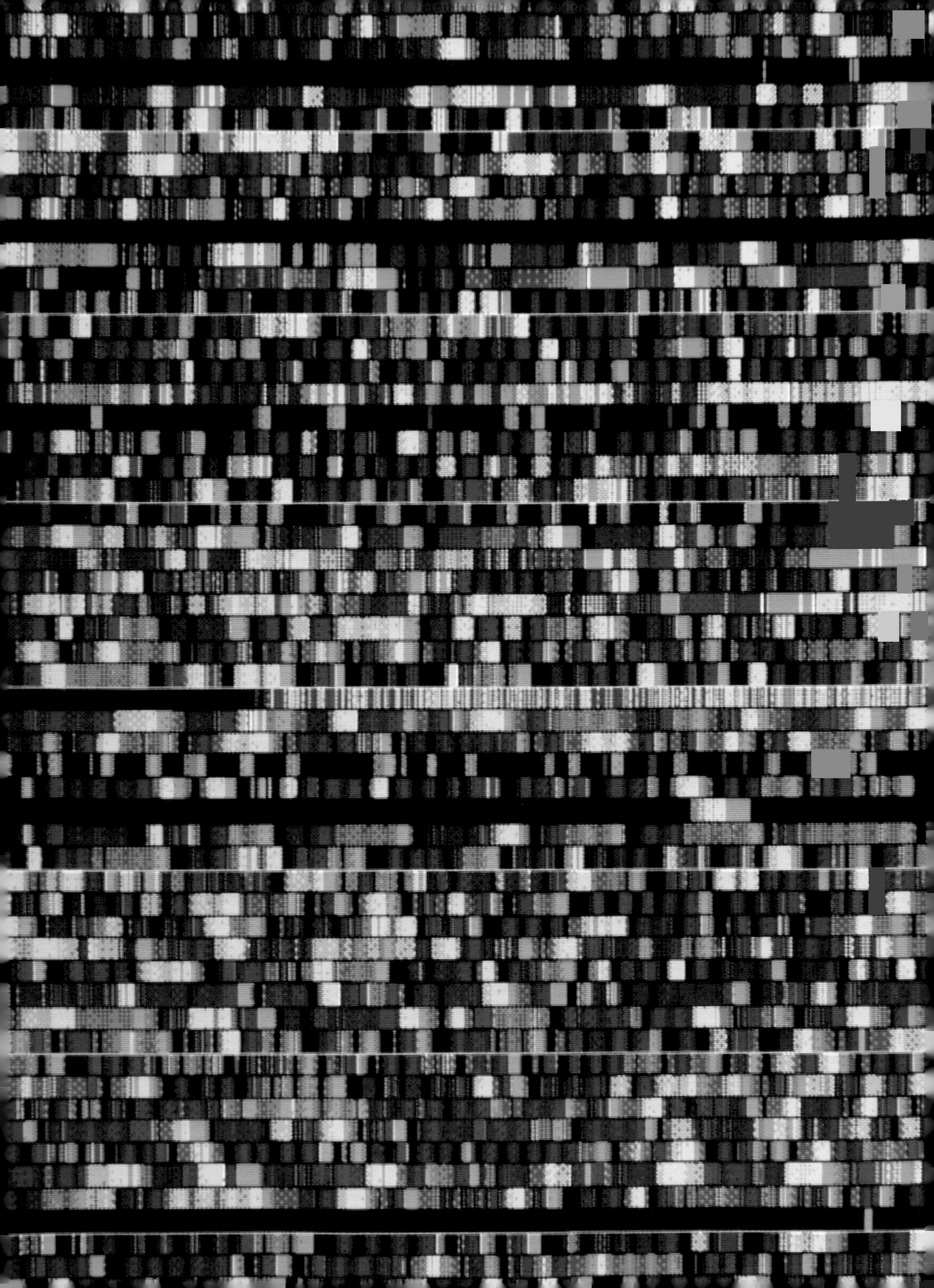

21 Combinatorics

BREAKTHROUGH THE SCIENCE OF COUNTING THE NUMBER OF ARRANGEMENTS OR SUBDIVISIONS OF A COLLECTION. A MILESTONE WAS THE PROOF OF THE BINOMIAL THEOREM.

DISCOVERER PINGALA (C. 200 BC), ABU BEKR IBN MUHAMMAD IBN AL-HUSAYN AL-KARAJI (AD 953–1029), BLAISE PASCAL (1623–62).

LEGACY PASCAL'S TRIANGLE AND THE BINOMIAL THEOREM ARE USED BY MATHEMATICIANS AND SCIENTISTS DAILY. MEANWHILE, MORE COMPLICATED COMBINATORIAL QUESTIONS CONTINUE TO PERPLEX MODERN THINKERS.

Combinatorics is the analysis of the number of ways a collection of objects can be arranged or subdivided. Combinatorial questions are useful, and often difficult, in their own right, occurring throughout mathematics and science. Through the famous binomial theorem, combinatorics is also central to an understanding of algebra.

If there are five people in a room and three are to be selected to play a game, how many possible choices are there? The answer is certainly not immediately obvious, but combinatorial questions like this appear in a great variety of situations, throughout science and in everyday life. Within mathematics, they are particularly important in the study of probability theory.

Factorials

The fundamental insight into the question involves what modern mathematicians know as factorials. If the five people in a room are going to stand in a line, how many different ways might they do it? Well, there are five choices for the person to stand at the front. Once that has been decided, there are four possible choices for the next person in the line, and then three, and so on. Once the first four positions have been filled, there is only one person left to stand at the back. This suggests that the number is $5 \times 4 \times 3 \times 2 \times 1$. Since the early 19th century, mathematicians have used an exclamation mark to represent this: 5! (This is read as "five factorial.") In this case, the answer comes out as $5! = 120$, which starts to illustrate that the factorial is an extremely fast-growing function: 10! is around

OPPOSITE A section from the human genome, with the colors yellow, green, red and blue representing different base pairs. To read it requires careful combinatorial analysis; with 3 billion base pairs in the genome, the number of possible subsequences grows out of control very quickly, making searching it a challenge.

3.6 million, while 60! (the number of different ways that 60 people can line up) exceeds the number of atoms in the observable universe. This illustrates an important moral: experimentation alone is not adequate to settle many combinatorial questions. A more abstract approach was required, and these topics began to be investigated by Indian mathematicians in around 800 BC.

Permutations and combinations

Factorials can be used to analyze more complicated situations. For instance, suppose that in a room of five people, three stand in a line. The number of ways this can happen is given by $5 \times 4 \times 3$. In terms of factorials, it is illuminating to write it as $\frac{5!}{2!}$, which is $\frac{5!}{(5-3)!}$. This illustrates the general principle at work. In modern terminology, the number of possible *permutations* of r people from a total group of n is expressed by $\frac{n!}{(n-r)!}$.

The number of different ways that 60 people can line up exceeds the number of atoms in the observable universe.

This does not answer the game players' question, however. Two lines of people may be different, even if they contain the same people; the exact ordering matters. If we are only interested in picking a collection of people, and orderings do not matter, it is not a permutation which is needed but a *combination*. The number of combinations of r people from a group of n is smaller than the number of permutations, since the same group of people can be ordered in different ways. In fact, $r!$ different orderings are possible, which suggests that the number of combinations should be the number of permutations divided by $r!$ This comes out as:

$$\frac{n!}{(n-r)! \times r!}$$

In the case of three people being selected from five, the answer is:

$$\frac{5!}{2! \times 3!} = 10$$

These combinations can be read off one of the most famous objects in mathematics, Pascal's triangle.

Pascal's triangle

Begin with a single 1. Below it write two more 1s. Each subsequent row begins and ends with a 1, and each number in between is calculated as the total of the two numbers above. This simple idea can be extended for as many rows as required, producing a beautiful pattern.

This triangle is named after the 17th-century French scholar Blaise Pascal, but it was known to several earlier thinkers. Its first recorded occurrence is in the work of Pingala, an Indian literary theorist who wrote his *Chandahsutra* around 200 BC. Yet its full implications were not investigated until much later. It was rediscovered in 1303 by the Chinese mathematician Szu Yuen Yu Chien.

ABOVE Many patterns lurk within Pascal's triangle. The diagonals, after the initial line of 1s, list the whole numbers in order, followed by the triangular numbers, tetrahedral numbers, followed by their higher-dimensional equivalents.

The wonderful thing about Pascal's triangle is that combinations can be read straight off it. The row 1, 5, 10, 10, 5, 1, for instance, lists the number of ways of selecting objects from a group of five. There are exactly five ways to select one object, then ten ways to select two, also ten combinations of three, five possible combinations of four, and just one way to choose all five. (The 1 at the beginning of the row can be interpreted as saying there is exactly one way to select zero objects from five.)

Many other beautiful patterns lurk within Pascal's triangle. One example is the class of triangular numbers which describe the number of balls that make up an equilateral triangle. These numbers lie along the third diagonal in Pascal's triangle: 1, 3, 6, 10, 15, and so on.

The binomial theorem

Pascal's triangle and the theory of combinations are not only useful for counting objects. They also play a critical role in algebra. Algebraists are often faced with expressions in which several brackets are multiplied together. For instance, the expression $(1 + x)^2$ means $(1 + x) \times (1 + x)$. Working this out fully gives $1 + 2x + x^2$, while

$$(1 + x)^5 = 1 + 5x + 10x^2 + 10x^3 + 5x^4 + x^5.$$

The numbers involved here are taken straight from a row of Pascal's triangle. This famous fact has saved mathematicians many, many hours of tedious calculation over the centuries. Evaluating $(1 + x)^5$ by hand would require 32 individual steps—running a high risk of error. But thanks to this binomial theorem, the answer can be written down immediately. It was first proved by the Arabic mathematician Al-Karaji, around the year 1000.

22 The Fibonacci sequence

BREAKTHROUGH THE GOLDEN SECTION, STUDIED BY THE PYTHAGOREANS, AND THE FIBONACCI SEQUENCE, FIRST CONTEMPLATED IN ANCIENT INDIA, WERE FOUND TO BE INTIMATELY RELATED.

DISCOVERER PYTHAGORAS (570–475 BC), PINGALA (C. 200 BC), LEONARDO FIBONACCI (1170–1250), JOHANNES KEPLER (1571–1630), JACQUES BINET (1786–1856).

LEGACY THE FAME OF THE FIBONACCI SEQUENCE AND THE GOLDEN SECTION NOW EXTENDS BEYOND MATHEMATICS AND INTO POPULAR CULTURE.

Mathematicians have always taken delight in the many beautiful objects that arise in geometry. But over time, one topic in particular has come to symbolize the ideal of geometrical perfection, and the phenomenon of the same mathematics appearing in strikingly different scenarios.

The symbol of the Pythagorean sect was a pentagram or five-pointed star, a shape they endowed with a deep mystical significance. Within the edges of the pentagram lurks a particular number, which would become synonymous with the idea of geometrical perfection.

The pentagram and the golden section

The point where one side of a pentagram crosses another divides each side into two parts. These two parts are not the same length: one is longer than the other. Nevertheless, there is a beautiful kind of symmetry within the numbers of the situation, which emerges when you look at the ratios of the various segments. It turns out that the whole side lies in the same ratio to the long segment as the long segment does to the short one. Calling the long segment a and the short one b, then the whole side has length $a + b$. So the statement about their ratios is that $\frac{a+b}{a} = \frac{a}{b}$.

This information is enough to calculate what the single number is that describes both these fractions. Today we call it the golden section. Usually denoted by the Greek letter ϕ (phi), it has an exact value of $\phi = \frac{1+\sqrt{5}}{2}$, and is an irrational (see page 29) though not transcendental (see page 197) number. Its approximate value is 1.618. In algebraic

OPPOSITE The number of spirals in a sunflower head often comes out as a Fibonacci number. In this case, counting the spirals reveals 21 in one direction, and 33 in the other. The same phenomenon occurs in other contexts, such as the spirals of fruitlets on a pineapple.

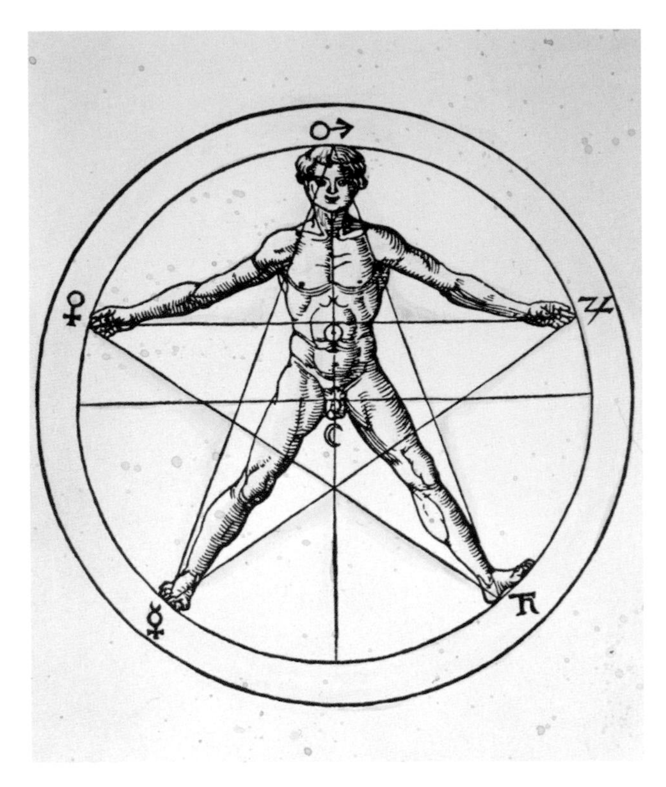

ABOVE The occultist Heinrich Cornelius Agrippa produced this picture of a man inscribed in a pentagram in the 16th century, one of a long line of attempts to connect the golden section to the human form. Pentagrams and the golden section have historically been a topic of considerable superstition.

terms, ϕ is interesting for satisfying the equation $1 + \phi = \phi^2$.

The Pythagoreans were not the only people to study the golden section. It is discussed in Euclid's *Elements*, too, where it is named the "extreme and mean ratio"; Euclid shows how to construct the golden section as a ruler-and-compass construction (see page 45).

The golden section in the arts

The artist, sculptor and architect Phidias, who lived in the fifth century BC, is said to be the first to have recognized the aesthetic possibilities of the golden section. Indeed, it is from his name that we get the symbol ϕ. Unfortunately, none of Phidias' works have survived so far as we know, though his 12-meter-high statue of Zeus at Olympia was considered one of the wonders of the ancient world.

Since Phidias' time, several artists have taken a serious interest in ϕ, including Leonardo da Vinci and Salvador Dalí. It is often claimed that a golden rectangle, that is to say, a rectangle whose sides are in a ratio of ϕ, is among the most aesthetically pleasing shapes. But attempts to test this psychologically have never been completely convincing.

A golden rectangle does have a very beautiful mathematical property, however. If you remove from it a square whose width is that of the shorter side, what is left is a smaller rectangle. But, unexpectedly, this new shape is also a golden rectangle. Repeating this process of sectioning off a square to leave a smaller rectangle produces a very elegant pattern of nested golden rectangles of ever decreasing sizes. By connecting the corners of these, a golden spiral is formed, which is a very close approximation to a logarithmic spiral (see page 142).

Fibonacci's sequence

The golden section is closely related to another, even more famous piece of mathematics, whose origins lie in India. Leonardo da Pisa, better known as Fibonacci, spent his youth traveling around the Arabic countries of the Mediterranean basin. The most important outcome of his travels was the introduction of the Hindu–Arabic numeral system into Europe (see page 73). Another of the finds presented in his book *Liber Abaci* was a certain sequence, first contemplated by the Indian literary theorist Pingala a thousand years before when analyzing possible meters for poems. The Fibonacci sequence, as it became known, runs as follows:

$$1, 1, 2, 3, 5, 8, 13, 21, 33, 54, \ldots$$

The pattern here is that each subsequent number is produced by adding up the preceding two.

A golden rectangle does have a very beautiful mathematical property, however. If you remove from it a square whose width is that of the shorter side, what is left is a smaller rectangle. But, unexpectedly, this new shape is also a golden rectangle.

This sequence has become emblematic of mathematical patterns emerging in different places in nature. Counting the petals on several species of flower often reveals a Fibonacci number (meaning a number from the Fibonacci sequence), as do the number of spirals on certain types of cacti, pine cones, pineapples and several other types of plant.

Binet's formula

The relationship between the Fibonacci sequence and the golden section is revealed by looking at the fractions of successive members of the sequence:

$$\frac{1}{1}, \frac{2}{1}, \frac{3}{2}, \frac{5}{3}, \ldots$$

It was shown by Johannes Kepler that, while the original Fibonacci sequence simply grows without limit, this new sequence gets ever closer to a fixed number. What is more, this limit is none other than ϕ.

For this reason, ϕ appears in the formula for calculating the Fibonacci numbers. If you want to calculate the 50th Fibonacci number, say, without working through the intermediate 49, this formula will do it. It was known to Leonard Euler and Daniel Bernoulli, among others, but takes its name from Jacques Binet, who discovered it in 1843. It says that the nth term of the Fibonacci sequence is:

$$\frac{1}{\sqrt{5}}(\phi^n + (1 - \phi)^n)$$

For $n = 50$, the answer comes out as 12,586,269,025.

23 The harmonic series

BREAKTHROUGH NICOLAS ORESME WAS THE FIRST TO RECOGNIZE THAT ADDING TOGETHER INFINITELY MANY NUMBERS IS A VERY DELICATE MATTER, WHICH CAN PRODUCE SOME SURPRISING RESULTS.

DISCOVERER NICOLE ORESME (1323–82), PIETRO MENGOLI (1626–86), LEONHARD EULER (1707–83).

LEGACY THIS WAS THE FIRST DEEP FACT IN THE THEORY OF INFINITE SERIES, WHICH REMAINS ONE OF THE SUBTLEST TOPICS IN MATHEMATICS.

What happens when you try to add together infinitely many numbers? This question was addressed by Nicole Oresme in the 14th century, and again by Leonhard Euler in the 18th century, with spectacular results.

Like many scholars of the Middle Ages, Nicole Oresme was not an academic specialist, but he studied almost everything he could. A philosopher and a physicist, he was also among the most influential of the early economists, as well as an outspoken skeptic of astrology and the occult. Almost two hundred years before Copernicus, Oresme realized that the shifting night skies could be explained by a rotating Earth (though he eventually conformed to the prevailing view of the times that it was the heavens, and not the Earth, that turned).

Buried within his manuscripts was the first proof of a great mathematical theorem, which was subsequently lost for several hundred years before being rediscovered by 17th-century mathematicians such Pietro Mengoli. Oresme's breakthrough concerned addition, one of the most ancient and familiar procedures in mathematics. But, Oresme asked, what happens when one tries to add up infinitely many numbers, instead of just a small collection?

OPPOSITE The numbers of the harmonic series are central not only to mathematics but to music, where they represent the harmonics of a base note. Overlaying different harmonics produces sounds of different timbres, giving different instruments their characteristic sounds.

Convergent and divergent series

In many cases, the question is meaningless. If you try to work out 1 + 2 + 3 + 4 + 5 + ..., the sum just grows larger and larger without limit. After 14 steps, the total exceeds 100. After 45, it has surpassed

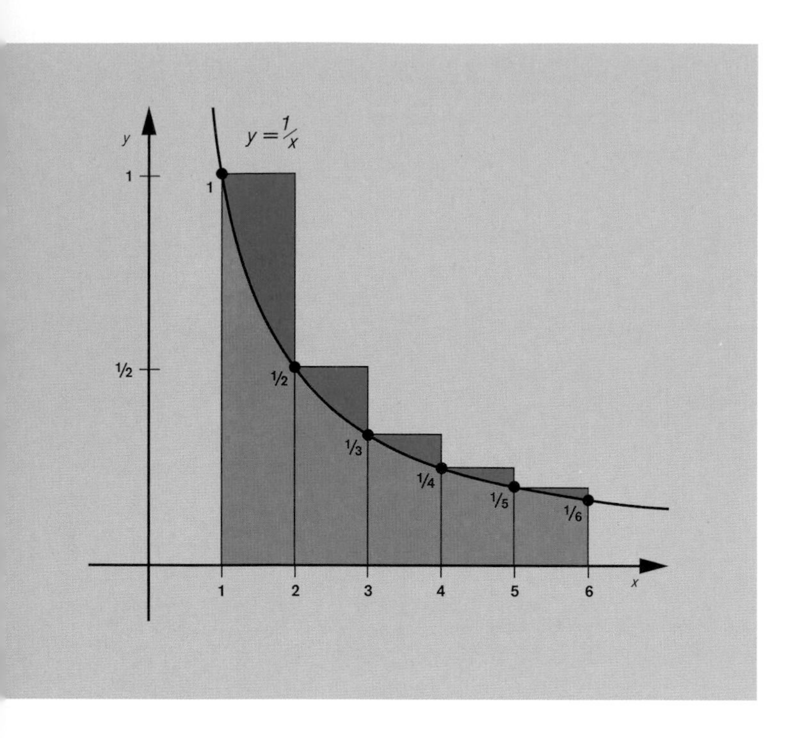

ABOVE The red region shows the area under the curve $y = \frac{1}{x}$, while the rectangles give the harmonic series. Although both areas are infinite, the difference between them (in blue) is a fixed, finite number known as the Euler-Mascheroni constant, with a value of around 0.557.

1000. As more and more numbers are added on, the total outgrows any boundary you might try to impose. This is what is known as a divergent series. Another example is 1 + 1 + 1 + 1 + 1 + ..., which simply counts. After one step, it has reached 1. After two steps, 2, and after 100 steps it has reached 100, and so on.

In these examples, trying to add together infinitely many numbers is a lost cause. In many cases this is true, but in some situations the picture can be very different. For instance, if you work out 0.9 + 0.09 + 0.009 + 0.0009 + ..., then the sequence of totals reads 0.9, 0.99, 0.999, 0.9999, ... As more and more numbers are added on, the total gets ever closer to 1, but never reaches or surpasses it. This is an example of a convergent series.

The harmonic series

The obvious difference between these examples is that the individual terms in the convergent example become smaller and smaller as the series progresses. It is not hard to see that this must be a necessary requirement for convergence. So, Oresme pondered, what happens to the series of fractions which begins $1 + \frac{1}{2} + \frac{1}{3} + \frac{1}{4} + \frac{1}{5} + \ldots$? This is called the harmonic series, because fractions like $\frac{1}{2}, \frac{1}{3}, \frac{1}{4}, \ldots$ are central to the mathematical theory of music. (Relative to a given root note, notes with frequencies expressed by these fractions are the harmonics: the octave, perfect fifth, perfect fourth, and so on.)

At first sight, it looks as if the harmonics should form the type of series that might close in on some fixed limit. After all, each term is smaller than the last. By the time one is adding on millionths, it is hard to imagine that the total will grow much further. But Nicole Oresme provided a simple and dramatic demonstration that this is illusory. In fact, the harmonic series grows and grows without limit. Contrary to every appearance, the harmonic series is divergent.

What made this discovery so surprising is that, after the first few terms, the series grows so slowly as to be almost imperceptible. For the total to reach just 10, over 12,000 terms have to be added together. To reach a total of 50 requires over 10^{20} terms. (That's a 1 with 20 zeroes after it.) Adding on a new term every second, this would take longer than the age of the Universe. Nevertheless, as Oresme showed, the series total will eventually exceed any number you can name, even if it takes an unimaginably long time to get there.

Contrary to every appearance, the harmonic series is divergent. What made this discovery so surprising is that, after the first few terms, the series grows so slowly as to be almost imperceptible.

Nicole Oresme had shown that a common-sense approach to infinite series is not enough, since their behavior can be far from obvious. When this insight was rediscovered in the 17th century, it opened up more questions than it answered. For example, what happens if the series is restricted to prime numbers: $1+\frac{1}{2}+\frac{1}{3}+\frac{1}{5}+\frac{1}{7}+\frac{1}{11}+ \ldots$? Or square numbers: $1+\frac{1}{4}+\frac{1}{9}+\frac{1}{16}+\frac{1}{25}+\ldots$ (that is, $1+\frac{1}{2\times2}+\frac{1}{3\times3}+\frac{1}{4\times4}+\frac{1}{5\times5}+\ldots$)?

The Basel problem

These questions fell to the greatest mathematician of the age: Leonhard Euler. In 1737, he proved that the prime series follows the same pattern as the harmonic series, and is divergent (though even more slowly).

The square series, however, is different. Pietro Mengoli, who had independently rediscovered Oresme's theorem, realized that the square series does not escape to infinity, but closes in on a certain finite number. But he was unable to answer the obvious question: what is this mysterious limiting value? In 1644, he threw this out as a challenge to his contemporaries in the mathematical world, not even the most brilliant of whom were able to meet it. This conundrum became known as the Basel problem. Quite simply, it asked: what is the value of $1+\frac{1}{4}+\frac{1}{9}+\frac{1}{16}+\frac{1}{25}+\ldots$?

Euler finally revealed the answer in 1735, employing new and imaginative techniques. The solution was very far from expected, proving that the limit of the series is the number $\frac{\pi^2}{6}$ (that is, $\frac{\pi\times\pi}{6}$ which is around 1.645).

Questions about infinite series are certainly delightful puzzles, but they are much more than that. The subtle and sophisticated methods that flowed from Oresme's breakthrough developed into the modern subject known as mathematical analysis. A central question in the area is the Riemann hypothesis (see page 206), where the Riemann zeta function is a generalization of the series of both Oresme and Euler.

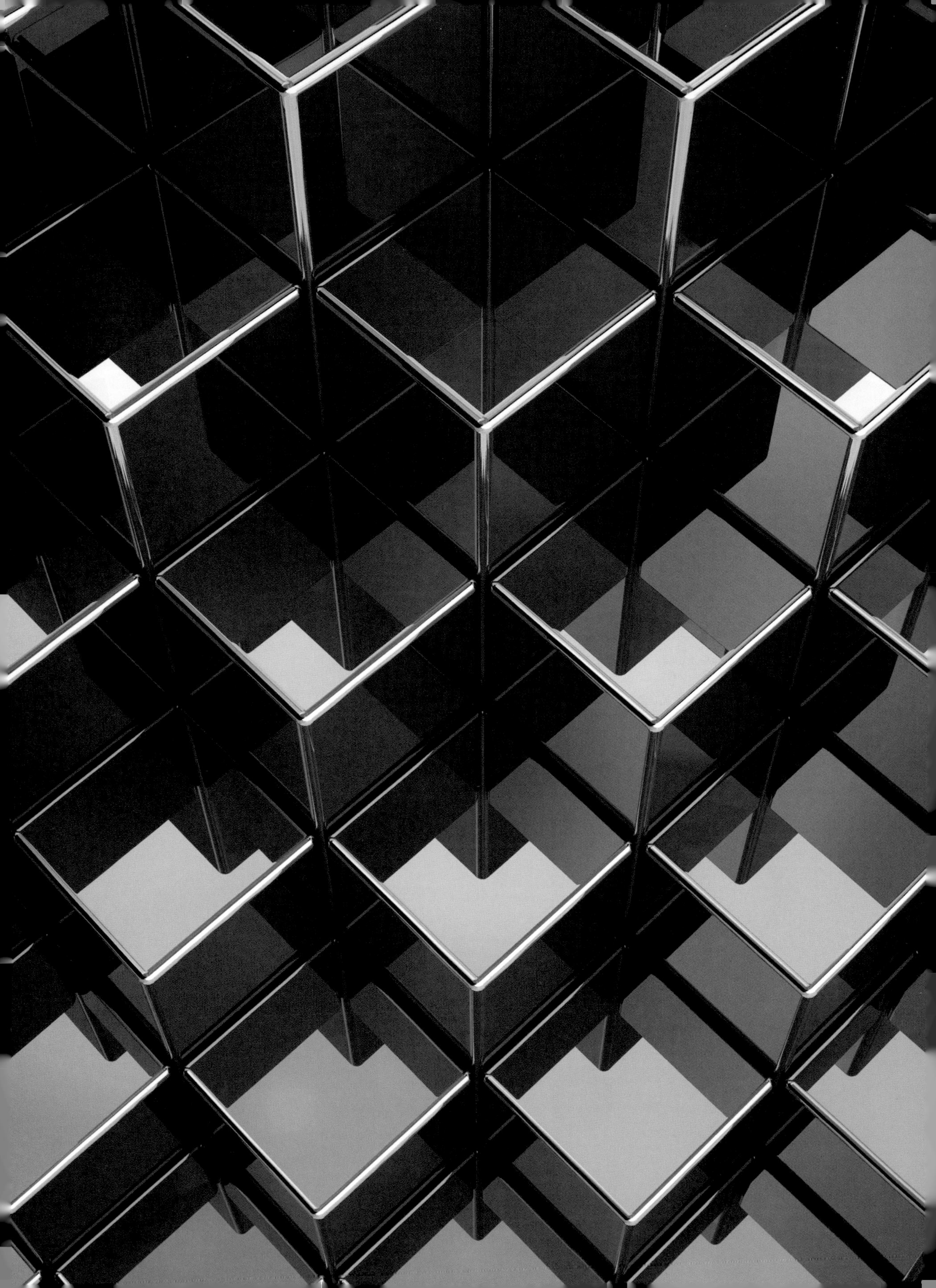

24 Cubic and quartic equations

BREAKTHROUGH CARDANO'S BOOK *ARS MAGNA* CONTAINED ADVANCED TECHNIQUES FOR SOLVING SOPHISTICATED EQUATIONS KNOWN AS CUBICS AND QUARTICS.

DISCOVERER GIROLAMO CARDANO (1501–76), LODOVICO FERRARI (1522–64), SCIPIONE DEL FERRO (1465–1526), NICCOLÒ FONTANA (1500–57).

LEGACY *ARS MAGNA* MARKED THE END OF AN ERA OF ALGEBRAIC INVESTIGATION. ITS METHODS (NOW LARGELY COMPUTERIZED) REMAIN INVALUABLE TODAY.

One of mathematicians' central preoccupations over the centuries has been solving equations. It was a particular obsession in 16th-century Italy. The great challenges of the era were to find techniques for solving especially fearsome types of equation known as "cubic" and "quartic." These were eventually conquered in Cardano's historic book, *Ars Magna*.

As progress was made, equation solving assumed the role of intellectual gladiatorial combat among the top theorists of the day. The likes of Girolamo Cardano, Lodovico Ferrari, Scipione del Ferro and Niccolò Fontana would challenge each other to public contests, staking everything on lists of fiendish puzzles for the other to solve. Terrible disputes arose when they caught glimpses of each other's work, or made public techniques they had sworn to keep secret. It was an intellectual crucible in which reputations were made and destroyed. The mathematical legacy of this extraordinary chapter in mathematics was laid out in *Ars Magna* in 1545. This provided, for the first time, definitive accounts of cubic and quartic equations.

OPPOSITE This picture of cubes is not a photograph, but was created on a computer using ray-tracing software. This technique typically requires numerous cubic equations to be solved; even today this is done essentially using the method described in Cardano's *Ars Magna*.

Equations and solutions

Solving an equation means identifying an unknown number from some indirect information about it. For instance, if a certain number when doubled is 6, then it is obvious that the original number must have been 3. In symbolic form, this equation would be written as $2 \times x = 6$, using the letter x to represent the unknown number. In fact, it is customary in algebra to omit the times sign: $2x = 6$. The solution to this equation, then, is $x = 3$.

In trickier cases, the unknown number undergoes more than just one procedure. If a number is tripled and then added to 4 to give 25, what was it? Questions like this can be answered quite simply by reversing the operations which are done. Begin by subtracting 4 from 25 to get 21. Then, instead of multiplying by 3, divide by 3, to get an answer of 7. This equation might be written as $3x + 4 = 25$ with solution $x = 7$.

Despite major progress by Omar Khayyam in around 1100, these equations still held enough ammunition for the mathematicians of the Italian Renaissance to do battle.

It has been understood for millennia that any equation that involves only adding, subtracting, multiplying and dividing the mystery number by other known quantities can, with practice, easily be solved this way. All that is required is to undo each of the steps in turn until the answer is revealed. But things became altogether more complicated when the unknown number was also multiplied by itself.

If a number is multiplied by itself and the answer is 9, what was the number? We might write this as $x^2 = 9$ (where x^2 is shorthand for $x \times x$). This is not too difficult to solve: obviously, the number 3 satisfies this condition. But a deeper understanding of mathematics produces a surprise: this equation admits a second solution. According to the theory of negative numbers, it is also true that $-3 \times -3 = 9$. So -3 is an equally valid answer. This illustrates the general principle that an equation involving x^2 typically has two separate solutions.

Using the notation of powers, the last equation would appear as $x^2 = 9$ with two solutions: $x = 3$ and $x = -3$. A more problematic equation is $x^2 - 5x + 6 = 0$. Here, the unknown number is first multiplied by itself, then 5 times the original number is subtracted and 6 is added, leaving zero. The trouble is that it is far from obvious how to "undo" this sequence of steps. Equations like this, which involve x^2 terms, are called quadratic equations. In this case, the two solutions are 2 and 3. Once you know this, it is easy to check that these are indeed correct. But how to find the solutions in the first place?

It was the ninth-century mathematician Abu Ja'far Muhammad ibn Musa Al-Khwarizmi, famous for his book on algebra, who first discovered a method for solving any quadratic equation (see page 85). Since then, his technique has been imprinted on the minds of generations of school students. It says that if you begin with the equation

$$ax^2 + bx + c = 0$$

where a, b and c are any three numbers, then the solutions are obtained via the equation

$$x = \frac{-b \pm \sqrt{b^2 - 4ac}}{2a}.$$

(The two answers are obtained by interpreting "±" first as "+" and then as "−.")

The conquest of the cubic and quartic

Although the quadratic formula continues to frighten students on first meeting, once you get used to it, it is a very friendly creature. In discovering it, Al Khwarizmi effectively closed the book on quadratic equations. Just by applying his formula, every quadratic equation can now be solved. Cubic equations, however, which additionally involve the term x^3 (that is, $x \times x \times x$) and typically have three different solutions, were another matter. And quartic equations, with their x^4 terms and four solutions, were even worse. Despite major progress by Omar Khayyam in around 1100, these equations still held enough ammunition for the mathematicians of the Italian Renaissance to do battle. Once found, the final results were hugely more complex than Al Khwarizmi's solution to the quadratic.

BELOW The red line shows a cubic curve and the blue one a quartic. Solving these equations involves finding the points where they cross the horizontal axis. Typically a cubic equation will have three solutions, and a quartic four.

In *Ars Magna*, Girolamo Cardano presented complete techniques for solving both forms of equation. He attributed the bulk of the work on the cubic to Niccolò Fontana (known as Tartaglia, meaning "the stammerer") and Scipione del Ferro, while it was Ludovico Ferrari who mastered the quartic.

The publication of *Ars Magna* was a momentous event in the history of algebra. But it by no means marked the end of mathematicians' fascination in equation-solving. The next challenge was the quintic, with its x^5 term and five solutions. But here a major surprise lay in store, which would not be revealed until 1820 (see page 178).

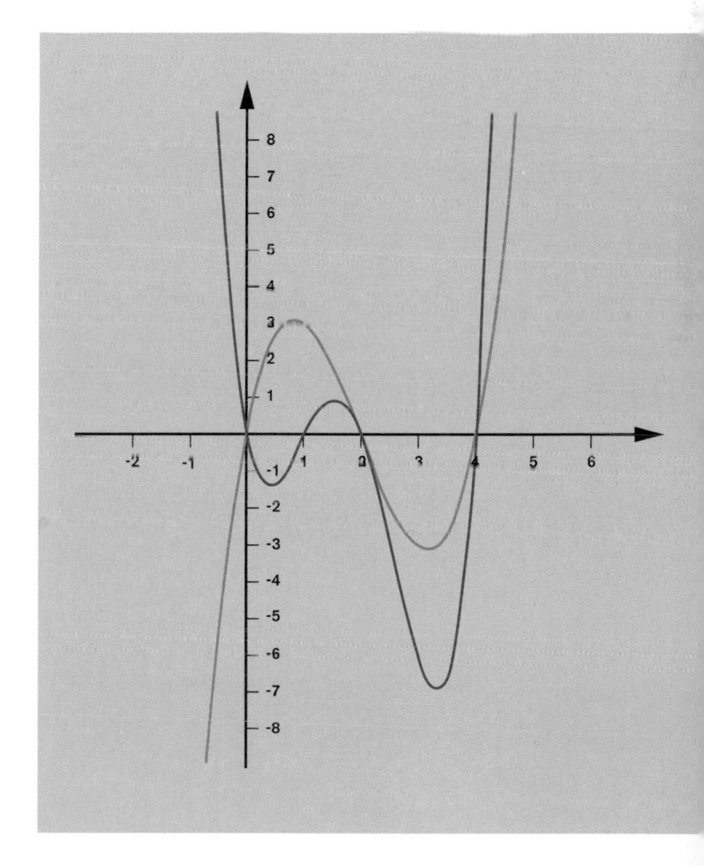

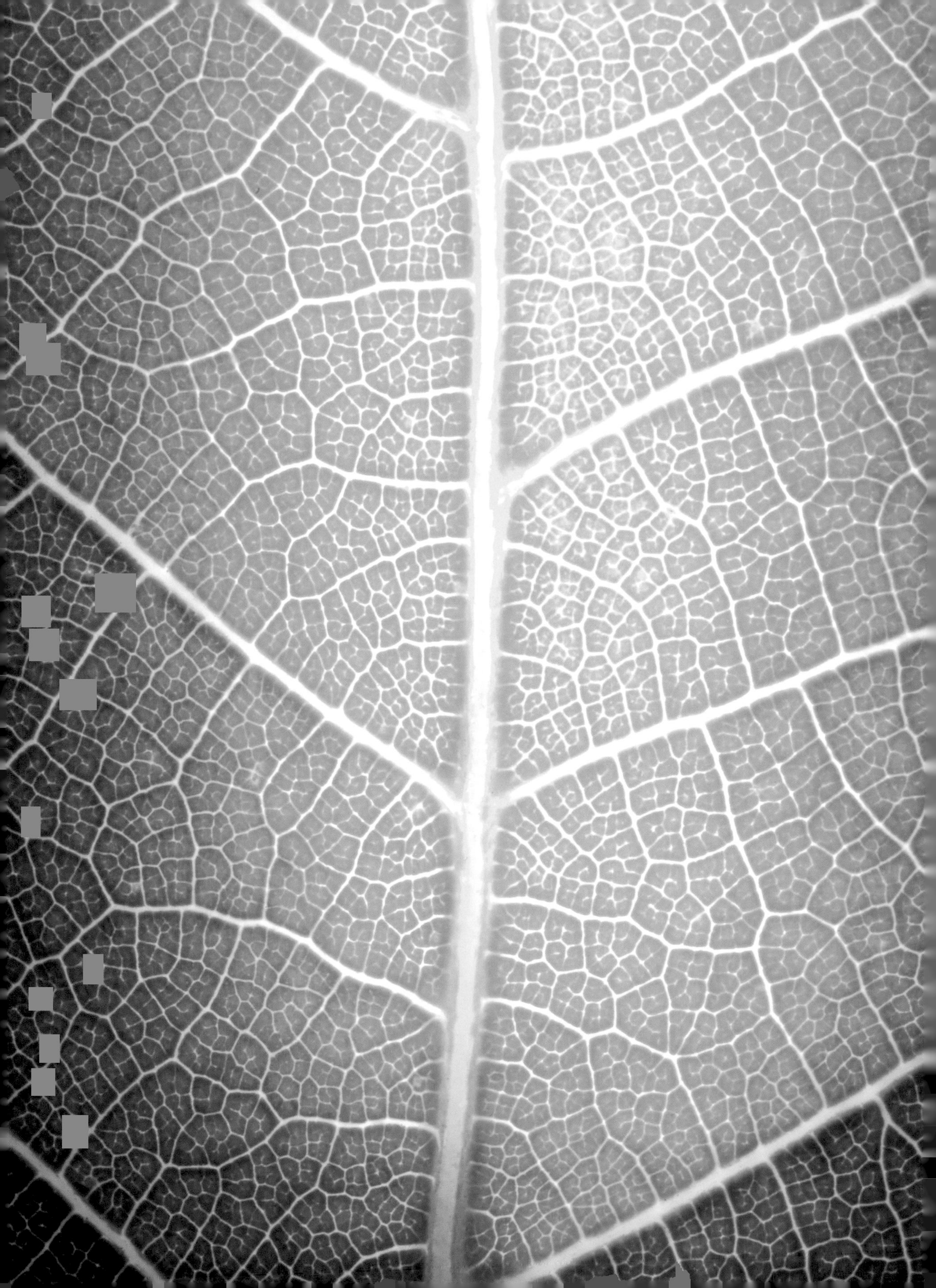

25 The complex numbers

BREAKTHROUGH BOMBELLI WROTE THE RULES FOR AN EXTENDED NUMBER SYSTEM THAT INCLUDED SO-CALLED IMAGINARY NUMBERS.

DISCOVERER RAFAEL BOMBELLI (1526–72).

LEGACY THE COMPLEX NUMBERS FORM THE PRINCIPAL BACKDROP FOR ALMOST ALL MODERN BRANCHES OF MATHEMATICS, AND HAVE PROVED PARTICULARLY SIGNIFICANT TO UNDERSTANDING THE PHYSICS OF QUANTUM MECHANICS.

At certain moments in history, mathematicians have been forced to completely re-evaluate their concept of number. One such juncture arrived in the 16th century, when researchers suddenly found themselves faced with "imaginary numbers" in their formulae. It was Rafael Bombelli who grasped this nettle, turning the problem into a magnificent breakthrough as he wrote the rules for an extended number system known as the complex numbers. Today, Bombelli's system remains the backdrop against which most modern mathematics is carried out.

A previous extension of the number system was the incorporation of negative numbers (see page 81). In fact, negative numbers already contain the seeds of the more abstract entities known as imaginary numbers. The clue lay in negative multiplication.

Arithmetic of negative numbers

It is no surprise that multiplying a negative number by a positive number should yield a negative result: $3 \times (-2) = -6$, for instance. But it is a common mistake to believe that multiplying two negative numbers together should also produce a negative answer. Not so, that will always give a positive result: $(-3) \times (-2) = 6$, for example. This has an important consequence when numbers are multiplied by themselves. Of course, a positive number multiplied by itself always produces a positive result: $2 \times 2 = 4$. At the same time, a negative number multiplied by itself also produces a positive result: $(-2) \times (-2) = 4$. The upshot was this:

OPPOSITE The process of photosynthesis, by which plants convert carbon dioxide into oxygen, relies on quantum effects at the atomic level. To describe such processes precisely requires grappling with complex numbers.

there was no number which yielded a negative result when multiplied by itself. If in the course of their research a mathematician needed a number which multiplied by itself to produce -1, say, then they were stuck. The expression $\sqrt{-1}$ corresponded to no known number. To put it another way, the equation $x \times x = -1$ (or $x^2 = -1$) seemed to have no solutions.

Bombelli's *L'Algebra*

In most situations, the absence of solutions to equations like $x^2 = -1$ did not cause a problem. It was just a fact of life, and mathematicians managed to cope perfectly well with it for hundreds of years. But in the algebraic hothouse of 17th-century Italy, it began to present a serious inconvenience. While researchers such as Girolomo Cardano were struggling to find the solutions to difficult types of equations known as cubic and quartic (page 101), they found expressions like $\sqrt{-1}$ appearing in their work with ever greater frequency. When this happened, these algebraists faced a dilemma. The obvious course of action was to give up the calculation; after all, it had seemingly degenerated into meaninglessness. But some found that if they persevered, the problems would sometimes resolve themselves. A little later, they might find an expression like $\sqrt{-1} \times \sqrt{-1}$, which could then be replaced with -1. Tellingly, although the line of argument had apparently passed through some realm of unreality, the answers that emerged at the end were completely correct.

Descartes was not the only person to dislike imaginary numbers. Cardano described working with them as "mental torture"–surely a sentiment echoed by many mathematical students in the subsequent centuries.

The breakthrough came in 1572, when Rafael Bombelli published his book *L'Algebra*. It contained the rules for a newly extended number system, which included quantities such as $\sqrt{-1}$. Later, René Descartes derisively termed $\sqrt{-1}$ an "imaginary" number, and regrettably the name stuck. Descartes was not the only person to dislike imaginary numbers. Cardano described working with them as "mental torture"—surely a sentiment echoed by many mathematical students in the subsequent centuries. Despite these reservations, mathematicians of this era gradually came to accept that Bombelli's rules were robust, and they became bolder at working with quantities like $\sqrt{-1}$. Yet this was done without fully recognizing them as genuine numbers.

i–the imaginary unit

It was not until one of history's greatest mathematicians embraced it that Bombelli's extended number system fully came in from the cold. In the 18th century, Leonhard Euler gave $\sqrt{-1}$ the name by which it is still known: i or the imaginary unit. Other imaginary numbers are multiples of i: objects like $2i$, $-3i$ and $\frac{2}{3}i$. But Bombelli's system does not only consist of real numbers (such as 2, -1 and π) and imaginary numbers. It also includes the result of combining the two: $2 + 2i$, $-1 -3i$ and $\pi + \frac{2}{3}i$. Taken together, this system of complex numbers provides the backdrop for most modern mathematics.

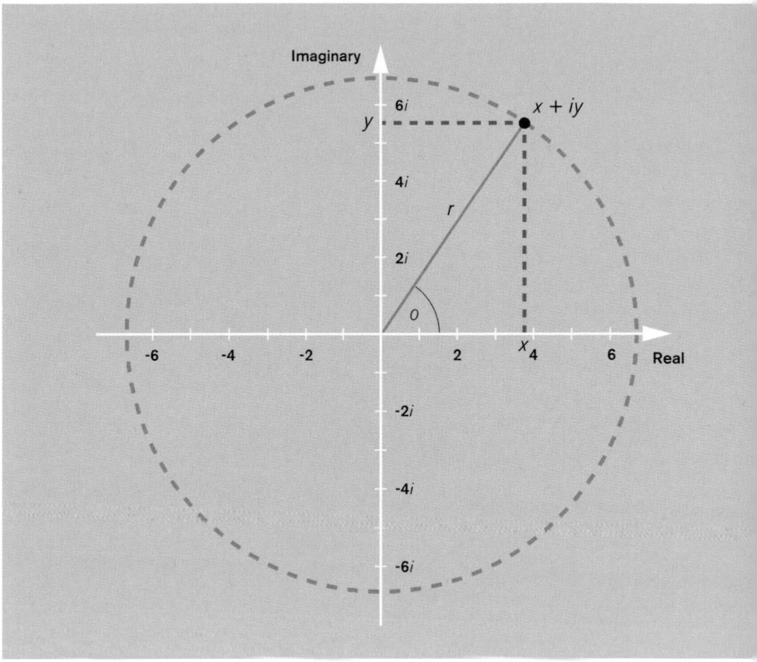

Euler probed the possibilities of the new system and realized that familiar objects, such as the trigonometric functions well known from centuries of triangular geometry, may take on dramatically new appearances when transported into the complex domain. This observation led him to his most famous discovery, Euler's formula (see page 155). The recently developed theory of calculus (page 133) would also find a very welcoming home among the complex numbers, in the work of Augustin-Louis Cauchy. Thus the potential of the complex numbers was gradually being realized. Their crowning glory was the discovery of the fundamental theorem of algebra in the early 19th century (see page 165).

ABOVE The Argand diagram depicts the complex numbers as a 2-dimensional flat plane. Any complex number such as $x + iy$ can be described geometrically by its angle θ and its magnitude r.

Complex geometry

Around the year 1800, at the same time that deep theorems about complex numbers were being proved, Caspar Wessel and Robert Argand finally found a way to relieve the "mental torture" associated with them. They realized that the number system has a geometric representation. If the real numbers are drawn along a horizontal line from left to right, then the imaginary numbers run perpendicularly, along a vertical axis, and the entire system of complex numbers corresponds to the whole plane. This approach to the complex numbers makes a wonderful setting for geometry. For example, multiplying something by i is equivalent to rotating it by 90° around the origin, while multiplying by -1 is a reflection through the vertical axis. This confluence of geometric and algebraic ideas would be of profound importance in the years ahead.

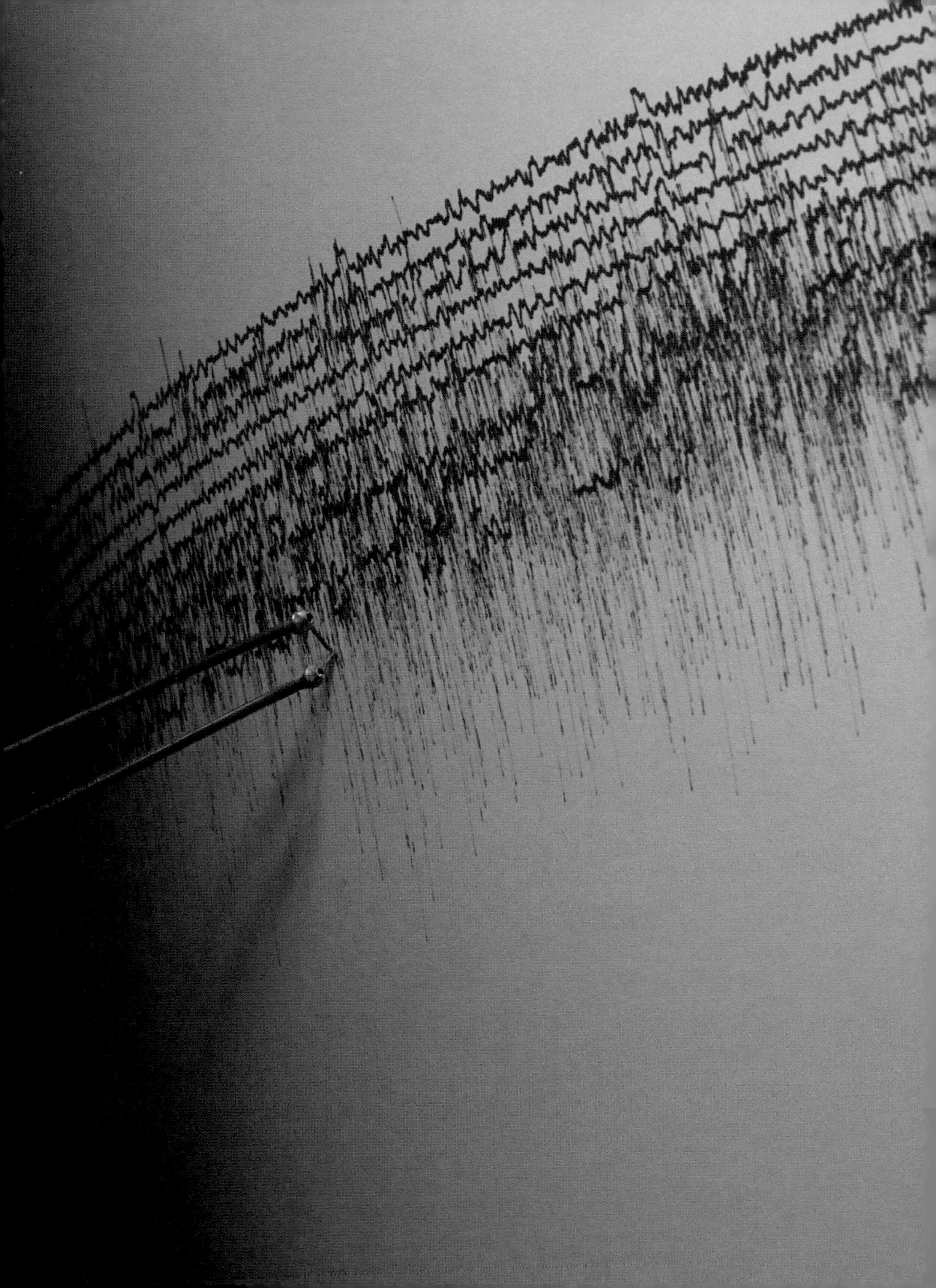

26 Logarithms

BREAKTHROUGH LOGARITHMS WERE DISCOVERED BY NAPIER IN 1594. THEY ARE THE OPPOSITES OF POWERS, AND WOULD GO ON TO ATTRACT INTEREST ACROSS THE WHOLE OF SCIENCE.

DISCOVERER JOHN NAPIER (1550–1617), HENRY BRIGGS (1561–1630).

LEGACY FOR CENTURIES, LOGARITHM TABLES WERE INVALUABLE TOOLS FOR SCIENTISTS AND ENGINEERS. NOWADAYS, THEIR DEEP CONNECTION TO THE EXPONENTIAL FUNCTION (SEE PAGE 155) ENSURES THAT THEY REMAIN OF GREAT MATHEMATICAL INTEREST.

In years gone by, many people routinely kept tables of logarithms on hand, to assist with multiplication and division. In the late 20th century, the pocket calculator finally consigned the log table to history. But fascinating discoveries in the deeper mathematical waters of series and calculus ensure that logarithms themselves will never go out of fashion.

In the late 16th century, John Napier began to study what he initially called "artificial numbers." He had discovered a technique for turning difficult multiplication problems into much simpler questions of addition. In order to multiply two numbers, such as 4587 and 1962, he would first calculate the two corresponding artificial numbers and *add* these together. Then he would de-artificialize the result, that is to say, calculate the ordinary number which corresponded to the artificial one. Even though this process had involved no multiplication, this answer would indeed be the result of multiplying together the two original numbers: 8,999,694.

OPPOSITE
A seismograph measuring the power of an earthquake. The Richter scale, which quantifies the strength of earthquakes, is logarithmic: an earthquake which registers 3 on the scale is ten times stronger than one which measures 2.

Napier's logarithms

Napier later settled on a better name for his artificial numbers: logarithms. Today, we understand that logarithms are nothing more than the opposites of powers, which in turn are repeated multiplication. So "2 to the power 3" means three 2s multiplied together: $2 \times 2 \times 2 = 8$. This is also written as $2^3 = 8$. Correspondingly, we say that "the logarithm of 8 to base 2 is 3," which is written $\log_2 8 = 3$.

One can take logarithms to any base. For instance, to base 10, the logarithm of 1000 is 3 (because $10 \times 10 \times 10 = 1000$). For Napier's multiplication procedure, one needs to fix a base throughout. So, to multiply 8 by 64, one first takes their logarithms to base 2, giving 3 and 6. These are then added: $3 + 6 = 9$, and the final step is to undo the logarithm process, which is to say calculate $2^9 = 512$. (You can check that $8 \times 64 = 512$.)

Briggs's log tables

It was not long after John Napier's invention of logarithms that Henry Briggs set about turning them into a practical tool. Because of our decimal system, Briggs settled on base 10 as the most convenient and began to assemble the first "log table": a book of logarithms for all the numbers from 1 to 1000. Over the years, Briggs and later thinkers extended the tables to much larger sets of numbers.

The extraordinary Tables du Cadastre amounted to 17 large folio volumes and included the logarithms of the numbers up to 200,000, given to 19 decimal places.

Of course, for most numbers, the logarithm is not a whole number. So researchers had to decide to what level of accuracy to give their logarithms. In the late 18th century, Gaspard de Prony supervised the construction of the extraordinary Tables du Cadastre, which amounted to 17 large folio volumes and included the logarithms of the numbers up to 200,000, given to 19 decimal places (24 for the larger numbers).

The natural logarithm

Logarithms have been of great use to scientists since their original discovery by Napier. As the eminent scientist Pierre-Simon Laplace said, "by shortening the labors, the invention of logarithms doubled the life of the astronomer." But their mathematical significance runs deeper than their function as calculating devices. This was first realized in 1650 by Pietro Mengoli, whose work on series (see page 97) unexpectedly converged with his interest in logarithms.

The harmonic series is the expression $1 + \frac{1}{2} + \frac{1}{3} + \frac{1}{4} + \frac{1}{5} + \ldots$. Mengoli realized that this expression, somewhat surprisingly, does not approach a finite limit but actually grows ever larger without bound. However, by subtly changing the question, he was able to find a related expression which does converge to a fixed number, namely $1 - \frac{1}{2} + \frac{1}{3} - \frac{1}{4} + \frac{1}{5} - \ldots$.

This alternating series has a fixed limit of around 0.693147. Mengoli showed that this limiting number is the natural logarithm of 2 (traditionally written "ln 2" though pronounced "log two"). The natural logarithm is a logarithm like any other, but with a very special choice of base, namely the number e (see page 155), which has a value of around 2.71828. Indeed, it was through the natural logarithm and results like Mengoli's that the exponential function, which would go on to become one of the most important objects in mathematics, first came to prominence.

Indeed, the purely pragmatic search for ever more accurate log tables was a powerful driver of the more abstract theory of series. In 1668, Nicholas Mercator published a work entitled *Logarithmo-technica,* in which he found a series formula for the natural logarithm:

$$\ln(1 + x) = x - \frac{x^2}{2} + \frac{x^3}{3} - \frac{x^4}{4} + \ldots$$

This beautiful theorem was an extension of Mengoli's result, which corresponded to the value of $x = 1$.

Gr. 0

+ | −

min	*Sinus*	*Logarithmi*	*Differentiæ*	*Logarithmi*	*Sinus*	
30	87265	47413852	47413471	381	9999619	30
31	90174	47085961	47085554	407	9999593	29
32	93083	46768483	46768049	434	9999566	28
33	95992	46460773	46460312	461	9999539	27
34	98901	46162254	46161765	489	9999511	26
35	101809	45872392	45871874	518	9999482	25
36	104718	45590688	45590140	548	9999452	24
37	107627	45316714	45316135	579	9999421	23
38	110536	45050041	45049430	611	9999389	22
39	113445	44790296	44789652	644	9999357	21
40	116353	44537132	44536455	677	9999323	20
41	119262	44290216	44289505	711	9999289	19
42	122171	44049255	44048509	746	9999254	18
43	125079	43813959	43813177	782	9999218	17
44	127988	43584078	43583259	819	9999181	16
45	130896	43359360	43358503	857	9999143	15
46	133805	43139582	43138686	896	9999105	14
47	136714	42924534	42923599	935	9999065	13
48	139622	42714014	42713039	975	9999025	12
49	142531	42507833	42506817	1016	9998984	11
50	145439	42305826	42304768	1058	9998942	10
51	148348	42107812	42106711	1101	9998900	9
52	151257	41913644	41912499	1145	9998856	8
53	154165	41723175	41721986	1189	9998811	7
54	157074	41536271	41535037	1234	9998766	6
55	159982	41352795	41351515	1280	9998720	5
56	162891	41172626	41171299	1327	9998672	4
57	165799	41006643	41005268	1375	9998625	3
58	168708	40821746	40820322	1424	9998577	2
59	171616	40650816	40649343	1473	9998527	1
60	174524	40482764	40481241	1523	9998477	0
						min
						Gr. 89

89

a

ABOVE One of the earliest log tables, taken from John Napier's *Mirifici Logarithmorum Canonis Descriptio* (Description of the Wonderful Canon of Logarithms) of 1614. While Napier worked with what would later become known as the natural logarithm, Henry Briggs investigated the logarithm to base ten.

Calculus and logarithms

Mercator's theorem hinted at the "naturalness" of the natural logarithm. But a fuller story had to wait for the implications of Newton and Leibniz' theory of calculus (see page 133).

The equation $y = \frac{1}{x}$ describes an important relationship called the reciprocal. It is this equation which relates the number two to a half, four to a quarter, a million to a millionth, and so on. Geometrically, it describes a curve called hyperbola. The natural logarithm appears, unexpectedly, as the area beneath this curve. This is a consequence of the fact that the natural logarithm is the inverse of the exponential function (see page 155). From this it follows that the differential of the natural logarithm ($y = \ln x$) is none other than the reciprocal ($y = \frac{1}{x}$). While log tables have now been replaced by computers, this deep fact ensures that logarithms continue to play a central role in mathematics.

27 Polyhedra

BREAKTHROUGH EACH GENERATION OF GEOMETERS HAS EXPANDED THE COLLECTION OF KNOWN SHAPES. A MAJOR DEVELOPMENT WAS THE DISCOVERY OF NEW REGULAR POLYHEDRA BY JOHANNES KEPLER.

DISCOVERER ARCHIMEDES (C. 287–212 BC), JOHANNES KEPLER (1571–1630), LOUIS POINSOT (1777–1859).

LEGACY THE FAMILIES OF POLYHEDRA FOUND SO FAR GIVE GREAT INSIGHTS INTO THE GEOMETRY OF 3-DIMENSIONAL SPACE.

A geometer's idea of perfection is symmetry. Throughout the history of the subject, geometers have sought to discover and classify the most symmetrical shapes they can find. Central to this quest are polyhedra: 3-dimensional objects built from flat faces and straight edges. The story of polyhedra began with Theaetetus' analysis of the Platonic solids (see page 37). But over the centuries that followed, many further families of polyhedra have been found.

Examples of symmetrical shapes are all around us. Perhaps the most familiar example is the square. This satisfies the geometers' notion of beauty, because a square's sides all have the same length and its angles are all equal. A triangle can also satisfy these criteria, but not every triangle qualifies—only the most symmetrical member of the triangle family, the equilateral triangle. Beyond the square is the regular pentagon with its five equal sides, followed by the regular hexagon with six, and so on. As the number of sides increases, the shapes grow ever closer to the apotheosis of geometrical perfection: the circle.

OPPOSITE
A small stellated dodecahedron decorates the floor of St. Mark's Basilica in Venice. The mosaic dates from around 1430, and predates Johannes Kepler's first analysis of the Kepler–Poinsot polyhedra in the 17th century.

This list of regular polygons has been well known since antiquity, and it is the starting point of many further mathematical investigations. One particular question has been revisited time and again: what 3-dimensional solid shapes can be built by gluing these flat shapes together? This question has fascinated scientists for thousands of years, sometimes to the point of obsession.

Archimedean solids

The first breakthrough was provided by Theaetetus. His five Platonic solids are the 3-dimensional analogs of the regular polygons. They are perfectly symmetrical: all the faces are identical, and each face is itself a regular polygon.

A century later, Archimedes pushed this line of investigation further. By weakening the amount of symmetry required, Archimedes found some beautiful polyhedra, the likes of which no one had ever seen before. Crucially, he dropped the requirement that all the faces should be identical, though they should still all be regular polygons. His shapes still maintained a high level of symmetry: the arrangement of faces at each corner is identical to every other. With these considerations, Archimedes arrived at a family of 13 exquisite new solids, all more complicated and intricate than Plato's. The most famous of Archimedes' shapes is the truncated icosahedron, better known as the soccer ball, built from a mixture of 12 pentagons and 20 hexagons. As well as Archimedes' 13 individual shapes, there are two infinite families of shapes which satisfy the criteria, known as the prisms and antiprisms.

Archimedes arrived at a family of 13 exquisite new solids, all more complicated and intricate than Plato's. The most famous of Archimedes' shapes is the truncated icosahedron, better known as the soccer ball.

Kepler–Poinsot polyhedra

Unfortunately, Archimedes' great work on polyhedra was lost. During the Renaissance period, European scientists and artists including Leonardo da Vinci and Johannes Kepler set about rediscovering the 13 beautiful Archimedean solids. However, this search also turned up a major surprise: some new highly symmetrical shapes which seemed to satisfy the characteristics of a Platonic solid, and yet were missing from Theaetetus' ancient list.

In 1619, Johannes Kepler discovered two shapes, the great and small stellated dodecahedra, whose faces were all identical and were all constructed from lines of equal length, with equal angles. The difference was that rather than having pentagonal faces, as the dodecahedron does, Kepler's two shapes each had faces which are pentagrams. So the edges of the shape pass through each other at certain points, giving the shapes a striking star-like effect.

In fact, Kepler had been beaten to the punch by the artist Paolo Uccello, whose depiction of the small stellated dodecahedron can be seen adorning the mosaic floor of St. Mark's Basilica in Venice. Almost 200 years later, Louis Poinsot found two more beautiful star polyhedra: the great dodecahedron and the great icosahedron. Together, these four can be considered to complete the list of regular polyhedra begun by Theaetetus.

ABOVE Vebjørn Sands' *Kepler Star* near Olso airport in Norway, completed in 1999, takes the form of an icosahedron and dodecahedron (two of the Platonic solids) nested within a great stellated dodecahedron, one of the Kepler–Poinsot polyhedra.

Johnson solids

The study of polyhedra continued throughout the 20th century as geometers steeled themselves to tackle less symmetrical shapes. One example was well known to the ancient Egyptians: the pyramid. This has one square face on which it usually sits, plus four triangular faces (equilateral triangles, of course). It is not an Archimedean solid, as the corner at the top has a different configuration from the others. The pyramid suggests that by lessening the amount of symmetry required, an array of new relevant shapes might be found. During the 20th century, geometers set themselves a challenge: to discover the full catalogue of 3-dimensional shapes that can be built from regular polygons. It was this question that Norman Johnson addressed in 1966 in publishing a list of the 92 Johnson solids.

The defining characteristic of his solids is not symmetry but convexity, which roughly means that the shape contains no holes or parts that stick out too far from the main body of the shape. (Kepler's star polyhedra fail this test.) Johnson believed that his list of convex polyhedra was complete, but he was not able to prove this for certain. Then, in 1969, Victor Zalgaller succeeded in finishing the job by showing that Johnson's list is indeed complete. Besides the solids described by Plato and Archimedes, there are no other convex solids that can be built from regular polygons.

The first Johnson solid is the familiar pyramid, and the second is its bigger brother: a pyramid with a pentagonal base. Beyond this, there is no pyramid with a hexagonal base, since six equilateral triangles can no longer fit together in the right way. Despite eschewing all questions of symmetry (except that of the individual faces), many of Johnson's solids have an undeniable beauty.

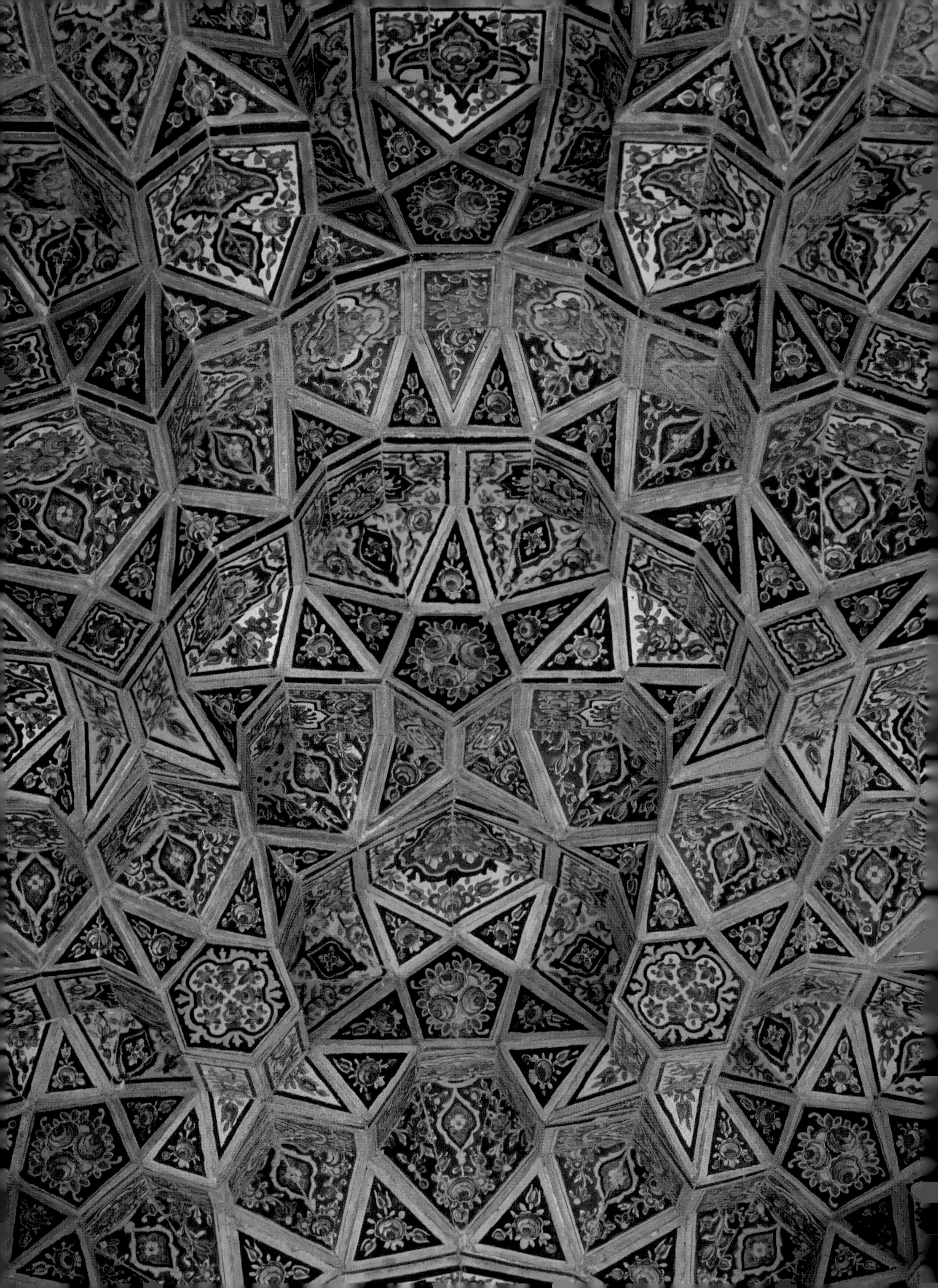

28 Tessellations

BREAKTHROUGH A TESSELLATION OCCURS WHEN TILES FIT TOGETHER TO CREATE A REPEATING PATTERN. AS WELL AS BEING AESTHETICALLY PLEASING, THESE POSE SEVERAL INTERESTING GEOMETRIC QUESTIONS.

DISCOVERER PAPPUS OF ALEXANDRIA (C. AD 290–350) JOHANNES KEPLER (1571–1630)

LEGACY WHILE TESSELLATIONS HAVE RECENTLY FOUND UNEXPECTED APPLICATIONS IN PHYSICS AND CHEMISTRY, THEY ARE BY NO MEANS THOROUGHLY UNDERSTOOD TODAY.

Which shapes can fit together to create a repeating pattern? This simple question has fascinated mosaicists and artists since time immemorial. It also has a long history within mathematics. Though the theory of tessellations seems simple at first, when the same question is posed in more complex settings, it leads to some profound and difficult mathematics.

A regular polygon is a perfectly symmetrical straight-edged shape. This means that its edges must be of equal length, and all its angles are the same. Perhaps the most famous example is the square. But every designer and DIY enthusiast knows something else about the square, namely that it tessellates.

Regular tessellations

Square tiles fit next to each other perfectly and can be used to cover as large a space as required, with no overlaps between them and no gaps left untiled. This fact has been exploited for thousands of years, but not every shape has this property. A regular pentagon, for instance, with its five equal sides, cannot tessellate. The interior angle of a pentagon is 108°, so there is no way to fit several together so that they add up 360°, as four angles of 90° do.

It has been known for many years, at least since the work of the geometer Pappus of Alexandria in around AD 325, that only three regular polygons tessellate: the square, the equilateral triangle and the hexagon. These are

OPPOSITE A colorful irregular tiling adorning the vaulted roof of a niche in Shiraz, Iran. The shapes are chosen so that every patch is covered, and no tiles overlap—a basic requirement for a tessellation.

ABOVE Square tiles tessellating. Along with the equilateral triangle and hexagon, the square is one of the only three regular shapes to tessellate. The overlaid pattern involving 8-pointed stars is more complex, and one of an infinite collection of irregular tessellations.

the only ones whose corner angles slot together perfectly: regular pentagons, heptagons, octagons or shapes with more sides simply cannot. These are known as the three regular tessellations.

Irregular tessellations

Regular pentagons do not tessellate: their corners don't slot together in the right way. But some irregular pentagons do. These are five-sided shapes whose sides are straight but of different lengths, and which have unequal angles. Despite their seeming simplicity, pentagonal tessellations remain rather mysterious today. Fourteen different ways are currently known for pentagons to fit together to tile a floor, and as yet no one has proven that there may not be other ways.

On the other hand, every triangle tessellates, whether equilateral or not. So, too, does every four-sided shape. Therefore parallelograms, trapezia and kites are all examples of tessellating figures. For shapes with more sides, such as irregular heptagons, the story is a little more complicated. There are no convex shapes, meaning shapes which contain no angle of more than 180°, which tessellate. But some interesting non-convex tilings are known, such as the Voderberg tiling, which is built from tiles made from a non-convex irregular 9-gon or nonagon.

Kepler's semiregular tessellations

In 1619, the astronomer and mathematician Johannes Kepler investigated tessellations involving tiles of more than one shape. Such patterns had been explored by artists for centuries, but Kepler had set his sights on something more thorough: a mathematical classification. Of course, the number of possible such tessellations is bewildering, so Kepler focused on those showing some nice overall symmetry. He insisted that every tile should be a regular polygon, and that each corner where tiles meet should be identical to every other. One common example is the pattern made from octagons and squares. Kepler was able to list eight different configurations overall, which are today known as the semiregular tessellations. They are also sometimes known as the Archimedean tessellations, in recognition of their relationship to the Archimedean solids (see page 114).

Hyperbolic tilings

In the 21st century, tessellations continue to fascinate today's geometers. Nowadays, we have several new mathematical spaces to tile. Instead of the familiar flat Euclidean plane, the discovery of hyperbolic space in the 19th century (see page 189) prompted the question of which regular shapes would tessellate here. The answer is utterly different from the Euclidean world, where only the triangle, square and hexagon tessellate. In fact, in hyperbolic geometry there are infinitely many regular tessellations, because every regular polygon can be used to fill the hyperbolic plane. Indeed, they each tessellate in many different ways. If an equilateral triangle is used to tile the ordinary Euclidean plane, it must be that six meet at each corner. But in the hyperbolic plane, there are many possibilities: triangles can meet in sevens, eights, or indeed any higher number.

Pentagonal tessellations remain rather mysterious today. Fourteen different ways are currently known for pentagons to fit together to tile a floor, and as yet no one has proven that there may not be other ways.

This may seem surprising, but the same is true in spherical geometry. We can think of the Platonic solids (see page 37) as being the regular tilings of a sphere. In this case, too, equilateral triangles can meet in threes (to give a tetrahedron), fours (an octahedron), or fives (an icosahedron).

Honeycombs

The question of tessellations can be asked just as well in higher dimensions. A cube is also a tessellating shape, since one can fill 3-dimensional space with cubes, with no overlaps or gaps. In fact, the cube is the only Platonic solid that can do this (see page 37), although Aristotle mistakenly believed that the tetrahedron tessellates. But there are plenty of other 3-dimensional shapes that can fill space on their own. Of the Archimedean solids, only the truncated octahedron fills space, its six hexagons and six squares perfectly meshing with each other. There are other space-filling polyhedra, however. Triangular and hexagonal prisms work well, as does a rhomboid dodecahedron (a pretty shape constructed from 12 rhombuses).

When tiles of more than one shape are allowed, the question becomes much harder to answer, and the topic is not yet fully understood. An example that has recently come to prominence, however, is the Weaire–Phelan foam, a famous counterexample to Kelvin's conjecture (see page 361).

29 Kepler's laws

BREAKTHROUGH KEPLER PROVIDED BEAUTIFUL GEOMETRICAL DESCRIPTIONS OF PLANETARY ORBITS. THESE PROVIDED THE IMPETUS FOR NEWTON'S THEORY OF UNIVERSAL GRAVITATION.

DISCOVERER JOHANNES KEPLER (1571–1630), ISAAC NEWTON (1642–1727).

LEGACY THE DISCOVERIES OF KEPLER AND NEWTON MARKED A NEW UNDERSTANDING OF OUR UNIVERSE AND THE BEGINNING OF A NEW TYPE OF PHYSICS, WHICH WOULD DEMAND EVER INCREASING MATHEMATICAL SOPHISTICATION.

For as long as people have been studying mathematics, the subject has provided an intellectual tool for understanding the physical Universe. A particular challenge has been to make sense of our Solar System, which is held together by a single force: gravity. A major breakthrough was the beautiful geometrical account of gravity provided by Johannes Kepler.

For millennia, people have been looking at the night sky, noting down their observations, in an attempt to understand the motions of the Earth, Moon, Sun and other celestial bodies. Nicolaus Copernicus' insight that the Earth orbits the Sun, rather than vice versa, was a huge step forward. All the same, the data collected by the great 16th-century astronomer Tycho Brahe seemed to contain anomalies.

Kepler's laws

The explanation was provided by Johannes Kepler in a wonderful piece of mathematical physics. Previous astronomers had assumed that orbiting bodies should always follow circular paths. Kepler's analysis told him that this was not correct. In fact, using Tycho Brahe's data, he made a convincing case that Mars travels around the Sun following a curve called an ellipse. This was one of the conic sections well known to geometers, studied by Apollonius centuries before (see page 57). What is more, the Sun is not at the center of the ellipse. Instead, it sits at a point near one end called the focus.

OPPOSITE Saturn's moon Dione. Dione's elliptical orbit is very close to a circle, with an eccentricity of just 0.002 (an eccentricity 0 representing a perfect circle). In contrast, Pluto's orbit around the Sun has an eccentricity of 0.248, while the Earth's is 0.017.

Kepler elevated this observation to the status of a rule: orbiting bodies should always travel in elliptical paths, with the Sun at a focus point. This was his first law of planetary motion, published in his influential 1621 work *Epitome astronomia Copernicanae* (The Epitome of Copernican Astronomy). The work also contained his other two laws, which further refined elliptical orbits.

Kepler's laws of planetary motion were a triumph of scientific observation combined with geometrical reasoning.

In his second law, Kepler addressed the question of how fast a planet travels. He realized that as Mars travels closer to the Sun, it appears to move more quickly. Why should this happen? Kepler revealed the answer with a beautiful geometric idea. He imagined that the planet was attached to the Sun with a straight rod. Over the course of a fixed period of time (an hour, say), the rod will sweep out a certain area. Kepler's second law states that the area of that region will always be the same, no matter the stage of the planetary cycle. It follows from this that the planet must travel faster when it is closer to the Sun than when it is more distant.

Kepler's third law was subtler still. It related the duration of a planet's entire orbit to the length of the resulting ellipse (that is to say the straight distance from one end of the orbit to the other). Kepler posited that the square of the duration (d^2 or $d \times d$) should be proportional to the cube of the length (l^3 or $l \times l \times l$).

Universal gravitation

Kepler's laws of planetary motion were a triumph of scientific observation combined with geometrical reasoning. Even today they are considered standard tenets of astronomy, though we now know that relativistic effects mean they do not apply with perfect accuracy. But the question Kepler failed to address satisfactorily was why the laws should be true.

It fell to Isaac Newton to answer this question. We know that the Moon (as well as ourselves) is attracted to the Earth by gravity. The Earth, meanwhile, is itself locked in orbit around the Sun by the same force. But there is a curious asymmetry here. In each case, the force only seems to apply in one direction. But why is the Sun not also attracted to the Earth? Or, for that matter, why is the Moon not drawn toward the Sun?

Newton addressed this question with his universal law of gravitation. He stated that a pair of heavy bodies should each feel the same mutual attraction. So the Earth is indeed attracted to the Moon, and the Sun to the Earth. The difference lies in their relative masses. The Sun is over

300,000 times heavier than the Earth, so the same gravitational force which keeps the Earth in orbit has little visible effect on the Sun. Nevertheless, it is an important point of principle that it must feel that force in some way. In other solar systems with only two stars, the principle of mutual attraction can be seen more clearly (see page 237).

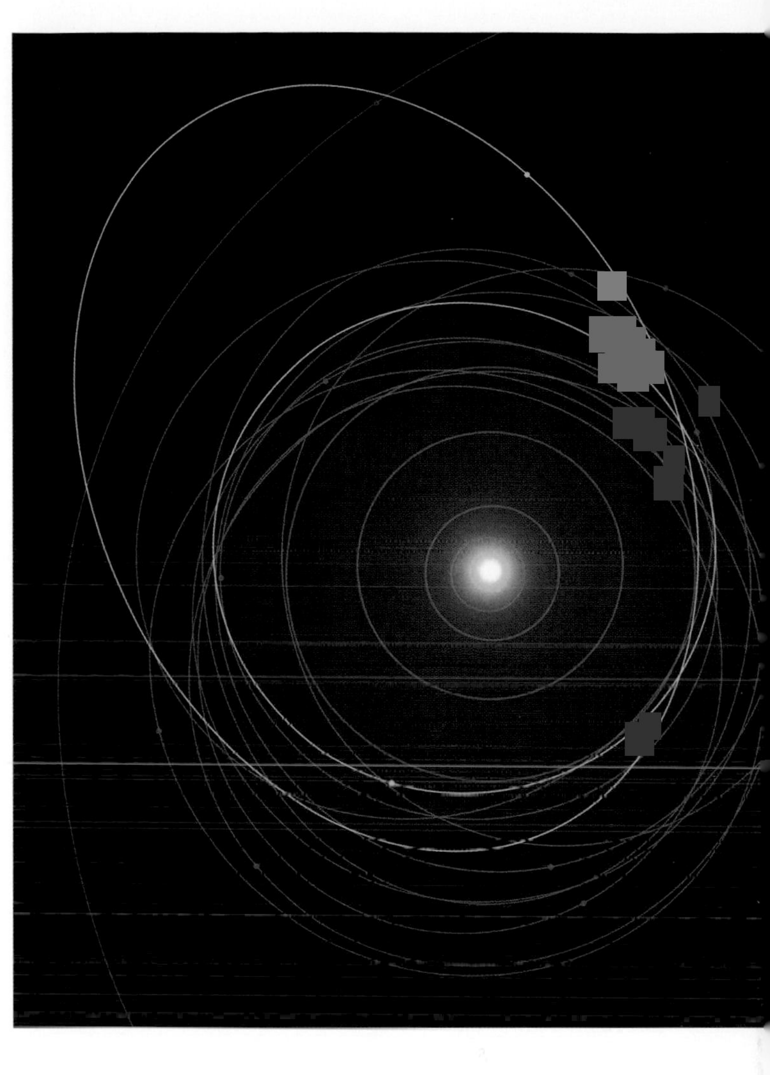

ABOVE The Solar System, showing the elliptical orbits of the four outer planets (Jupiter, Saturn, Uranus and Neptune) in green, three dwarf planets (Ceres, Pluto and Eris) in yellow and ten candidate dwarf planets in brown.

Newton's inverse square law

What determines the mutual attraction of two objects? An obvious answer is the mass of each of them. Gravity only becomes significant at large masses, which is why we do not notice the gravitational pull between people or everyday objects such as books. The second part of the answer is more challenging. If the Sun is far heavier than the Earth, then why do we not fall off the Earth and tumble toward the Sun? The answer is that the Sun is further away from us than the Earth. Gravitational attraction diminishes over distance. Newton contemplated exactly how this happens. If a person travels four times further from the center of the Earth, will the gravity they feel be reduced by four times? Kepler had believed so, but this conclusion did not mesh as well with his three laws.

Newton corrected this error. In fact, gravity decreases with the square of the distance between the objects. So, when a person is four times further from the center of the Earth, their gravity will be reduced by 16 times. This inverse square law was a major discovery, and Newton was able to use it to derive Kepler's laws mathematically.

30 Projective geometry

BREAKTHROUGH DESARGUES CONQUERED THE MATHEMATICS OF PERSPECTIVE BY INCORPORATING ARTISTS' VANISHING POINTS INTO GEOMETRY.

DISCOVERER FILIPPO BRUNELLESCHI (1377–1446), GIRARD DESARGUES (1591–1661).

LEGACY DESARGUES' ANALYSIS PROVIDED VALUABLE TECHNIQUES FOR VISUAL ARTISTS; HE ALSO FOUNDED THE NEW MATHEMATICAL SUBJECT OF PROJECTIVE GEOMETRY.

When you look at something far away, it appears smaller than an identical object nearby. This simple phenomenon has troubled artists for millennia. But when it was contemplated by the mathematician Girard Desargues, his conclusions went on to revolutionize both geometry and visual art.

It requires skill and experience to produce a picture that looks natural when drawing a scene containing objects of different sizes at different distances. Harder still is the phenomenon of foreshortening, where the shape of an object seems to change with distance. To an artist sketching a man lying on a bed, his feet, being nearer to the viewer, might appear larger than his head. At the same time, his legs will seem shorter when viewed along their length rather than front on.

The problem of perspective

Artists of every era have grappled with the challenges of representing the 3-dimensional world on a 2-dimensional canvas. Many early works of art, such as those from ancient Egypt, tend to look "wrong," a testament to the difficulty of this task. There is a strong mathematical component to this artistic problem. The primary technique for perspective is the vanishing point, first systematically employed by Renaissance artists Filippo Brunelleschi and Paolo Uccello. The idea stems from observing parallel lines, such as train tracks, as they disappear off into the distance. To the naked eye it seems as if these lines become closer and closer together until they eventually converge at some faraway point. Of

OPPOSITE Rays of light converging on a vanishing point. This apparent convergence is one of the methods by which humans gauge distance, and is essential in representing perspective in two dimensions.

course, this is an illusion: the tracks always remain a fixed distance apart and never actually meet. But their representations on the artist's canvas do meet, and this place is called the vanishing point.

Desargues' new geometry

In the early 17th century, Girard Desargues great idea was to incorporate vanishing points into geometry, in order to settle an ancient conundrum. Mathematicians had long noticed a beautiful symmetry between the geometry of points and straight lines. If you begin by marking any two points on a page, a famous law of Euclidean geometry says that there must be exactly one straight line which connects them. The *dual* of this statement comes from interchanging the roles of points and lines. It says that if you start with two straight lines on the page, then they will meet at exactly one point. This dual statement is usually true. But, inconveniently, it is not quite universally valid. There are special pairs of lines, parallel lines, which never meet.

However, in the world of art, parallel lines do meet: at a vanishing point. So Desargues had the bold idea of incorporating the same idea into mathematics, where the vanishing point goes by the name of a "point at infinity." His new subject of projective geometry took place in this extended setting. Just as Euclid had first formalized the study of lines and circles on a flat page, so Desargues began the study of shapes on the projective plane, meaning the Euclidean plane augmented with additional points at infinity. The immediate payoff for this increased abstraction was that the duality between points and lines now worked perfectly. On a projective plane, it was finally true that every pair of lines meets at a single point, so there are no such things as parallel lines.

In the world of art, parallel lines do meet: at a vanishing point. So Desargues had the bold idea of incorporating the same idea into mathematics, where the vanishing point goes by the name of a "point at infinity."

Much of modern geometry still takes place in projective spaces. Although these are harder to describe than traditional Euclidean settings, many geometric techniques work more smoothly there. For example, in a projective situation the conic sections (see page 57) cease to be three different shapes, but are revealed as different perspectives on one single curve.

Desargues' theorem

Vanishing points remain an essential technique for today's artists. But they do not completely resolve all the difficulties of representing depth. Suppose an artist is drawing a room in which the carpet is decorated

with a triangular design. How might she accurately represent that pattern? Of course, the canvas must contain a triangle too. But, because of foreshortening, it will not appear as the one on the floor seen from directly above.

ABOVE A spiral staircase and study in perspective, by Hans Vredeman de Vries (1604). The use of multiple vanishing points provided artists with newly accurate ways to represent 3-dimensional spaces on paper.

The painter has two triangles to consider: one on the carpet and another on the canvas. The aim is for the two to be "in perspective." This means that if each corner of one triangle were joined by a laser beam to the corresponding corner of the other, then the resulting three beams should all meet at a single point. But how can this be achieved (especially without the use of lasers)? Again, Girard Desargues provided an answer. Desargues' theorem asserts that two triangles are correctly in perspective precisely when another, simpler criterion is met. If you extend one edge of one triangle and the corresponding edge of the other triangle, the two extended edges should meet at a point. Following this procedure with all three pairs of edges produces three points. Desargues' requirement was that these three points should lie along a straight line. If they do, then the two triangles are in perspective. If not, then they are not. It is far from obvious that Desargues' criterion is equivalent to the two triangles being in perspective. For this reason, his theorem has been of great practical value to painters ever since, as it yields a convenient method for telling whether or not two shapes are in perspective.

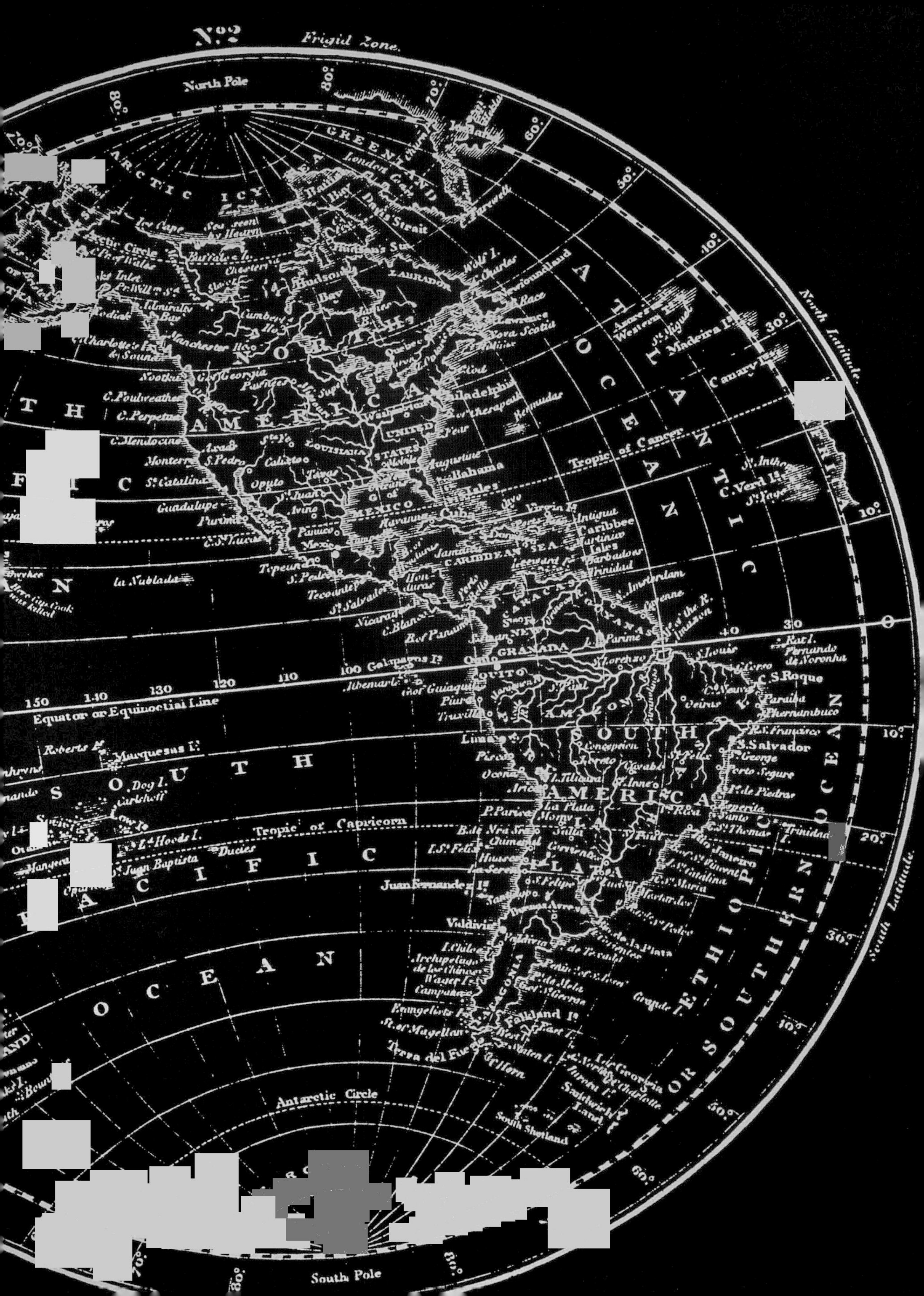

31 Coordinates

BREAKTHROUGH COORDINATES ARE NUMBERS USED TO IDENTIFY THE LOCATION OF POINTS. THEIR DEVELOPMENT BY DESCARTES WAS A MILESTONE IN THE STUDY OF GEOMETRY.

DISCOVERER RENÉ DESCARTES (1596–1650).

LEGACY CARTESIAN COORDINATES ARE USED EVERY DAY, NOT ONLY BY MATHEMATICIANS, BUT BY GEOGRAPHERS AND MAPMAKERS. WITHIN MATHEMATICS, THEY HAVE ALLOWED NUMERICAL AND ALGEBRAIC TECHNIQUES GREATLY TO ENHANCE OUR UNDERSTANDING OF GEOMETRY.

René Descartes is primarily famous as a philosopher. Yet the lines between philosophy, mathematics and science were much fuzzier in his times than they are today, and Descartes' mathematical legacy is of equal significance. His most important innovation was Cartesian coordinates, opening the way for the study of modern geometry.

Since humans first contemplated mathematics, the subject has consisted of two major branches: numbers and shapes. That the two are closely related was well understood by Euclid, who fruitfully applied numbers to analyze shapes. Geometrical questions have sparked major developments in number theory, as seen in Archimedes' analysis of circles (see page 53), or the emergence of irrational numbers (see page 29). Nevertheless, the two subjects remained fundamentally distinct. Geometry took as its starting point objects that were constructed not from numbers, but from other basic ingredients such as points, lines and planes. For geometry to achieve a higher level of sophistication, it was necessary to find a way to incorporate geometric objects fully into a purely numerical system.

OPPOSITE A map of the Earth, with coordinates given as latitude and longitude. The equator is 0° latitude, with the North pole 90° north, and the South pole 90° south. Longitude is traditionally measured from the Greenwich meridian, so Mexico City has coordinates of 19° North, 99° West.

René Descartes

Cartesian doubt was Descartes' observation that he could not be certain of the world outside his own mind. For all he could tell, his consciousness and sensory perceptions might be the product of a fiendish scientific experiment in an alien universe. Yet, Descartes argued, the one thing he could be sure about is that he, in some form, exists. Otherwise what was

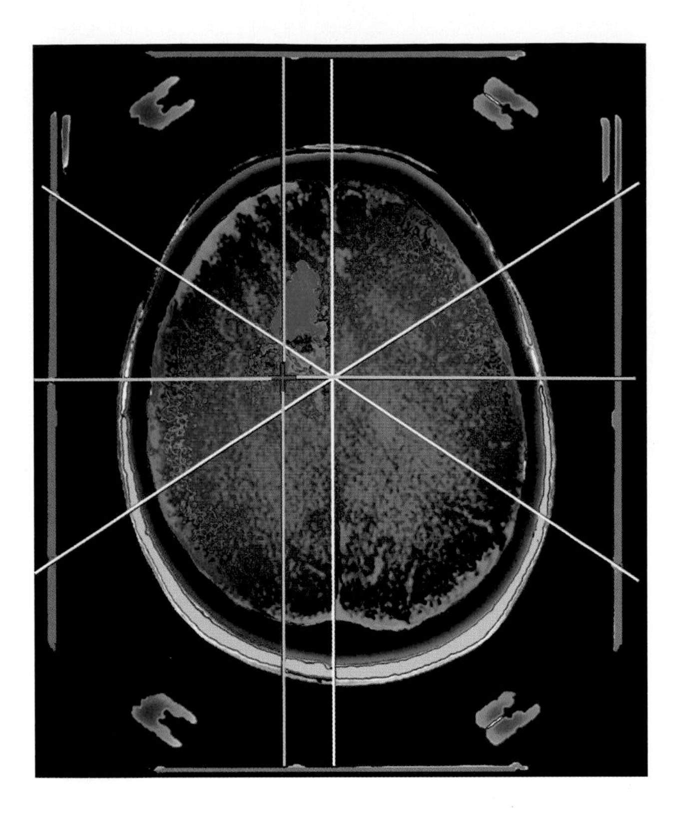

ABOVE Medical imaging is just one of the many applications of coordinates today. In this colored computed tomography (CT) scan of a human brain, the white lines and purple cross specify the coordinates for a biopsy on a tumor (shown in green).

experiencing his thoughts? He expressed this insight in the pithy phrase "*Cogito ergo sum*" or "I think, therefore I am" (in fact, he first wrote it in French: "*Je pense donc je suis*").

These ideas were fleshed out in his work *Discourse on Method* (first published in French as *Discours de la méthode*) of 1637. An appendix, entitled "*La Géométrie,*" was where Descartes forged his mathematical legacy. He realized that two numbers were enough to identify any location on a page. He could begin by saying how far the point is from the left-hand edge. Perhaps it is 3 inches away. This identifies it as lying on a certain vertical line, and he only needed to say how far up the page it is. If it is 4 inches from the bottom edge, then the pair of numbers (3, 4) specify the point exactly. (To distinguish (3, 4) from (4, 3) we need to uphold the convention that left–right always comes before up–down.)

What works on a piece of paper works equally well in the more abstract setting of the plane, as studied by geometers since Euclid. After all, a plane is nothing more than an idealized page, extending infinitely in all directions. Into this situation Descartes added some artificial edges, called axes. The first is a horizontal line which functions as the bottom edge. The other is a vertical line, standing in for the left-hand edge. A point can then be identified by its distances from each axis. Of course, there are also points to the left of the vertical axis. Descartes employed negative numbers (see page 81) to identify these points. Such a point might be written (−2, 5), while (3, −7) is a point below the horizontal axis. The center of the plane is the origin, where the two axes cross. It has coordinates (0, 0).

Coordinates are not limited to 2-dimensional spaces. A position in 3-dimensional space is specified by three spatial coordinates, while incorporating a fourth coordinate for time allows the position of an event in spacetime to be specified. These methods are central to technologies from air-traffic control to medical imaging, while geometers use coordinates to navigate through still higher-dimensional spaces.

Cartography

Descartes' coordinates would transform mathematics, and they remain vitally important in geography. For thousands of years, humans have used maps to begin to understand and describe their environments. Indeed, maps are among the oldest pictures we have. In Pavlov, in the modern Czech Republic, an engraving on a rock from around 25,000 years ago appears to be a depiction of the surrounding area. It has also been suggested that the most famous of all prehistoric artworks, the cave paintings from Lascaux in France, may include maps of the night sky. Drawn around 17,000 years ago, these caves include nearly 2000 images, including figurative paintings of animals, as well as more abstract designs. Some of these are believed to represent constellations of stars.

As humans traveled ever more widely, mapmaking became an increasingly important activity. But how do you know exactly where you are on a map? To address this question, all modern maps come with a superimposed grid, whose lines are marked with numbers. Using this innovation, it is easy to describe the location of any place using just two numbers: one to represent how far east or west the position is, and one to represent how far north or south it is.

We usually think of maps as being flat; this is certainly the most convenient format for carrying when we travel around. Yet flat maps pose a serious geometrical challenge.

Map projections

We usually think of maps as being flat; this is certainly the most convenient format for carrying when we travel around. Yet flat maps pose a serious geometrical challenge. The Earth, after all, is not flat, but spherical. To represent the geography of the Earth is bound to entail some distortion. Might Descartes' coordinates be of assistance here? Could a place's latitude and longitude simply be plotted as Cartesian coordinates, to make a flat, accurate map of the Earth?

Such a map should have the property that the distance between two points on the map scales up to give their distance on the surface of the Earth. Unfortunately, such a map is geometrically impossible. The fundamental obstacle is a consequence of Carl Friedrich Gauss's analysis of curved surfaces (see page 185). Gauss showed that curvature is intrinsic to a shape, meaning that the only way to obtain a truly accurate picture of the Earth's surface is on a globe.

32 Calculus

BREAKTHROUGH NEWTON AND LEIBNIZ WERE ABLE TO IDENTIFY SIMPLE ALGEBRAIC LAWS TO DESCRIBE SEEMINGLY COMPLEX CHANGING SYSTEMS.

DISCOVERER ISAAC NEWTON (1643–1727), GOTTFRIED LEIBNIZ (1646–1716), KARL WEIERSTRASS (1815–1897).

LEGACY ARGUABLY THE SINGLE MOST IMPORTANT DISCOVERY IN THE HISTORY OF MATHEMATICS, CALCULUS IS NOW A CRUCIAL TOOL FOR ALL SCIENTISTS AND ENGINEERS.

The invention of calculus was a defining moment in the history of mathematics, and one which dramatically increased the subject's value to non-mathematicians. Innumerable branches of science seek to understand systems which change over time or vary across space. In the 17th century, for the first time, the technical tools were assembled to make such problems amenable to mathematical analysis.

As well as being one of the most important moments in mathematics, the invention of calculus was also among the ugliest. Two of the greatest thinkers of the day each claimed the discovery as their own, resulting in a bitter dispute which took on an international dimension.

The feud between Newton and Leibniz

In England, Isaac Newton was a brilliant young researcher, whose discoveries in optics, gravitation and astronomy revolutionized the science of his day and propelled him to fame and fortune. In later life, he went on to hold several influential positions in British public life, as a Member of Parliament, Master of the Royal Mint, Knight of the Realm and President of the country's most prestigious scientific establishment, the Royal Society.

Living in Germany, Gottfried Leibniz commanded huge respect from his contemporaries, but he was never a public figure like Newton. He was one of the earliest pioneers of computation, designing the first machine capable of addition, subtraction, multiplication and division.

OPPOSITE Calculus is essential throughout modern engineering, for calculating the stresses different components of a building will come under, and understanding in detail how curved pieces will fit together.

Leibniz also wrote with great perspicacity on mathematical logic, long before that subject gained recognition, as well as on history and law. Philosophically, he is remembered for his astonishing optimism in arguing that our Universe is the best of all possible worlds. Leibniz was also the first to understand the potential power of representing number in binary, the language spoken today by computers the world over.

Rates of change

The problem over which Newton and Leibniz clashed was the mathematics of change. From the Earth's orbit around the Sun to the flow of water in a river, our world is rarely static; it is full of objects which evolve, rotate or grow with time. But mathematics had always been hindered in its ability to model such systems. Even a single moving object posed problems, whether it was an animal chasing its prey or a comet hurtling through space. As time ticks by, many aspects change: the object's position, its speed, its acceleration. What is the relationship between these?

From the Earth's orbit around the Sun to the flow of water in a river, our world is rarely static; it is full of objects which evolve, rotate or grow with time. But mathematics had always been hindered in its ability to model such systems.

The simplest case is easy to understand. If a bicycle travels at exactly 10 miles per hour, then after 1 hour it has traveled 10 miles; after 2 hours it has traveled 20 miles, and so on. But most situations do not fit this neat pattern. With an accelerating and decelerating bicycle, it is difficult to pin down the relationship between its speed at any given moment and the distance it has covered overall.

The essence of the problem is geometric. It amounts to this: given a curve on a piece of paper, can we work out its steepness at a certain point? If the curve represents the bicycle's position, the curve's steepness represents its rate of change of position, which is to say its speed. This fundamental relationship has been known for thousands of years, since the work of Archimedes. What were missing were the mathematical rules for measuring this steepness, from an algebraic description of the original curve. This was the problem that Newton and Leibniz solved.

Gradients and limits

It is easy to work out the steepness of the graph approximately. Simply replace the curve with a straight line whose steepness roughly matches it at the point being measured. Although this is a rather clumsy approach, Newton and Leibniz saw that it contained the seed of an exact method. Connecting two points with a straight line produces an approximation to the curve between them. If the two points are far apart, the curve may significantly deviate from the straight line. But as the two points become closer, the resulting approximation is more and more accurate. The brilliant idea Newton and Leibniz had was to ask what would happen if the two points were infinitely close together?

ABOVE The relationship between the position, speed and acceleration of a cyclist form elementary examples of calculus at work. More sophisticated techniques are needed to understand flow of air around the cyclist, and help design aerodynamic equipment.

To make this work, they exploited a weird idea: infinitesimal numbers. This allowed them to derive a formula of a straight line that was a perfect fit to the given curve at the point of interest. What was more, the algebraic rules that emerged from this analysis were easy to apply. Just by mastering these rules, rates of change could be calculated by purely algebraic means, with no need to draw graphs or contemplate infinitesimals. The rules included: if the formula for a vehicle's position is x^2, then its speed is $2x$. If the position is x^3, then the speed is $3x^2$. Generally, if the position is x^n, then the speed is nx^{n-1}.

Infinitesimals later fell out of favor as being insufficiently rigorous. It was not until the 19th century that Karl Weierstrass put calculus on a truly firm footing. Nevertheless, the laws that Leibniz and Newton had discovered did survive and have demonstrated their worth in countless scenarios.

The Royal Society's verdict

The dispute between Newton and Leibniz was an acrimonious affair. While friends of the two scholars launched accusations against each other, in Britain the Royal Society set up an investigation into the matter. It came down unambiguously on Newton's side, and its damning conclusion accused Leibniz of outright plagiarism. Perhaps this result was inevitable, however, since Leibniz's account of events was never sought; and what was more, the report's summary was written by the society's president, Newton himself.

33 Differential geometry

BREAKTHROUGH TECHNIQUES OF CALCULUS PROVIDED GEOMETERS WITH THE TOOLS TO SOLVE HARD PROBLEMS.

DISCOVERER JOHANN BERNOULLI (1667–1748), JAKOB BERNOULLI (1654–1705), CHRISTIAAN HUYGENS (1629–95).

LEGACY THE EARLY OUTINGS OF DIFFERENTIAL GEOMETRY ALL INVOLVED CURVES. LATER, WORK ON HIGHER-DIMENSIONAL OBJECTS (SEE PAGE 201), WOULD MAKE DIFFERENTIAL GEOMETRY INDISPENSABLE TO MODERN PHYSICS.

The discovery of calculus by Newton and Leibniz provided geometers with exciting new tools. Within a few years, several previously intractable problems had been solved, showcasing the great practical power of the new technique. This new type of mathematics became known as differential geometry.

If you hold both ends of a chain, not pulling it taut but allowing it to droop in the middle under its own weight, a curve is formed. This curve should be mathematically describable, but what is it?

Galileo contemplated this question in his final work of 1638, *The Discourses and Mathematical Demonstrations Relating to Two New Sciences*. Galileo observed that a parabola was a reasonably close fit. Geometers had been studying parabolas for millennia (see page 57), so this was perhaps the natural first guess. But a chain curve was not a parabola, as the closer analysis of Joachim Jungius showed in 1669. So what was this chain curve, or catenary as it came to be called?

The catenary

Jakob Bernoulli was one mathematician to contemplate the nature of the catenary. Unable to provide any answers himself, he threw the problem open in 1690 in the scientific journal *Acta Eruditorum*.

One person to take up the challenge was Jakob's brother, Johann Bernoulli. He took the greatest pleasure in succeeding where his elder

OPPOSITE The Gateway Arch in St. Louis, Missouri, USA, takes the form of an inverted catenary, the same shape that a chain will form under its own weight if left to hang from its two ends.

brother had failed, later writing to a friend: “The efforts of my brother were without success; for my part, I was more fortunate, for I found the skill (I say it without boasting, why should I conceal the truth?) to solve it in full … the next morning, filled with joy, I ran to my brother, who was still struggling miserably … like Galileo, thinking that the catenary was a parabola. ‘Stop! Stop!’ I said to him, ‘don’t torture yourself any more, trying to prove the identity of the catenary with the parabola, since it is entirely false.’”

What makes Huygens’ curve so unusual is that if you place an object on it and allow it to slide down the curve, it will always take the same time to reach the bottom, no matter how high up the curve it starts.

Johann Bernoulli had exploited the newly developed science of calculus to come up with his answer (see page 181). While the parabola is described by a very simple equation such as $y = x^2$, the catenary involves the far subtler exponential function (see page 155), and can be expressed as $y = \frac{1}{2}(e^x + e^{-x})$.

The Bernoulli dynasty

Jacob and Johann Bernoulli were the first generation of the Bernoulli dynasty, a Swiss family who would enliven European mathematics for the best part of two centuries. Johann’s son, Daniel Bernoulli, would go on to be an early influential figure in fluid mechanics (see page 181). The two are said to have fallen out when Johann shamefully attempted to claim credit for his son’s discoveries in the field of hydraulics. Although the eight or so mathematicians of the Bernoulli family had diverse interests, including making major advances in probability and number theory, their greatest legacy was in the early investigation of calculus, and especially in discovering its marvelous potential for applications in geometry and physics.

The tautochrone problem

Johann Bernoulli was not the only person to solve Jakob’s catenary problem. Gottfried Leibniz reached the same conclusion, as did another European expert on curves, the Dutch scientist Christiaan Huygens.

In 1659, Huygens had already discovered another extraordinary curve. Like a parabola or catenary, it begins steeply, then flattens out into a dip at the bottom. What makes Huygens’ curve so unusual is that if you place an object on it and allow it to slide down the curve, it will always take the same time to reach the bottom, no matter how high up the curve it starts. This curve became known as the tautochrone curve (“tautochrone” meaning “same time”). But what was it?

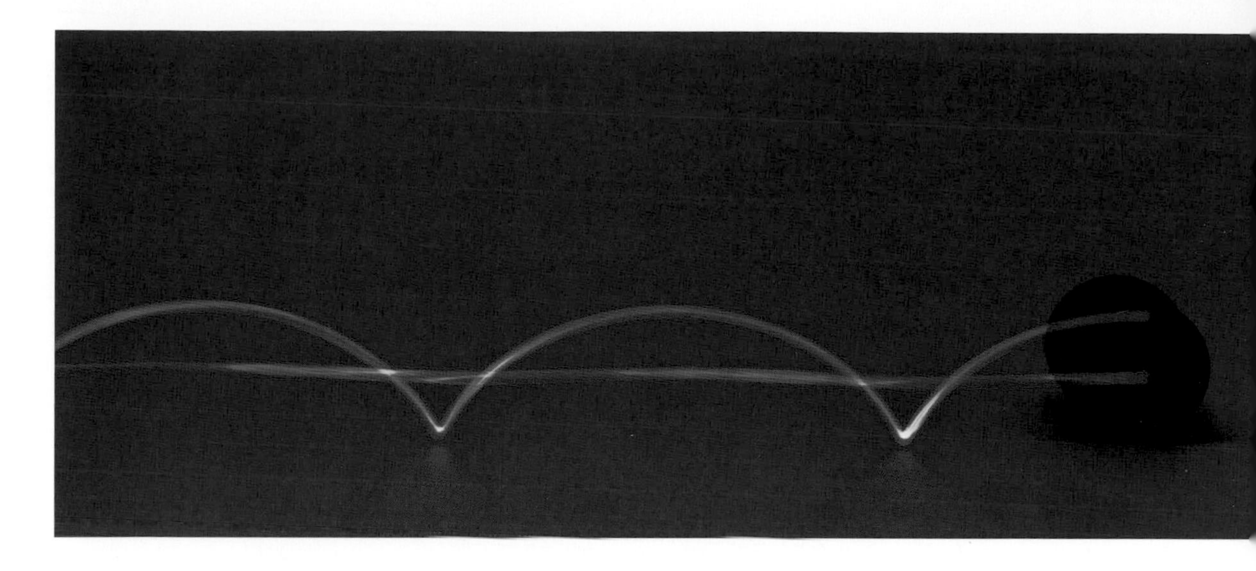

In one of the earliest applications of calculus, Huygens was able to provide the answer: it was a curve called a cycloid. This shape has a very elegant description. If you start with a circle, and mark a point on its rim, and then roll that circle along a horizontal line like a wheel, the path traced out by the specified point is a cycloid.

ABOVE A cycloid produced by a slow-exposure photograph of a cylinder rolling along. A red LED on its rim produces the cycloid curve. The same curve, when turned upside-down, is the solution to the tautochrone and brachistochrone problems.

Outside mathematics, Huygens is principally famous for his invention of the pendulum clock, which revolutionized time-keeping. In fact, he later went on to combine his interests, designing a clock whose pendulum swung in the path of a cycloid, rather than in the arc of a circle. This was a less successful innovation, however.

The brachistochrone problem

In 1696, it was Johann Bernoulli's turn to pose a problem to the readers of *Acta Eruditorum*. The question he asked was: suppose you want an object to slide along a ramp, starting from a certain point on the wall and ending at a specified place on the floor. What shape should the ramp be, for the object to descend quickest? This question became known as the brachistochrone problem, meaning "shortest time." The obvious answer was a straight line, but Galileo had already established that using a segment of a circle was actually quicker. Was that the answer?

Several people were able to apply the new methods of differential geometry to deduce the answer to the brachistochrone problem. They included Bernoulli himself, his brother Jakob, and both Gottfried Leibniz and Isaac Newton. The answer, they showed, was the same as for Huygens' tautochrone problem: the cycloid.

34 Polar coordinates

BREAKTHROUGH POLAR COORDINATES ARE A WAY OF DESCRIBING THE POSITION OF A POINT ON A PLANE, USING A DISTANCE AND AN ANGLE.

DISCOVERER ARCHIMEDES (C. 287–212 BC), JACOB BERNOULLI (1654–1705).

LEGACY FOR MANY SHAPES, SUCH AS THE FAMOUS SPIRALS OF ARCHIMEDES AND BERNOULLI, POLAR COORDINATES ARE FAR MORE CONVENIENT THAN THE CARTESIAN KIND.

The history of geometry has repeatedly shown the benefits of analyzing shapes using the more abstract approach of algebra. But there are various different ways to convert geometrical objects into numerical ones. The most common system is Cartesian coordinates (see page 129), but no less important is the system of polar coordinates devised by Jacob Bernoulli.

In his book *On Spirals*, Archimedes describes one of his most famous geometrical discoveries. In fact, he attributes what we know as the Archimedean spiral to his late friend, the astronomer Conon of Samos, who is thought to have been the first to consider the object.

Archimedean spiral

Archimedes' spiral is a supremely elegant figure, beginning at the center of a page and then gradually cycling outward. The defining characteristic of Archimedes' spiral is that subsequent turns of the curve are always the same distance apart.

An Archimedean spiral is a very natural object. If you start with a piece of string and start coiling it around itself, the result is an Archimedean spiral. In fact, our Solar System contains an enormous such spiral, namely the magnetic field of the sun, which spirals outward from the center.

Archimedes' main interest in the spiral was as a tool for constructing other shapes. In common with the geometers of his period, Archimedes

OPPOSITE The Whirlpool Galaxy, at around 23 million light-years away, is a spiral galaxy. Its arms unfold outward along a logarithmic spiral, a shape that fascinated Jacob Bernoulli.

was fascinated by ruler-and-compass constructions such as squaring the circle and trisecting an angle (see page 194). We now know that these problems cannot be solved by ruler and compass alone. However, Archimedes was able to prove that with an extra tool in the shape of an Archimedean spiral, they can indeed be accomplished.

Logarithmic spiral

Many centuries later, the Renaissance thinker Jacob Bernoulli also became fascinated by a beautiful geometric spiral. But the shape that Bernoulli called "*spira mirabilis*" (or wonderful spiral) was not an Archimedean spiral, but what we now call a logarithmic spiral. While the turns of the Archimedean spiral are always the same distance apart, the logarithmic spiral becomes ever sparser as it cycles outward. Equivalently, it becomes ever more tightly raveled as it spirals inward. Indeed, starting at any point and traveling along the curve toward the center, the spiral crosses the horizontal axis infinitely many times, at ever tinier distances from the center.

Since their discovery, logarithmic spirals have become famous for appearing in nature in a diverse range of scenarios, from storm formations and spiral galaxies to nautilus shells and the flight paths of certain animals.

The logarithmic spiral has several very beautiful attributes. The one that especially bewitched Bernoulli was its self-similarity, a trait shared by fractal patterns (see page 253). If you expand the entire shape by a fixed amount, the resulting spiral is indistinguishable from before; as each turn expands, it simply leads into the next one.

Since their discovery, logarithmic spirals have become famous for appearing in nature in a diverse range of scenarios, from storm formations and spiral galaxies to nautilus shells and the flight paths of certain animals. The reason for this ubiquity comes from another beautiful criterion they satisfy. A circle is a curve with a unique property. If you join any point on the curve to the center, the resulting straight line will always be at an angle of exactly 90 degrees to the curve. A logarithmic spiral has the same property, but the defining angle may be different, in the simplest case 45 degrees, although it is possible to construct logarithmic spirals of any angle. For this reason, this curve is also known as an equiangular spiral.

Polar coordinates

Since Descartes' introduction of Cartesian coordinates (see page 129), they have been the standard way of describing geometrical objects in algebraic language. However, objects like Archimedean and logarithmic spirals are not easily described in this language. Bernoulli realized that a far more natural and intuitive description is in terms of polar coordinates. Indeed, in thinking about his spiral, Archimedes was essentially reasoning in the same terms, but it was not until the 17th century that polar coordinates became a standard geometric tool.

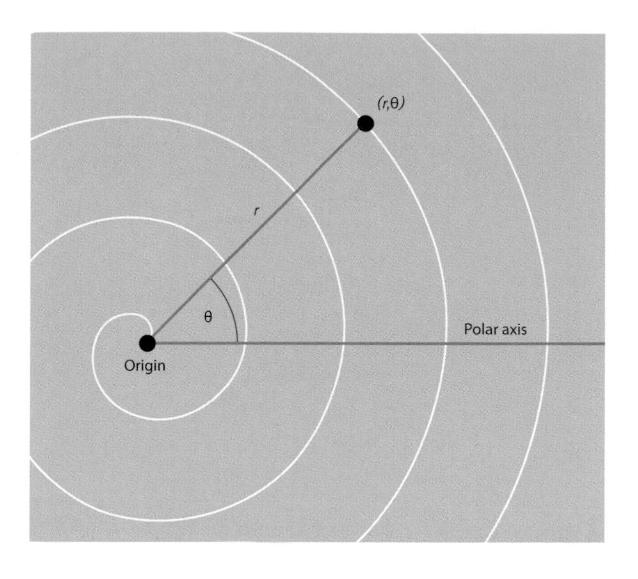

ABOVE Polar coordinates are specified by a distance r from the origin, and an angle θ, similar to a geographer's bearing. Certain shapes, such as this Archimedean spiral, admit simple descriptions in polar coordinates.

Like Cartesian coordinates, polar coordinates specify every point on the plane using two numbers. While Cartesian coordinates amount to defining a point by its grid reference on a map, polar coordinates are akin to instructions such as "Travel 2 miles north-west." The Cartesian system measures a point's position along horizontal and vertical axis, but polar coordinates measure the direct distance, usually denoted r, from the point to the center (or origin) of the plane. If $r = 2$, say, this information places the point on a circle of radius 2, centered at the origin.

To fix the point exactly, its second coordinate is the angle of elevation above the horizontal, usually written as θ. So if $\theta = 45°$ (which might also be written $\theta = \frac{\pi}{4}$ in radians, mathematicians' favorite measure of angle), then the point is the unique position on that circle at a bearing of 45 degrees, as it were.

Polar curves

Some curves have simpler descriptions in polar form than in Cartesian coordinates. The Archimedean spiral, for example, can be described by the beautifully simple equation $r = \theta$. As the spiral cycles round and round, the total angle through which it has rotated grows ever larger, as does its distance from the center. What is more, these two measurements grow at exactly the same rate.

The logarithmic spiral can also be represented elegantly in polar coordinates as $r = e^{\theta}$, or equivalently $\theta = \log r$. This is incomparably simpler than its Cartesian equation. This equation defines the curve with an angle of $\frac{\pi}{4}$, but another angle can be represented by the equation $r = e^{a\theta}$, for a suitable value of a.

35 Normal distribution

BREAKTHROUGH THE NORMAL DISTRIBUTION IS THE MOST IMPORTANT TOOL IN THE THEORY OF PROBABILITY. IT WAS FIRST DISCOVERED BY ABRAHAM DE MOIVRE DURING INVESTIGATIONS INTO COIN TOSSES.

DISCOVERER ABRAHAM DE MOIVRE (1667–1754), CARL FRIEDRICH GAUSS (1777–1855), PIERRE-SIMON LAPLACE (1749–1827).

LEGACY ALMOST EVERY PIECE OF MODERN STATISTICS AND DATA ANALYSIS USES THE NORMAL DISTRIBUTION OFTEN VIA THE CENTRAL LIMIT THEOREM

The mathematical theory of probability is designed to model random events, such as the outcome of a coin-toss. But even in such a simple situation, there are surprises in store. Tossing a coin 100 times, one would expect to get around 50 heads and 50 tails. Of course, this is unlikely to be exactly right. Yet, as more and more tosses are performed, the overall proportion of heads will get closer and closer to $\frac{1}{2}$. This fundamental fact is known as the *law of large numbers,* and it is the first of a series of results which retrieve order from even the most unpredictable situations. Central to this theory is a fundamental object known as the normal distribution.

The modern theory of probability began in a 1654 correspondence between Pierre de Fermat and Blaise Pascal about the so-called problem of points, a dilemma in a gambling game. Two people are playing a simple game of chance, which simply involves tossing a coin. Every time the coin lands heads up, Pierre wins a point. Every time it lands tails up, Blaise gets a point. The winner is the first person to ten points, and that person wins all the money in the pot. The difficulty is this: suppose something happens to stop the game—maybe the coin gets lost. The current score is 6–4 to Pierre. The two decide to abandon the game and split the pot as fairly as possible so as to best reflect the state of the game. But how should this be done?

OPPOSITE Games of chance, involving dice and coins, have been enjoyed by games-players and gamblers for at least 5000 years They also formed the first experiments for investigating probability, and even in this simple setting the sophisticated mathematics of the normal distribution quickly arises.

RIGHT The central limit theorem guarantees that if you repeatedly toss 100 coins and plot the number of heads on a graph, the normal distribution's tell-tale bell curve will be revealed.

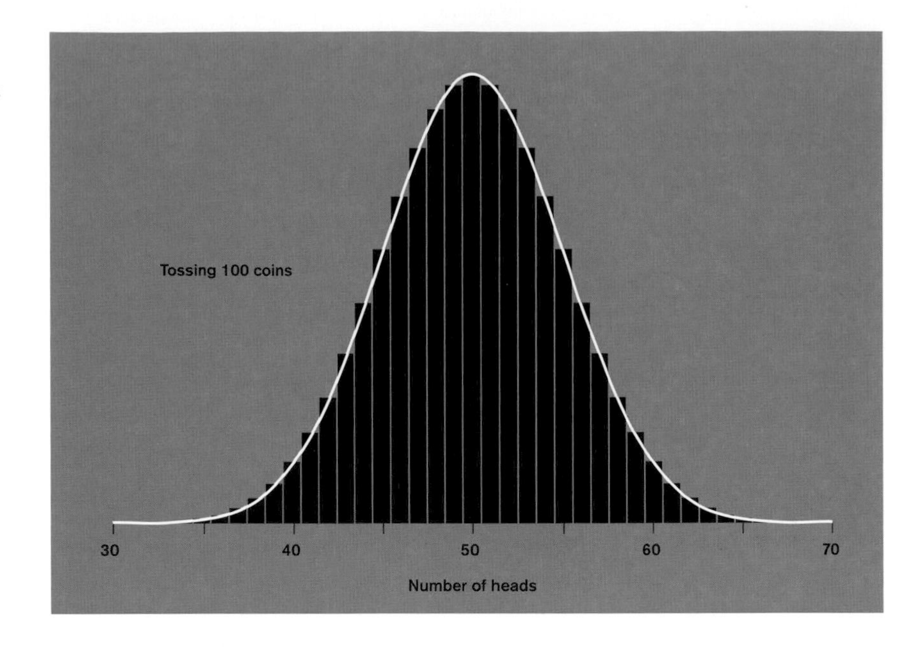

The problem of points

The naïve solution is for Pierre to take 60% of the pot and Blaise 40%. But the question is much subtler than it seemed. First, they had to decide what should count as an equitable arrangement. In doing this, Fermat and Pascal laid the platform for modern probability theory.

The method they derived hinged on formulating the probability that each player would win from the current situation, should they decide to play on. In this example, the probability of Pierre winning is actually $\frac{191}{256}$, while the probability of Blaise winning is $\frac{65}{256}$. These are the proportions into which the pot should be divided. These ideas would be central to the science of probability in the coming centuries.

The normal distribution

Abraham de Moivre was one of the first to try to understand the deeper mathematics underlying probability theory, in the early 18th century. His most important discovery was the normal distribution, which would go on to become one of the most ubiquitous objects in the whole of science. De Moivre did not fully flesh out his work, nor did he appreciate its profound importance. It was the later research of Carl Friedrich Gauss and Pierre-Simon Laplace which brought the normal distribution to prominence. For this reason, it is also known as the Gaussian distribution, or the bell curve. Today, the normal distribution is the primary tool used to describe how data is spread out.

The normal distribution occurs in a wide variety of scientific and statistical scenarios. For instance, if a marine biologist measures the lengths of Atlantic croaker fish, she is likely to see some variety—some fish will be longer than others. The so-called "expected" length here is 10 inches but the terminology is slightly unfortunate. It does not mean we should literally expect each fish to be 10 inches long. But if we took the average of a lot of fish, the answer would indeed be very near this value. When the lengths of the individual fish are compared with that expected value, some will be longer and some will be shorter. This spread is described by a normal distribution.

One of the reasons that the normal distribution is so important is that it arises in innumerable contexts—even where it is not immediately visible.

A normal distribution is determined by two numbers. The first is the expected value. The second is the standard deviation, which quantifies the spread: a small standard deviation means that the fish are all concentrated very near the expected value; while a larger standard deviation implies a broader spread. In the case of croaker fish, the standard deviation is 2 inches.

The central limit theorem

One of the reasons that the normal distribution is so important is that it arises in innumerable contexts - even where it is not immediately visible. Coin tosses seem to require a totally different type of mathematics from that of measuring fish. After all, the lengths of fish take possible numbers between some maximum and minimum. So it seems intuitively reasonable that they should form a bell curve.

A coin, on the other hand, can only produce only two results: heads or tails. Nevertheless, de Moivre found hints of a fundamental fact which would later become known as the central limit theorem. It says that here, too, there is a normal distribution lurking, which will emerge after enough time. If a coin is tossed 100 times, the average score for heads is the number of heads divided by the total number of tosses. The law of large numbers tells us that we should expect the answer to be round about $\frac{1}{2}$. But, of course, it is unlikely to be exactly that. So how are the possible values distributed around this expectation? The central limit theorem gives the answer: they are approximately normally distributed. What is more, the approximation becomes ever better, the more tosses are performed. Tossing a coin 1000 times will give a result very close indeed to a true normal distribution.

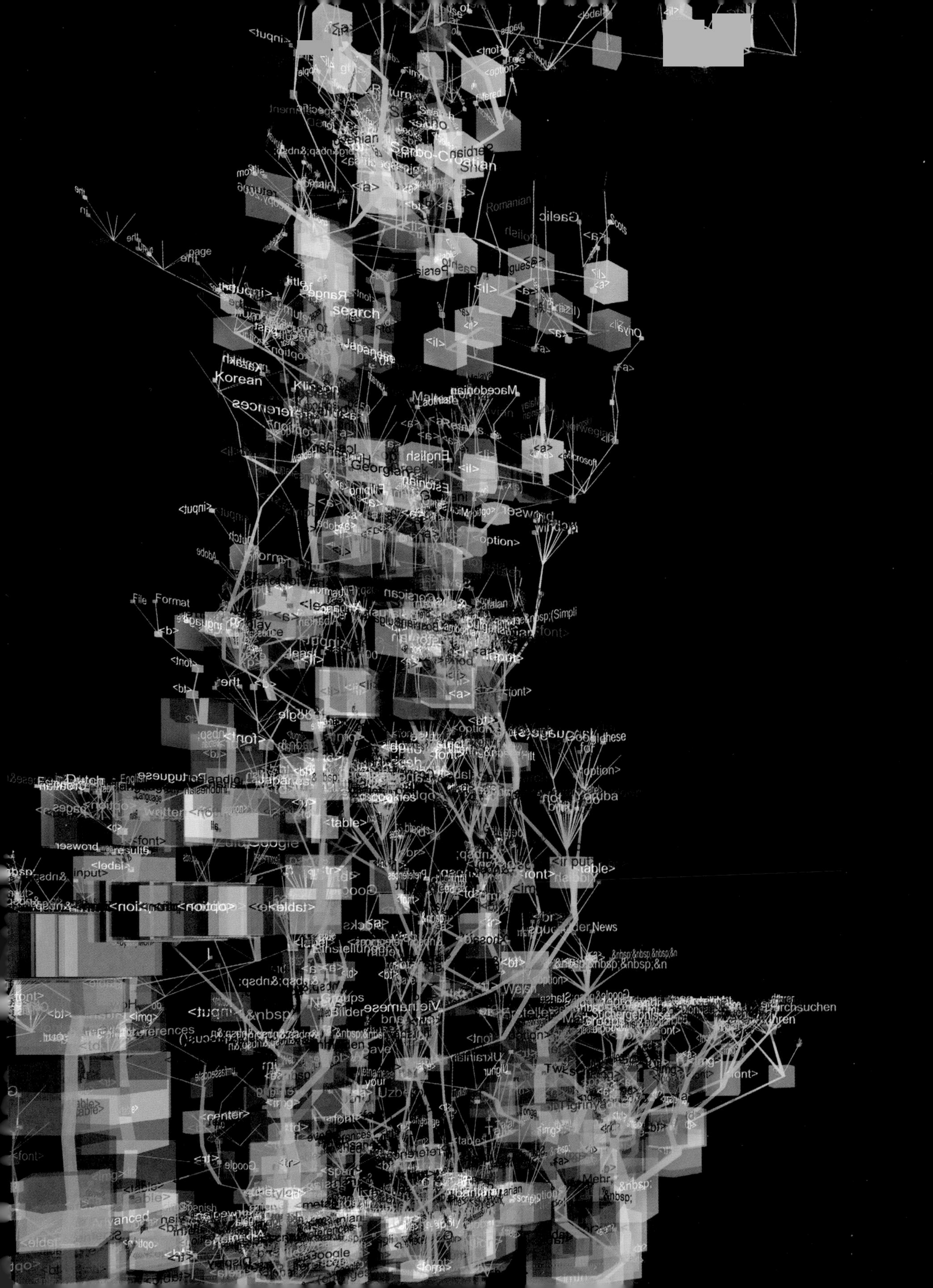

Serbo-Croatian
Romanian
Gaelic
Scots
search
Range
Japanese
Korean
Kazakh
Macedonian
English
Georgian
Norwegian
Microsoft
<option>
File
Format
Catalan
Corsican
Google
these
Portuguese
Dutch
Yoruba
<table>
browser
results
Einstellungen
Groups
Bilder
Vietnamese
Ukrainian
Welsh
News
Suchergebnisse
durchsuchen
Mehr
Advanced
Spanish
<center>
<input>
<font>
Display

36 Graph theory

BREAKTHROUGH GRAPHS ARE NETWORKS OF POINTS CONNECTED WITH EDGES. SUCH SIMPLE OBJECTS CAN EFFICIENTLY CAPTURE THE CRITICAL INFORMATION IN A GIVEN GEOMETRIC SITUATION.

DISCOVERER LEONHARD EULER (1707–83).

LEGACY GRAPHS DISTIL A PROBLEM TO ITS ESSENCE. THEY APPEAR THROUGHOUT MATHEMATICS, FROM THEORETICAL TOPOLOGY TO THE MOST PRACTICAL COMPUTATIONAL PROBLEMS.

Some mathematical breakthroughs usher in a new era of sophistication, where advanced techniques make it possible to settle technically demanding problems. But other discoveries go the other way. By stripping down a problem to its bare bones, a seemingly complex problem may be reduced to a simple one.

A famous case was the birth of graph theory, which is in constant use today to analyze networks of all kinds. Graph theory began with Leonhard Euler's solution to a strange riddle about the European town of Königsberg. Königsberg is now known as Kaliningrad and is part of Russia. In previous centuries, however, it was the capital of the German state of Prussia, and was a celebrated center of European intellectual life.

The bridges of Königsberg

Königsberg sat on the River Pregel, which split it into four districts. The districts were connected with seven bridges. These bridges presented the residents with a puzzle: was it possible to navigate the entire city, crossing each bridge exactly once?

All attempts to find this perfect city tour failed, leading to the conclusion that it couldn't be done. But might it not be that the right route simply hadn't yet been found? To rule that possibility out once and for all required more than experimentation; it needed a proof. When Carl Ehler, the mayor of nearby Danzig, mentioned the riddle to the great mathematician Leonhard Euler, Euler was initially dismissive, saying

OPPOSITE The source code of www.google.com illustrated as a mathematical graph. Source codes for creating websites are written in HTML, which is formatted by tags, each of which may contain several sub tags. Tags are represented as nodes, and connected to their sub tags with edges.

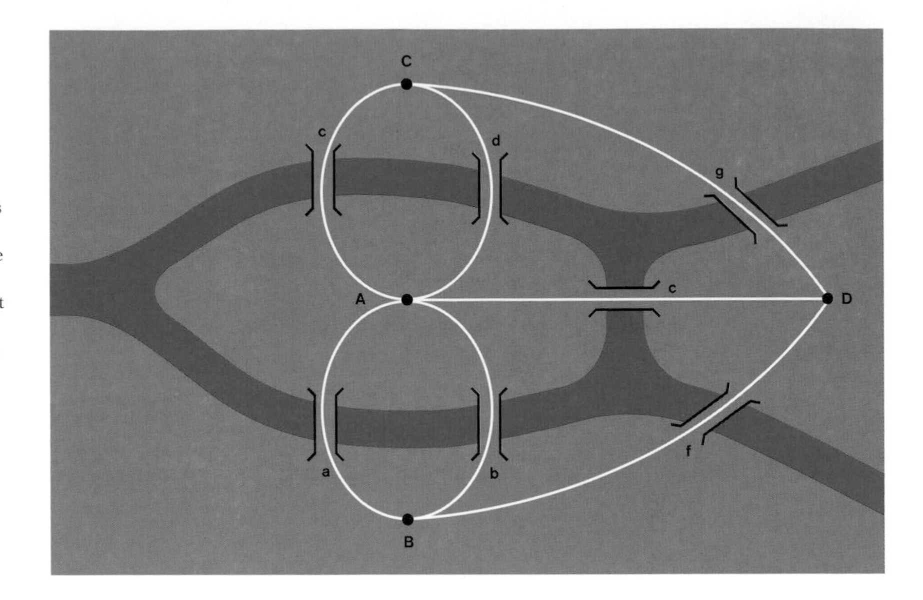

RIGHT The riddle of the bridges of Königsberg. In a graph-theoretic approach, the five landmasses are represented nodes, and the seven bridges become edges connecting them. The solution to the riddle hinges on the fact that every node has an odd number of edges emanating from it.

that the puzzle "bears little relationship to mathematics ... the solution is based on reason alone, and its discovery does not depend on any mathematical principle." Nevertheless, Euler solved the riddle quickly, establishing as expected that no perfect circuit of the town was possible. Although he initially considered the puzzle a triviality, he later wrote that it "seemed to me worthy of attention in that neither geometry, nor algebra, nor even the art of counting was sufficient to solve it." Indeed, what Euler had used was the rudiments of a brand-new type of mathematics, known as graph theory.

Graph theory

A graph, in this setting, is a network of nodes, some of which are connected with edges. Although he did not express it in these terms, Euler's solution to the problem of the bridges amounted to the realization that almost all the geographical details of the city were irrelevant. All that really mattered was the underlying graph: four nodes, representing the four city districts, and seven edges, representing the bridges. Euler called this approach "*geometria situs*," the geometry of place, ascribing the approach to Gottfried Leibniz. Their new geometry was a foretaste of topology (see page 213) where graphs play a major role.

Once reduced to a graph, the Königsberg bridge problem was easy to solve. The degree of a node is the number of edges that emerge from it. Euler established that to be navigable without repeating any edges, the degree of every node must be an even number. But the Königsberg graph has three nodes of degree 3 and one of degree 5.

Graphs and geometry

Graphs are extremely simple objects but encapsulate some very difficult problems. For example, there is no requirement when drawing a graph that edges should not cross. In fact, it is impossible to draw many graphs without edges crossing. This is the basis of a long-standing puzzle in which three houses each have to be connected to three utility companies: gas, water and electricity. However you try to organize the pipes, it is impossible to avoid at least one crossing.

Graphs that can be drawn on a page without edges crossing are known as planar. The three utilities puzzle centers on the fact that a graph in which three points are connected to each of three other points is a non-planar graph. Another is the complete graph on five vertices. This consists of five points, each connected to every other. (Four points can be connected without edges crossing.)

Euler's solution to the problem of the bridges of Königsberg amounted to the realization that almost all the geographical details of the city were irrelevant. All that really mattered was the underlying graph.

In 1930, Kazimierz Kuratowski established a surprising result: that these two special graphs are decisive in determining planarity. Kuratowski's theorem states that every non-planar graph must encode either the complete five-point graph or the three utilities graph.

Graphs and algorithms

In recent years, graph theory has become particularly relevant to deep and difficult questions of computer science. For example, imagine two large graphs, which appear different. But in graph theory, the exact positions of the nodes and the lengths of the edges are irrelevant. All that matters is whether two nodes are connected. So it might be that the two graphs are actually identical, despite appearances. But how would one check? This is the graph isomorphism problem.

In principle, it is easy to solve: one has to keep comparing nodes and edges of one graph against the other until all the possibilities are exhausted. The difficulty with this, however, is that it is likely to be an extremely lengthy procedure. In the jargon of computational complexity theory, the graph isomorphism problem is in *NP* (see page 339), but not known to be in *P*, which suggests it is highly unlikely to be tractable in the real world. Another famous example of a computationally intensive graph problem is the traveling salesman problem (see page 341).

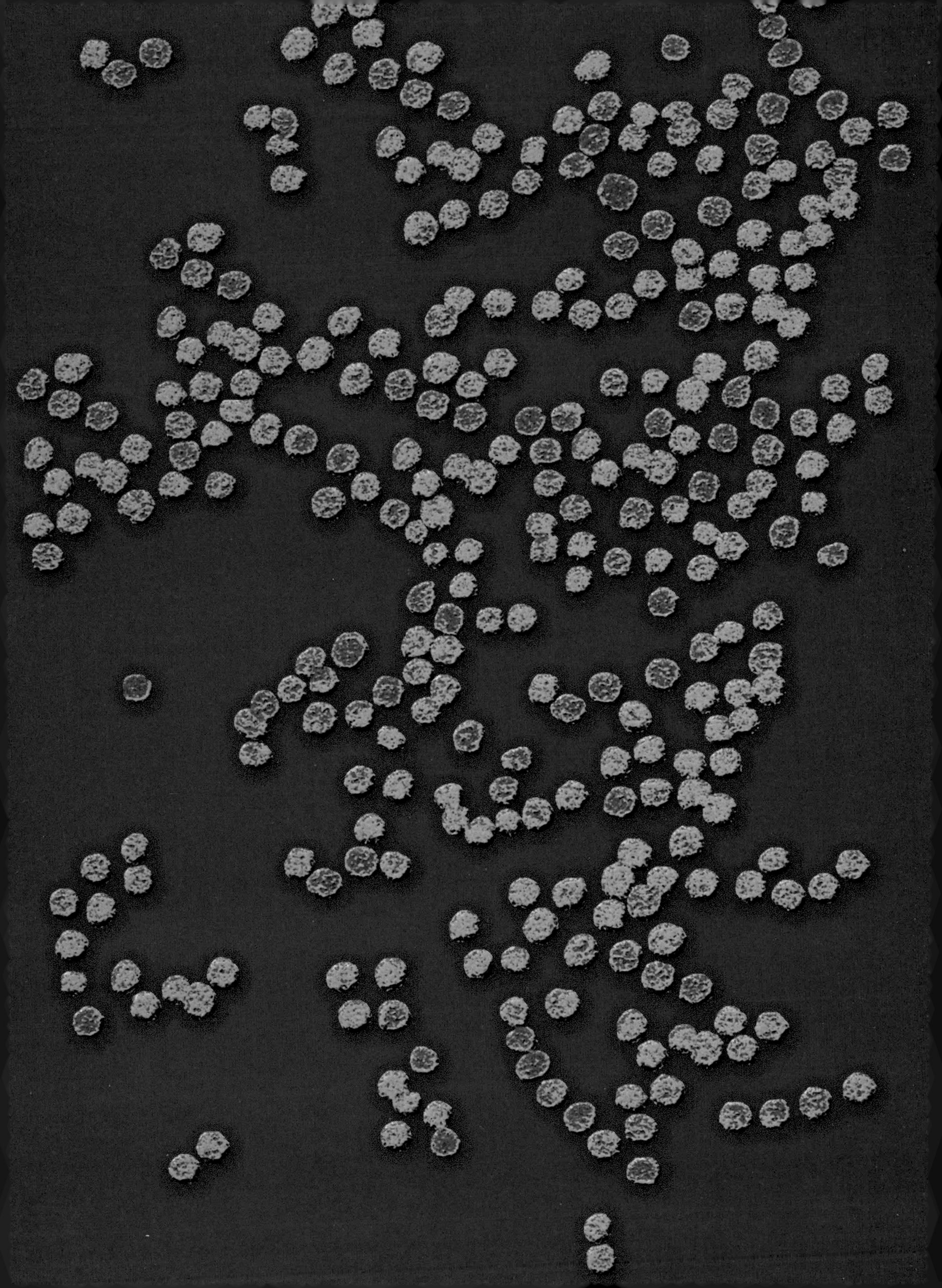

37 Exponentiation

BREAKTHROUGH IN THE 18TH CENTURY, THE NEW MATHEMATICAL TECHNOLOGY OF POWER SERIES ALLOWED THE FIRST EXTENSIVE ANALYSIS OF EXPONENTIAL GROWTH.

DISCOVERER JAMES GREGORY (1638–75), BROOK TAYLOR (1685–1731), LEONHARD EULER (1707–83).

LEGACY EULER'S WAS THE FIRST SERIOUS WORK IN THE SUBJECT OF COMPLEX ANALYSIS, WHICH CONTINUES TO BE A MAJOR TOPIC OF RESEARCH TODAY. OUT OF IT ALSO CAME THE MOST BEAUTIFUL EQUATION IN MATHEMATICS: EULER'S FORMULA.

For as long as mathematicians have contemplated numbers, they have understood a fundamental relationship between addition and multiplication, namely that multiplication is repeated addition. It is only a small step up to consider powers, meaning repeated multiplication. This perspective works well among the whole numbers, but how do powers sit within the sophisticated modern framework of the complex numbers? The answer to this question involved one of the most prized techniques in mathematical analysis: power series. It also resulted in the single most beautiful theorem in the subject: Euler's formula.

Every child learns that 4×3 is the number you get when you add three fours together: $4 + 4 + 4$. Expressed this way, it is by no means obvious that 4×3 and 3×4 should be the same thing. That they are equal can be seen in an array of objects arranged in three rows and four columns. Depending on whether you count the objects in rows or columns, this can either be thought of as three sets of four objects, or four sets of three.

This type of reasoning is likely to be some of the earliest abstract mathematics performed. A little more recently, mathematicians also began to consider exponentiation, or, as it is also known, raising to a power. Essentially, this is repeated multiplication. So $4^3 = 4 \times 4 \times 4$. The first suggestion that this is a trickier procedure is that $4^3 \neq 3^4$. Powers are central to several of the greatest mathematical stories, including Fermat's Last Theorem (see page 369) and Waring's problem (see page

OPPOSITE SARS viruses reproducing. Many organisms, viruses included, will reproduce at an exponential rate if left unchecked, which is one of the reasons that even a mild infection can be so dangerous.

241). But in these cases the exponent is usually a whole number.

Complex exponentiation

With the emergence of the new system of complex numbers (see page 105), it was necessary to extend arithmetical operations to work in this new, unfamiliar domain. For addition and multiplication, this was no great problem, the obvious approach worked smoothly. Exponentiation, on the other hand, was a much more delicate topic. What might it mean to raise a number such as 2 to a complex power: 2^i? Of course, the answer might be nothing at all. There is no reason to believe that every possible combination of mathematical symbols need have a sensible interpretation, any more than every collection of letters spells a meaningful word.

ABOVE The test of a 61-kiloton atomic bomb, in Nevada, 1953. Atomic bombs work by nuclear fission. As atoms of plutonium or uranium split into smaller atoms, they release energy and neutrons. These neutrons continue the chain reaction. It is their exponential growth which gives the bomb its power.

Yet it would turn out that 2^i can be endowed with meaning. Doing so would lead to the first serious work in the emerging subject of complex analysis. The idea came from the work of two British mathematicians: James Gregory and Brook Taylor, who explored what would come to be known as power series. In fact, the Indian astronomer Madhava had had similar insights hundreds of years earlier, contemplating the trigonometric functions sine and cosine (see page 63). These objects would acquire a new significance in the world of complex analysis.

Power series

A power series is the result of adding together increasing powers of the same number. The simplest case is just to add up all the powers of some number x, like this: $1 + x + x^2 + x^3 + x^4 + \dots$. It is not obvious that this will converge to the number $\frac{1}{1-x}$ (so long as $0 < x < 1$).

Like Madhava before him, James Gregory investigated sine and cosine, the functions that people had used for millennia to derive facts about the geometry of triangles. No formulae were previously known for these objects, but Gregory discovered that they could be elegantly expressed as power series:

$$\cos x = 1 - \frac{x^2}{2} + \frac{x^4}{4 \times 3 \times 2} - \dots$$

$$\sin x = x - \frac{x^3}{3 \times 2} + \frac{x^5}{5 \times 4 \times 3 \times 2} - \dots$$

Brook Taylor took this observation further. Mathematics is full of functions, meaning rules which take one number as input and spit out another as output. Taylor proved the hugely important theorem that most—or even all—important mathematical functions can be expressed as suitable power series.

There is no reason to believe that every possible combination of mathematical symbols need have a sensible interpretation, any more than every collection of letters should spell a meaningful word.

The exponential function

If every reasonable mathematical function can be expressed as a power series, Leonhard Euler reasoned that this might yield a way to make sense of complex exponentiation. He derived the expression for the most important power series of all, the exponential function:

$$e^x = 1 + x + \frac{x^2}{2} + \frac{x^3}{3 \times 2} + \frac{x^4}{4 \times 3 \times 2} + \frac{x^5}{5 \times 4 \times 3 \times 2} + \dots$$

The best-known instance of this series happens at the input value $x = 1$. Here, the output is the number e at around 2.7183.

This function has several properties which made it of unique importance to mathematicians in the centuries that followed. In particular, Euler realized that this was the right way to make sense of expressions like 2^i (it turns out to have a value of around $0.77 + 0.64i$).

Euler's formula

Euler's new exponentiation on the complex numbers was compatible with the usual "repeated multiplication" approach on the whole numbers. But he also made a further wonderful observation. He noticed that the power series of e^x looked rather like those for $\sin x$ and $\cos x$. More specifically, when Euler put an input value of i times z into the exponential function, he got a series which was exactly the same as the series for $\cos z$ added to i multiplied by the series for $\sin z$. What Euler had shown is that $e^{iz} = \cos z + i \sin z$.

This formula is true for all values of z. But something delightful happened when he picked the value $z = \pi$. In radians (mathematicians' preferred way of measuring angles), π represents half a revolution, or 180°. It is a fundamental fact of trigonometry that the sine of π is 0, while its cosine is -1. Hence, $e^{i\pi} = -1 + i \times 0$. Rearranging this slightly gives Euler's formula, by common consent the most beautiful in mathematics:

$$e^{i\pi} + 1 = 0$$

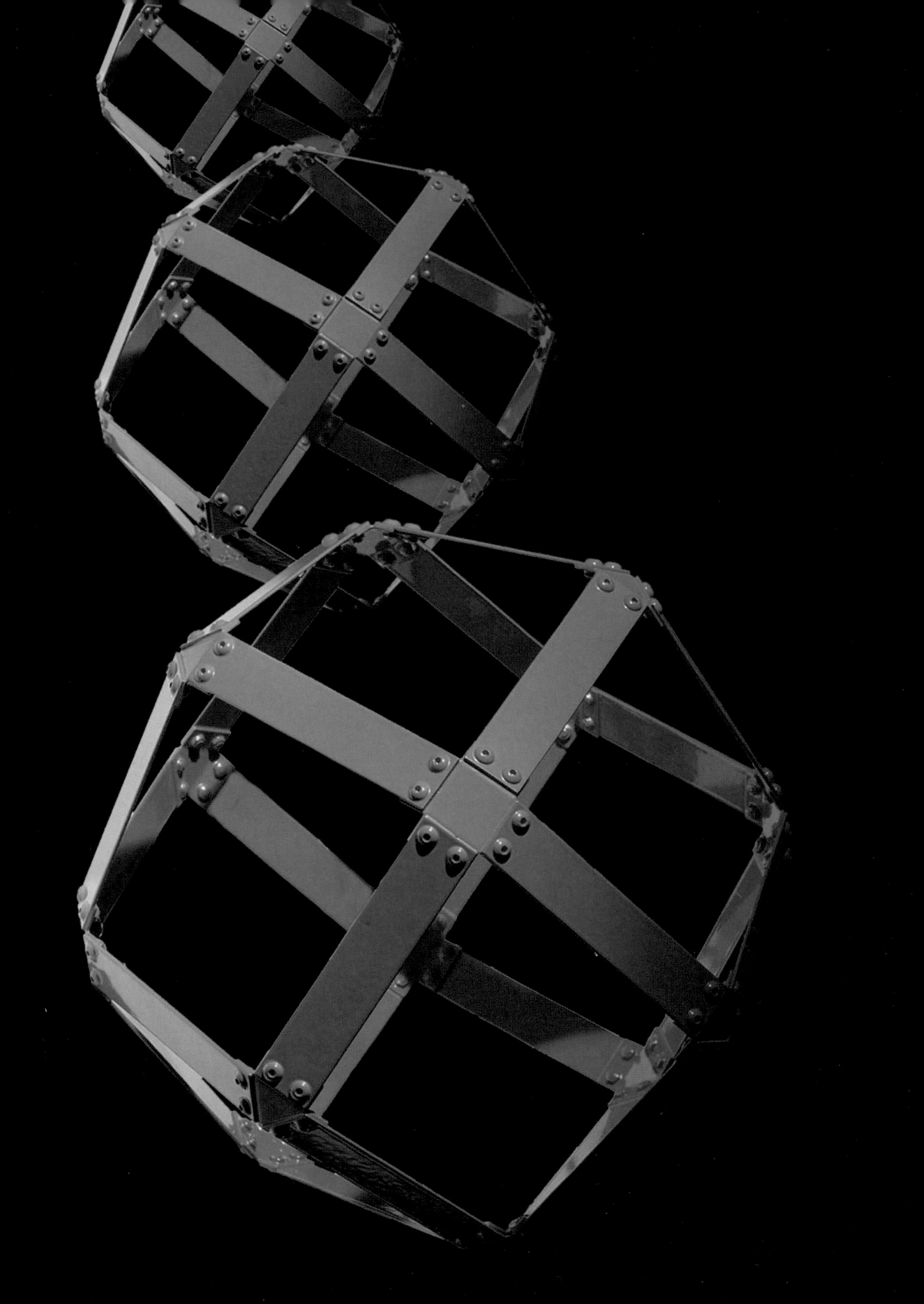

38 Euler characteristic

BREAKTHROUGH EULER DISCOVERED A REMARKABLE RELATIONSHIP BETWEEN THE NUMBER OF FACES, EDGES AND CORNERS ON A POLYHEDRON.

DISCOVERER LEONHARD EULER (1707–83).

LEGACY EULER'S OBSERVATION REPRESENTS A FUNDAMENTAL FACT ABOUT 3-DIMENSIONAL SHAPES, WHICH HAS BEEN A VITAL TOOL TO SCIENTISTS EVER SINCE. EFFORTS TO GENERALIZE THE EULER CHARACTERISTIC HAVE PRODUCED DEEP DEVELOPMENTS IN THE SCIENCE OF SHAPE.

Polyhedra are 3-dimensional shapes constructed from flat faces and straight edges meeting at corners. In 1750, Leonhard Euler wrote to his friend Christian Goldbach describing a beautiful pattern which elegantly relates the number of faces, edges and corners on any polyhedron.

The list of possible polyhedra is extensive, from a simple cube, to the truncated dodecahedron (which is to say a soccer ball), to magnificent geodesic spheres, such as that at the Epcot theme park in Florida. Yet, as Euler discovered, all these various shapes have something in common. Euler began with a polyhedron, such as a cube or truncated dodecahedron, and first counted its number of faces. In the case of the cube there are six, while the soccer ball has 12 pentagonal and 20 hexagonal faces, making a total of 32. In general, call the number of faces F.

Next, count the edges, and call that number E. A cube has 12, while for a soccer ball $E = 90$. Finally, count the number of corners, and denote that number by V (standing for vertices). For a cube $V = 12$; while on a soccer ball $V = 60$.

At first sight, there doesn't seem to be much pattern in these numbers. It is not surprising that shapes as different as a cube and soccer ball have correspondingly different values of F, E and V. Of course, another polyhedron will have different values again. But Euler noticed an astonishing similarity beneath the surface. Something remarkable happened when he calculated the quantity $V - E + F$ for each shape. A

OPPOSITE These polyhedra divide the surface of a sphere into twelve faces each with four sides. Known as a rhombic dodecahedron, this shape has 12 faces, 24 edges, and 14 vertices.

cube produces $8 - 12 + 6 = 2$, and a truncated dodecahedron produces $32 - 90 + 60 = 2$. Different though these two shapes are, in each case $V - E + F$ comes out as the same value: 2.

This is no coincidence. The underlying geodesic sphere at Epcot has $F = 11{,}520$, $E = 17{,}280$, and $V = 5762$, for which also $V - E + F = 2$. If you try the same thing with a dodecahedron, a pyramid or a hexagonal prism, the same result will hold.

Euler's result about polyhedra has been of enormous value for deepening our understanding of polyhedra and related shapes, as it imposes strict limits on what is possible. For example, as Euler wrote to Goldbach, it eliminates the possibility of any polyhedron having exactly seven edges.

Euler characteristic

There is a sense in which the scope of Euler's polyhedral formula extends even beyond polyhedra. It does not rely on the faces being flat, or the edges straight. If you start with a sphere and place a single dot on it, with a single "edge" circling round the equator of the sphere joining the dot to itself, this divides it into two hemispherical "faces." In this case, $V = 1$, $E = 1$ and $F = 2$. So the result is valid even here.

The question of why Euler's formula is true and what exactly it is telling us has been a major driver of mathematics for several centuries, especially in the field of topology.

This was one of the first great theorems of topology, long before that subject acquired its name, or was even recognized as a branch of mathematics in its own right. Today, topologists judge seemingly different shapes to be fundamentally the same if one can be pulled and stretched into the shape of the other. Crucially, the cube, pyramid, hexagonal prism and truncated dodecahedron are all nothing more than spheres from a topological perspective. Euler's theorem says that any way of dividing up a sphere into faces will always give a result where $V - E + F = 2$.

Yet, despite its wide applicability, there are polyhedra for which Euler's result fails. If the underlying shape contains a hole, such as a ring of four cuboids, each built from six rectangles, the answer will appear different. For this shape, $F = 16$, $E = 32$, and $V = 16$, giving $V - E + F = 0$. What is more, the same will apply for any polyhedron containing a single hole. Such shapes are topologically equivalent not to a sphere, but to a torus (or donut shape). Similarly, any way of dividing up a shape with two holes (a double torus) will always produce $V - E + F = -2$. These numbers, 2, 0, -2, describe the shape at a fundamental level. They are known as the Euler characteristic of the sphere, torus and double torus respectively.

Algebraic topology

From the start, the Euler characteristic hinted at a deep theory of space. It was far from clear why adding the number of faces to the number of corners and subtracting the number of edges should result in a meaningful quantity. The question of why Euler's formula is true and what exactly it is telling us has been a major driver of mathematics for several centuries, especially in the field of topology. Because the Euler characteristic can distinguish a sphere from a torus or a double torus, it played a pivotal role in the proof of the classification of closed surfaces (see page 213), perhaps the first true topological theorem.

ABOVE Spaceship Earth at Disney World's Epcot Center in Florida. The mathematics of geodesic domes was developed by Richard Buckminster Fuller in 1954. This example has 11,520 faces, 17,280 edges and 5762 vertices.

It was Henri Poincaré in the late 19th century who pushed the idea even further by lifting it into higher dimensions. Just as a polyhedron is built from vertices, edges and faces, so a higher-dimensional polytope (see page 201) is constructed from cells of lower dimension. Poincaré realized that, for a 3-dimensional polytope, the quantity "$C - F + E - V$" must be fixed, depending only on the underlying topology, not on the specific decomposition into cells (C).

Ultimately, Euler's observation about the faces of a polyhedron would lead to the modern subject of algebraic topology, in which structures inside a shape can be added and subtracted from each other, producing subtle algebraic objects that describe the underlying shape on a deep topological level.

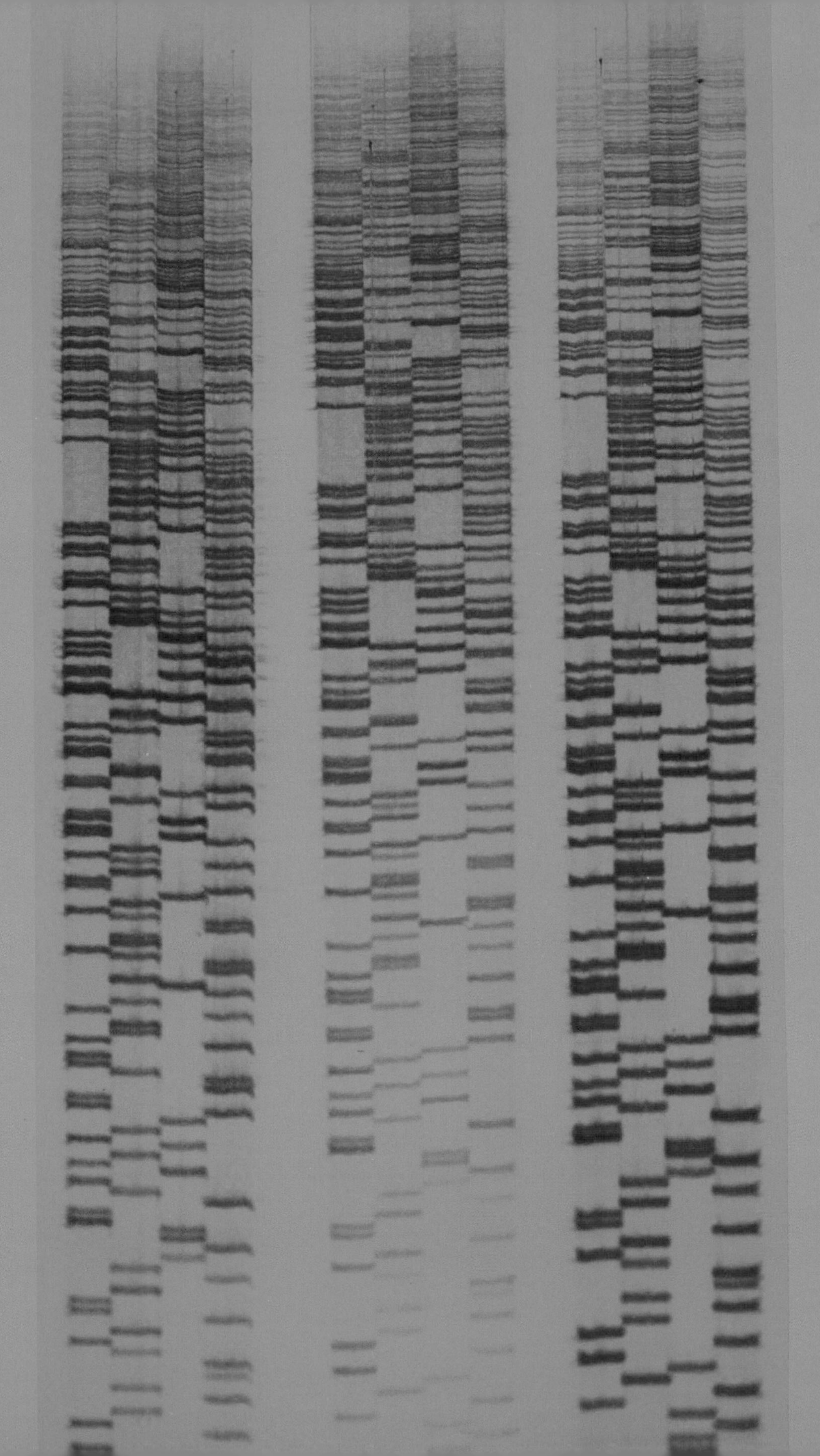

39 Conditional probability

BREAKTHROUGH CONDITIONAL PROBABILITY PERMITS THE ANALYSIS OF EVENTS WHOSE CHANCE OF HAPPENING DEPENDS ON THE OUTCOME OF OTHER EVENTS.

DISCOVERER THOMAS BAYES (1701–61).

LEGACY BAYESIAN REASONING IS USED BY MANY PEOPLE TODAY, FROM ANALYZING DATA OF ALL TYPES, TO ARTIFICIAL INTELLIGENCE RESEARCH.

Probability is typically illustrated using dice, coins and packs of cards. While these simple tools can communicate the basic ideas, the wider world is a messier, more complex place. For probability to be useful in real life, techniques had to be found to take into account events whose likelihood is not fixed, but may depend on the outcomes of other events. The breakthrough was the emergence of a new way to quantify uncertainty: conditional probability.

When coins are tossed and dice are rolled, it is tempting to think of the probabilities being set in stone: a 50% chance of getting a head, 1 in 6 for a six, and so on. Of course, this is the situation before the toss. As soon as the coin lands, this uncertainty is transformed into 100% certainty: a definite yes for a head, or no for a tail. But out in the wider world, things are not so clear-cut. Even with the coin, it is highly unlikely that the odds are exactly 50:50. The details of the coin's design may cause a slight bias, as may the person tossing it, if they have a flawed or somehow predictable technique. So the exact odds of the toss may depend on the specific coin selected, as well as the person chosen to perform the toss, and perhaps ambient conditions, such as the wind.

OPPOSITE An autoradiogram, showing a sequence of DNA bases. Our genetic make-up is determined by that of our parents, with chance determining how their DNA combines. To understand this process requires insights from probability theory. The chance of a hereditary disease being passed on is a question of conditional probability.

Bayes' theorem

What the theory of probability lacked was a method for analyzing events whose probability is not set in stone but depends on the

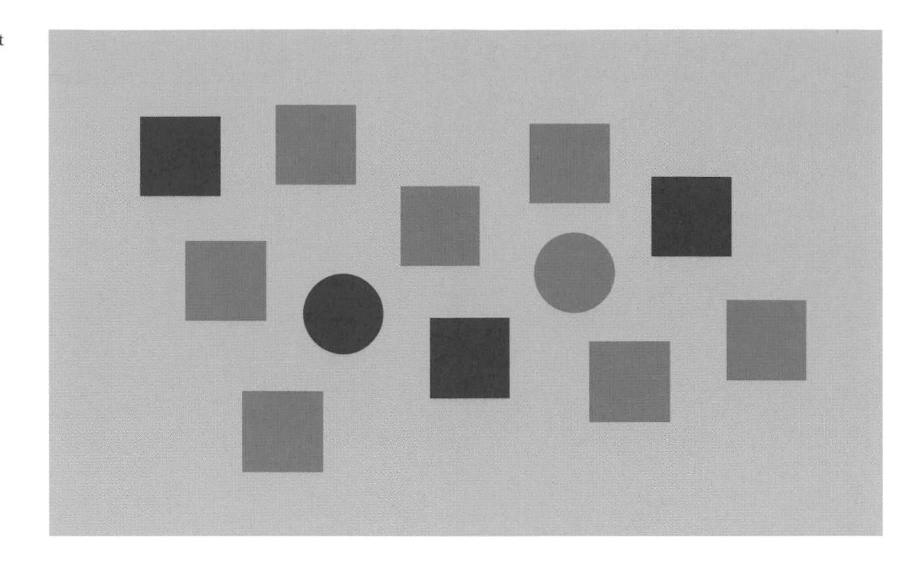

RIGHT The chance that a randomly chosen shape is red, given that it is a circle, is 50%. But the chance that it is a circle, given that it is red, is only 12.5%. People have a tendency to confuse these two measurements.

outcome of other events. This would entail some way of modifying probabilities to take into account new data. In real life, we are used to events affecting each other, but it was not until the early 18th century that the appropriate way to model this mathematically was found. The breakthrough came in a paper by the Reverend Thomas Bayes entitled "Essay Towards Solving a Problem in the Doctrine of Chances" which was published in 1763, two years after his death. In his paper, Bayes contemplates two events, which we might call A and B. In traditional probability theory, each has a certain probability of happening, which are often written as $P(A)$ and $P(B)$. Each of these will be a number somewhere between 0 and 1, with impossible events having a probability of 0, and a certainty having a probability of 1.

Bayes' innovation was to allow the two events to affect each other, that is, if A happens, this may alter the likelihood of B occurring. In an extreme case, it might rule it out completely, or guarantee it. Similarly, if A doesn't happen, B may be affected. This introduces two new quantities, now known as the conditional probabilities. Today, conditional probability is written as $P(A|B)$, which is interpreted as the probability of A given B, and similarly, $P(B|A)$ is the probability of B given A. So how are these four numbers related? In his celebrated theorem, Thomas Bayes was able to answer that question: $P(A|B) \times P(B) = P(B|A) \times P(B)$.

Conditional probability

Conditional probability and Bayes' theorem have become foundation stones for the modern study of uncertainty. They are particularly important in the theory of Markov processes (see page 245). Indeed, the definition of a Markov process is expressed in terms of conditional probability.

Yet conditional probability can have some very counterintuitive consequences. Humans, it seems, have a natural tendency to confuse the quantities $P(A|B)$ and $P(B|A)$, even when the two numbers are entirely different. This is a particular problem in the medical profession. What does it mean to say that a test for a certain disease is accurate? To start with, we should hope that a person who is genuinely suffering from the disease should have a high probability of testing positive. The other factor is that people without the disease should have a low probability of testing positive. So there are really two numbers that measure the accuracy of any test: the probability of a positive result given the presence of the disease (which we hope to be high), and the probability of a positive result given its absence (which we hope to be low).

Suppose, then, that the test has a 99% chance of coming up positive if the patient is genuinely infected, and a 5% chance of testing positive if they are not. What a patient will really want to know is the probability that the disease is present, given a positive test result. Unfortunately, the numbers we have looked at so far are wholly inadequate to determine the answer to this question!

The answer depends on the prior probability that the person had the disease. If they were tested completely at random, then the prior probability is simply the percentage of the population who have the disease. Let's suppose that the disease has an occurrence rate of 0.1%. If 100,000 people are tested, then on average 100 of them will have the disease, of whom 99 will test positive. But the remaining 99,900 uninfected people each have a 5% chance of testing positive, too, so on average 4,995 of them will do so. It is now clear that these false positives outnumber the true positives by a large degree. In more detail, if I test positive, the chance that I am one of the 99 true positives rather than one of the 4,995 false ones is $\frac{99}{4995 + 99}$, which is around 2%.

Conditional probability and Bayes' theorem have become foundation stones for the modern study of uncertainty.

This misunderstanding continues to cause endless problems today, from hospitals to law courts, but with the Reverend Bayes' insight, we at least have the right mental tools to understand these phenomena better.

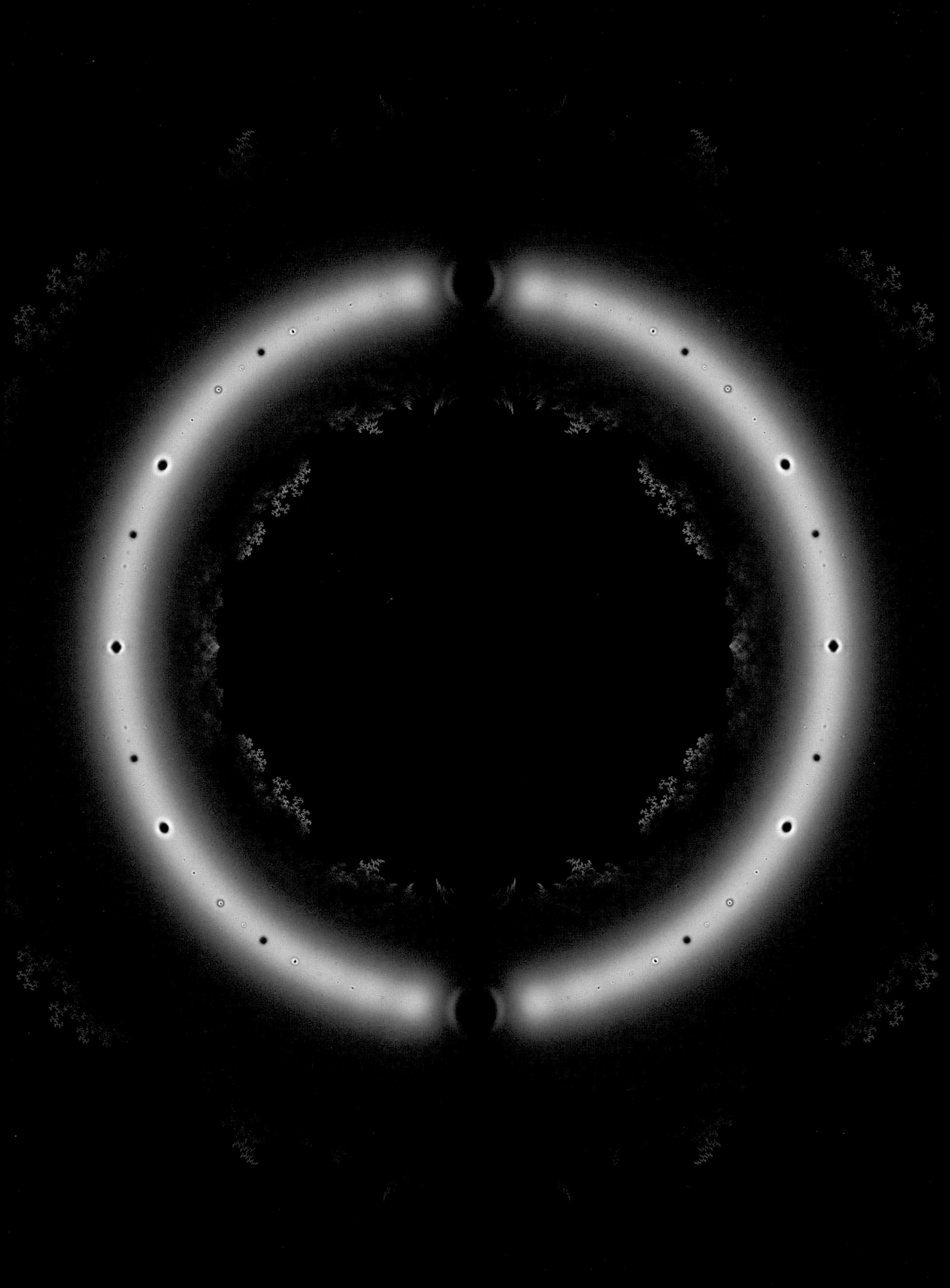

40 Fundamental theorem of algebra

BREAKTHROUGH THE FUNDAMENTAL THEOREM OF ALGEBRA ASSERTS THAT, FOR THE PURPOSES OF SOLVING EQUATIONS, THE SYSTEM OF COMPLEX NUMBERS IS EVERYTHING MATHEMATICIANS WILL EVER NEED.

DISCOVERER CARL FRIEDRICH GAUSS (1777–1855), ROBERT ARGAND (1768–1822).

LEGACY ONE OF THE MOST IMPORTANT FINDINGS IN MATHEMATICS, IT IS THE CHIEF REASON THAT THE COMPLEX NUMBERS ARE SO WIDELY USED IN MATHEMATICS TODAY.

Beginning with the tablet-makers of Babylon, one of the longest-running stories in mathematics has been the effort to solve equations. Yet at various moments, mathematicians have found their stock of numbers inadequate to the task, and they have had to move to extended number systems. At the beginning of the 19th century, a historic result was proved which showed that, with the system of complex numbers, this process had reached its end.

An equation is nothing more than the assertion that one quantity is equal to another: $3 + 4 = 7$. Often, though, there is an unknown number involved, commonly written as x. Solving the equation involves finding the value of x. So, if an equation reads $5 + x = 9$, then its solution is $x = 4$

This idea is nothing complicated. But at various moments in history, people have believed certain equations to be insoluble. An example is $5 + x = 2$. Diophantus (see page 69) considered such equations to be "absurd." But with the emergence of negative numbers (see page 81), its solution became clear: $x = -3$. This was one of the first cases of mathematicians moving to an extended number system and thereby finding solutions to more equations. With the system of positive and negative fractions, any equation of the form $ax + b = 0$ could now be solved (so long as $a \neq 0$). Yet this system of rational numbers cannot solve all equations. When he was calculating the length of one side of a triangle (see page 29), Hippasus of Metapontum needed to solve the equation $x \times x = 2$, or, written more briefly, $x^2 = 2$. Try as he might, however, he could

OPPOSITE The solutions of Littlewood equations, which have all their coefficients as either 1 or −1. The fundamental theorem of algebra guarantees that such an equation will have a solution in the complex numbers, which are plotted in this image.

find no fractional value of x that satisfied this equation. In fact, he proved that there can be no such fraction. Within the confines of the rational numbers, then, the equation $x^2 = 2$ simply has no solutions. Again, Diophantus considered this an "absurd" equation, but, in time, mathematicians became more comfortable working with irrational numbers. In this extended system, $x^2 = 2$ does have a solution, namely $\sqrt{2}$, which is around 1.41421356.

To solve equations built from complex numbers, would it be necessary to extend the number system yet again? This would surely be bad news. The complex numbers seemed abstract and difficult enough.

Equations and real numbers

Together, the collection of all positive and negative, rational and irrational numbers would come to be known as the real numbers. This was a powerful number system, but there was more bad news to come regarding equation-solving. As Girolomo Cardano (see page 101) and others had noticed, there still remain equations which cannot be solved. One such is $x^2 = -1$.

The arithmetic of negative numbers (see page 81) is very clear on one point: if you multiply two negative numbers together, you get a positive number. Similarly, if you multiply together two positive numbers, the result is of course positive. Therefore neither a positive nor a negative number when squared can give a result of -1.

This was bothersome for Cardano and the other Italian algebraists, who kept finding equations like this in their work. This led Raphael Bombelli to extend the number system once more, by incorporating a new number: i. This was defined as a solution to the equation $x^2 = -1$, which is to say $i = \sqrt{-1}$. This brought him to the enlarged number system that today we call the complex numbers (see page 105).

Equations and complex numbers

The complex numbers were nothing short of a mathematical revolution. But in terms of solving equations, exactly how much progress did they represent? They certainly brought the previously insoluble equation $x^2 = -1$ in from the cold, along with $x^2 = -2$ and $x^2 = -3$, and indeed $x^2 = a$ for any real number a. But what of more complicated equations, involving powers higher than 2? The answer was not immediately obvious. However, it was certainly difficult to point to any remaining "absurd" equations. Might any exist, or were the complex numbers enough for all possible purposes?

In 1797, Carl Friedrich Gauss announced the proof of a momentous fact: the complex numbers were enough to solve any equation built from real numbers. Whether solving $x^4 + x^3 + x^2 + x + 1 = 0$ or $\sqrt[5]{2}x^2 + \sqrt[3]{3}x = -\sqrt{5}$, Gauss's theorem guaranteed that there would be some value of x among the complex numbers that would make the equation true. This meant that someone interested in solving equations built from real numbers would have their needs fully met in the complex numbers. It was also a very surprising fact, since the definition of the complex numbers was drafted only to take into account one extra equation: $x^2 = -1$.

ABOVE The Oriental Pearl TV and Radio Tower in Shanghai. In modern communications, messages are typically sent as complex numbers (the real and imaginary parts being encoded in different signals). To decode these messages exploits the algebra of complex numbers.

Gauss must be considered one of history's greatest mathematicians, and the fundamental theorem of algebra as one of its greatest achievements. All the same, his theorem had two flaws. Although it was not spotted at the time, Gauss's argument actually contained a gap, relating to the geometry of curves in the complex plane. The second problem was deeper still: Gauss did not address the obvious next question. His theorem asserted that every equation built from real numbers must have a solution among the complex numbers. What, though, of equations built from complex numbers? An example is $x^2 = i$. To solve equations such as this, would it be necessary to extend the number system yet again? This would surely be bad news. The complex numbers seemed abstract and difficult enough; not many people would be prepared to step up another level.

In 1806, Robert Argand addressed both of these difficulties. Using a different approach to that of Gauss, Argand provided the first fully rigorous proof of the fundamental theorem. What was more, his version went even further. Argand proved that every equation built from complex numbers must already have a full complement of solutions among the complex numbers. For instance, the number $\frac{1}{\sqrt{2}} + \frac{i}{\sqrt{2}}$ is a solution to $x^2 = i$. After millennia of extending the number system, Argand's result marked the final accomplishment.

41 Fourier analysis

BREAKTHROUGH FOURIER ANALYSIS IS THE SCIENCE OF WAVES. JOSEPH FOURIER'S GREAT BREAKTHROUGH WAS TO PROVE THAT ALL WAVEFORMS CAN BE CONSTRUCTED FROM THE VERY SIMPLEST TYPES.

DISCOVERER JOSEPH FOURIER (1768–1830).

LEGACY FOURIER ANALYSIS IS CENTRAL TO MODERN TECHNOLOGY, FROM SYNTHESIZING SOUNDS FOR ELECTRONIC MUSIC TO SATELLITE COMMUNICATIONS.

The world contains many types of wave: from the vibrating string of a ukulele to the terrifying tremors of an earthquake, not to mention the electromagnetic radiation from the Sun on which we all depend. The mathematics of waves is a hugely important subject in science today, with countless applications from acoustics to telephone networks. Although it has fascinated thinkers since Pythagoras, in its modern form the theory of waves was sparked by the work of the French scientist and revolutionary Joseph Fourier.

Fourier's original motivation was the study of heat. Imagine you take a long iron rod, stick one end into a fire and put the other on ice, before setting it down in a temperate room. It is clear that one end of the rod will cool down, while the other warms up, until eventually the whole thing settles into the temperature of its surroundings. But if you monitor the temperature of a single point in the middle, what will you find? It was to answer tricky questions like this that the scientist developed what later became known as Fourier analysis.

OPPOSITE An abstract surface with a sine wave rippling through it. The mathematics of such waves is known as Fourier analysis, and is essential to a range of science and technology, from quantum physics to mobile phones.

Technically, heat is a flow rather than a wave, since its patterns don't repeat as light and sound waves do. Nevertheless, the techniques that Fourier developed proved equally valuable to the study of waves. Heat was not just a scientific concern to Fourier: it was something of a personal obsession. After living in the desert climate of Egypt as scientific assistant to Napoleon, Fourier developed a lifelong belief in the medical benefits of extreme heat. Back in Europe he would keep a roaring fire in his house in all seasons, and his friends would enter the sweltering rooms to find Fourier pondering mathematics while wrapped in blankets.

Waves and harmonics

The science of waves predated Fourier by thousands of years. Pythagoras is said to have devoted a great deal of time and energy to understanding the connection between the sonic properties of a musical note and the physical attributes of the corresponding wave. Pythagoras' waves were vibrations on the strings of a musical instrument, such as a lyre. But the same reasoning applies to the repeating patterns we see on today's oscilloscopes, such as on a hospital heart monitor. Each of these is a repetitive pattern—the same cycle repeated over and over again. The wavelength is the length of time it takes to complete one cycle.

There are as many different waveforms as there are different types of sound: from the roar of a motorbike to the wail of a saxophone. Of all these possible waves, the mathematician's favorite is the sine wave.

Pythagoras understood the relationship between wavelength and the pitch of a musical note. This is easiest to see on a stringed instrument. Shortening the string amounts to shortening the wavelength, and the sound that results when plucked has a correspondingly higher pitch than before. The most striking instance is when the string is reduced to exactly half its original length. Visually, the peaks and troughs of the resulting wave appear doubly squashed together compared to the original. This has a fascinating musical impact: the resulting pitch is exactly one octave above the first note. If the first is C, the second is also C, but an octave higher. This is what is known as the second harmonic of the first, original wave.

A similar thing happens when you compress the original wave threefold, to arrive at the third harmonic (which sounds an octave plus a fifth higher), and so on. These harmonics have been central to the theory of music since Pythagoras first turned them into a harmonious tuning system for instruments.

Harmonics were equally important to Joseph Fourier's work. It is not only the pitch of a note that derives from its waveform. The timbre of a sound is determined by the shape of its sound wave. So there are as many different waveforms as there are different types of sound: from the roar of a motorbike to the wail of a saxophone. Of all these possible waves, the mathematician's favorite is the sine wave. (Geometrically,

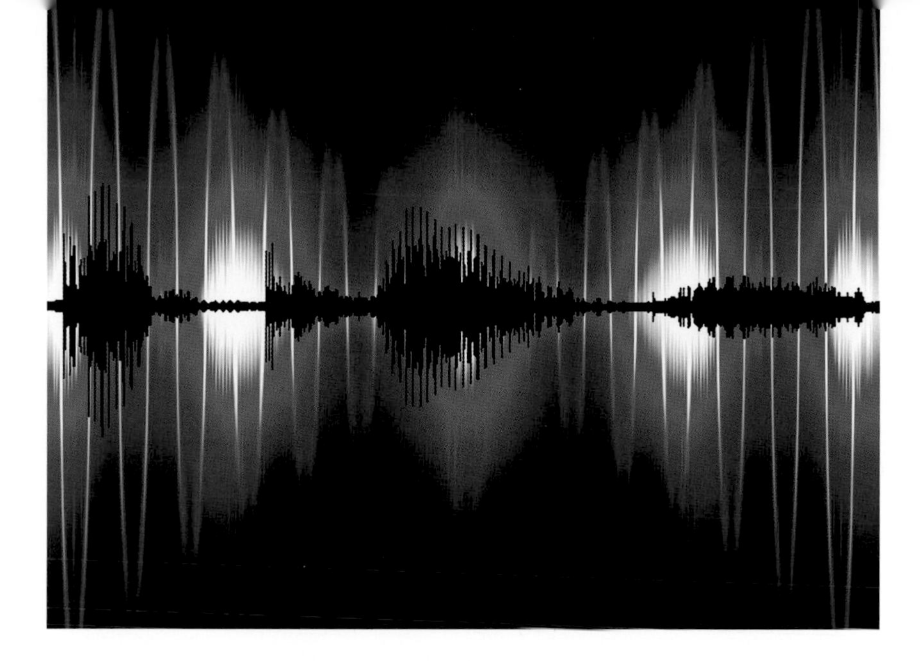

LEFT Separating out and analyzing different components of a sound is one of many technologies that relies on Fourier analysis. This artwork depicts computerized voice recognition.

it derives from an object rotating around a circle at a constant speed. Plotting its height on a graph produces a sine wave.)

Interference and Fourier's theorem

Of course, heartbeats and musical instruments have more complicated forms than the elegant sine wave. Yet thinkers such as Leonhard Euler and Daniel Bernoulli discovered how to build more sophisticated waveforms, using sine waves as the basic ingredients. If you take two waves and add them together, they interfere: they will reinforce each other at some places and cancel each other out at others. To create new waves, mathematicians began with a sine wave, and then added its various harmonics, after suitably adjusting their volumes.

Adding together more and more harmonics, and also including the corresponding cosine waves (which are identical to the sine waves, but delayed by a quarter of a cycle), Euler and Bernoulli could create a wide variety of waveforms. But Fourier's masterstroke was to prove that in fact all reasonable waveforms can be built this way. It is an astonishing fact: by incorporating all the harmonics of the base note, any sound whatsoever can be created just from sine waves! Fourier was even able to provide a formula for the volume needed for each harmonic. This remarkable theorem has been exceptionally useful. The technique of signal analysis which underlies today's mobile phone technology, radio communication and speech recognition software rely on Fourier's theorem. Indeed Fourier analysis now extends far beyond waves, with applications from the mathematics of prime numbers to the physics of quantum mechanics.

42 The real numbers

BREAKTHROUGH THE REAL NUMBERS SATISFY THE INTERMEDIATE VALUE PROPERTY, WHICH ASSERTS THAT A CURVE CONNECTING TWO POINTS MUST SOMEWHERE CROSS THE MIDLINE CONNECTING THEM.

DISCOVERER BERNARD BOLZANO (1781–1848), AUGUSTIN-LOUIS CAUCHY (1789–1857).

LEGACY THE REAL NUMBERS ALLOW GEOMETRICAL IDEAS TO BE PROVED WITH COMPLETE RIGOR. ALONG WITH THE COMPLEX NUMBERS, THEY FORM THE BACKDROP OF ALL MODERN MATHEMATICS.

Mathematics contains several types of theorem. Some are surprising and imaginative, or even shocking. Others are deep and technically demanding even to state, let alone to prove. But there are other theorems, where the surprise is that they require proof at all, for they seem at first sight little more than statements of the obvious. One such is the intermediate value theorem proved by Bernard Bolzano in 1817. Yet this was no triviality. It meant that, for the first time, mathematicians had developed a number system that fully matched their long-held geometrical intuition.

Take a page and divide it in half with a horizontal line. If someone picks one point above the line and one point below, it seems obvious that any curve connecting the two must cross the central line at some point (leaving aside trickery such as folding the paper over). Yet it was not until 1817 that mathematicians had the tools necessary to prove this. The famous result, the intermediate value theorem, was the moment that mathematical techniques and the conception of number finally caught up with geometric intuition. This fact went on to become a bedrock of modern mathematical analysis.

OPPOSITE Tailing pools in Canyonlands National Park, Utah, used to store the by-products of mining. To deposit material, miners are faced with the incontrovertible geometric fact that getting from one-side of a line to the other entails crossing it somewhere, a fact which took thousands of years to prove rigorously.

Euclid's lines

When Euclid wrote his great geometric treatise *The Elements*, he began with his famous laws: Euclid's postulates (see page 46). But before he could even state these rules, he needed to set out something even

more primitive: definitions of the concepts involved. Central to Euclid's work were straight lines. But what exactly is a straight line? Like most people, Euclid knew one when he saw it. But actually writing down a definition was another matter. He settled on: "A line is breadthless length." This conveyed the right idea: a line is a purely 1-dimensional object, along which you can only move left or right. Of course, in the real world, any line drawn on a page or displayed on a computer screen will have a certain thickness too, determined by the width of the pen nib or the resolution of the screen. Euclid's definition conceives of an idealized line whose breadth really is zero.

ABOVE Long-exposure photographs of star trails at the Arches National Park in Utah. The brightest track inside the arch represents the North star. The trails the stars leave in the sky are archetypal continuous curves, with no jumps or gaps. Understanding the geometry of continuity has taken mathematicians many years.

This was adequate for elementary geometrical purposes, and it remained the standard for over a thousand years. But in time, and especially with the discovery of calculus (see page 133), mathematicians began to become dissatisfied with having to rely on bare intuition for their fundamental definitions. They longed for something more robust. The key was to find a way to interpret the points, lines and curves of geometry purely numerically.

The first step in this direction was Descartes' innovation: Cartesian coordinates (see page 129). In Descartes' system, every point became identified with a pair of numbers—its coordinates, such as (1, 2). Similarly, straight lines went from being "breadthless lengths" to being equations, such as $y = x$. As a result, geometric notions were translated into algebraic relationships between numbers, and various commonsense assertions about elementary geometry became statements to prove or disprove. One such was the intermediate value property, which says that a curve connecting two points must necessarily cross a midline between them.

Functions and continuity

The intermediate property concerns curves. Indeed, it makes an assertion about every possible curve. But in the new system of Cartesian coordinates, where points are represented by pairs of numbers, what are curves?

It was Leonhard Euler who introduced the terminology that is now standard. The crucial notion is that of a function. Essentially, a rule that accepts numbers as inputs gives out other numbers as outputs. A curve can be considered as a picture of a function, where the horizontal position (which is to say the first coordinate) represents the input, and vertical position (or second coordinate) represents the corresponding output. But not every function produces a sensible curve. An unregulated function may jump around with all manner of gaps and jumps. To qualify as a curve, rather than a mess, the function must flow from one point to another. What this means is that small changes in input should result in correspondingly small changes in output. The term for a function obeying this law is *continuous*.

The problem was that the rational numbers themselves are riddled with holes. An enlarged number system was needed, one which incorporated every possible irrational number too, and left no remaining holes.

The intermediate value property

Expressed in numbers and functions, the intermediate value property could finally be expressed in precise mathematical language. The cost was that it became far less obvious than the naïve version. Indeed, the first suggestions were that it was not even true. If the underlying number system is the rational numbers (meaning fractions), it was easy to concoct a function which was below zero at an input value of 1, and above zero at input value of 2, but with no intermediate value between 1 and 2, producing an output of exactly zero. The problem was that the rational numbers themselves are riddled with holes. These holes appear at points corresponding to irrational numbers (see page 29). An enlarged number system was needed, one which incorporated every possible irrational number too, and left no remaining holes.

The efforts to formulate a number system for which the intermediate property was true resulted in the real numbers. In 1817, Bernard Bolzano realized that the real numbers would allow the intermediate value property finally to be proved. This was the first great theorem in the new discipline of real analysis, which merged notions from geometry and number theory.

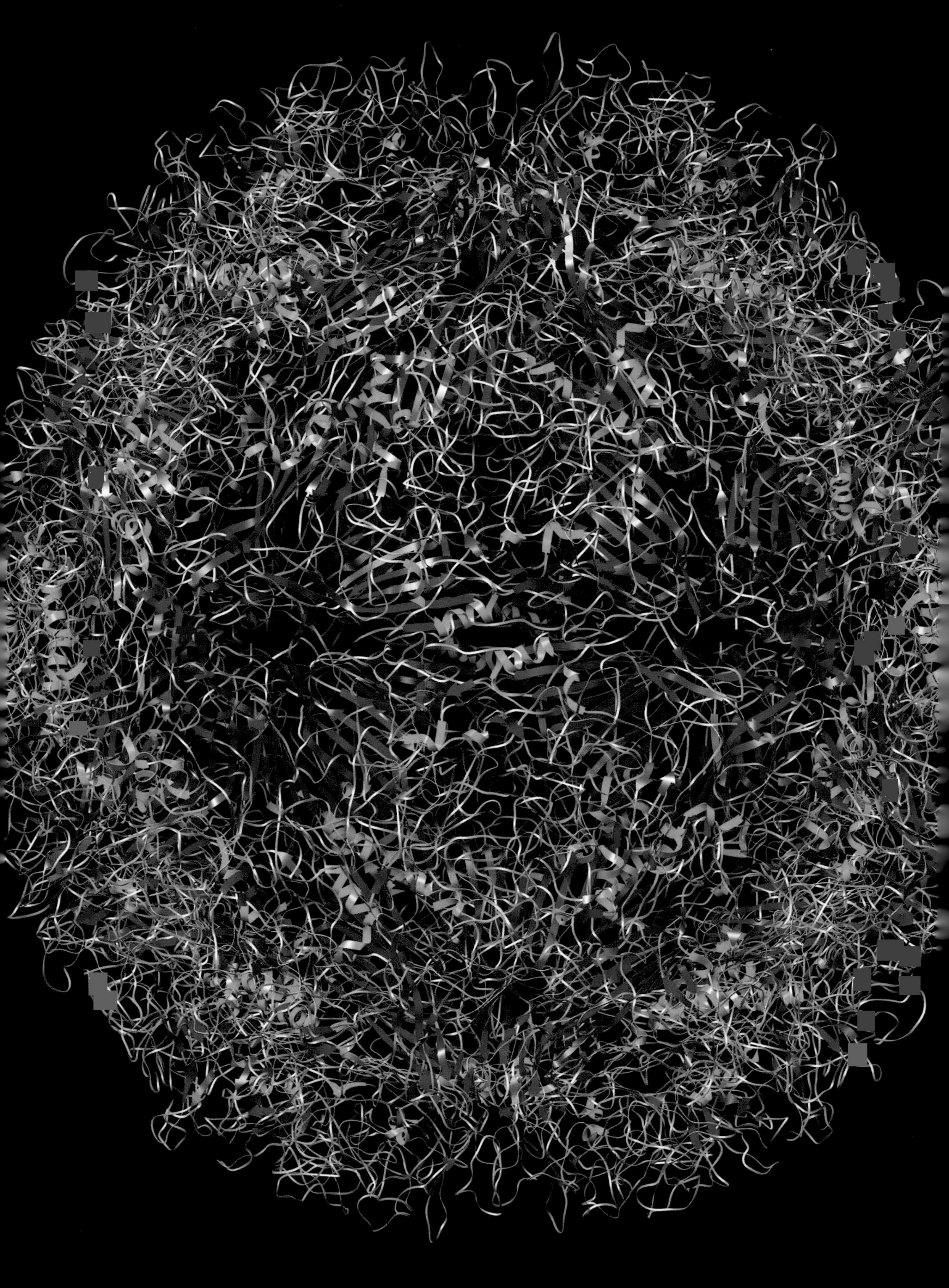

43 The unsolvability of the quintic

BREAKTHROUGH THERE IS NO SIMPLE PROCEDURE FOR SOLVING A CERTAIN TYPE OF EQUATION KNOWN AS A QUINTIC. THIS WAS PROVED BY NIELS ABEL, USING A NEW THEORY OF ABSTRACT SYMMETRY DEVELOPED FURTHER BY ÉVARISTE GALOIS.

DISCOVERER NIELS ABEL (1802–29), ÉVARISTE GALOIS (1811–32), PAOLO RUFFINI (1765–1822).

LEGACY THE WORK OF ABEL AND GALOIS REVOLUTIONIZED THE STUDY OF ALGEBRA THROUGH THE INVENTION OF THE THEORY OF ABSTRACT GROUPS.

Humanity's first use of mathematics was counting. But as we became more sophisticated, we started to solve equations. This remains a priority for today's mathematicians, and the desire to solve ever more complex equations has been a major driver of progress in the subject. Never was this truer than in the revolutionary work that Niels Abel and Évariste Galois performed on the quintic equation.

To solve an equation means to discover the value of a number (typically called x) from some indirect information about it. If a field is twice as long as it is wide, with an area of 800 square meters, how wide is it?

Complex equations

In 16th-century Italy, researchers had devoted all their efforts to two particularly tricky types of equation. The cubic is an equation of degree 3: it contains an x^3 term (that is, $x \times x \times x$). The quartic is more complex still, including degree 4, meaning that it contains an additional x^4 term. The period was a successful one for mathematics, and in Cardano's great work *Ars Magna*, a complete set of instructions were laid out which would allow any cubic or quartic equation to be solved (see page 101).

OPPOSITE A computer model of a human parvovirus. Like many viruses, its outer capsid exhibits icosahedral symmetry, closely related to the symmetry of a quintic equation.

The challenge for the next generation of algebraists was obvious: the quintic equation, one which also contains an x^5 term. But little did these mathematicians realize that the quintic equation concealed deep mathematical truths not hinted at by its lesser cousins.

It was clear, at any rate, that the quintic would be no trifling matter; the formula for the cubic was already complicated, and the quartic was even more involved. Doubtless the quintic would be even worse, and would require the sustained efforts of the top mathematicians to crack it. No lesser figure than Leonhard Euler set it in his sights, seeking a foolproof method for solving any quintic equation. Try as he might, however, the quintic resisted every such attempt. In the mid-18th century, Euler noted ruefully, "All the pains that have been taken in order to resolve equations of the fifth degree, and those of higher dimensions ... have been unsuccessful; so that we cannot give any general rules for finding the roots of equations which exceed the fourth degree."

Unsolvable equations

This unsatisfactory situation endured until 1820, when a young researcher from Norway revealed a stunning twist in the tale. The reason that Euler had failed to find a formula for the quintic is that there isn't one.

The formula for the quadratic, cubic and quartic equations are built from the simplest ingredients: addition, subtraction, multiplication, division and taking roots. The last of these is the opposite of raising to powers. Just as 9 is the square of 3, that is $9 = 3^2$, so 3 is the square root of 9, written $3 = \sqrt{9}$. All powers have their corresponding roots: just as 8 is 2 cubed (2^3 or $2 \times 2 \times 2$), so 2 is the cube root of 3, written as $2 = \sqrt[3]{8}$. It is no surprise that to solve equations which involve powers requires taking roots.

Just as rotating a square by 90 degrees swaps around its edges but leaves its overall appearance the same, so the solutions of an equation could be interchanged in such a way that its overall appearance remained the same.

Niels Abel's revelation was that there could be no formula for solving the quintic equation that uses only such simple tools. In Italy, Paolo Ruffini had arrived at the same conclusion two decades earlier, but had struggled to get his 500-page manuscript taken seriously by the mathematical community. In fact, on closer examination, Ruffini's lengthy argument contained a critical gap. Abel, on the other hand, penned a punchy six-page paper, setting out an argument that was complete and compelling. Their Abel–Ruffini theorem says that the general quintic equation cannot be solved just with arithmetic and roots.

Their theorem does not say that quintic equations do not have solutions. The fundamental theorem of algebra, proved a few years earlier by Carl Friedrich Gauss, guarantees that equations of every degree must be satisfied by some numbers (see page 165). The difficulty arises in trying to find those numbers.

The beginnings of group theory

Why should it be that the quintic equation, and those of all higher degrees, cannot be found with a single formula, while the quadratic, cubic and quartic can? Tragically, Abel, whose life had been plagued by grueling poverty and ill health, died at the age of 26, before he had completed his investigations into this question.

The problem would be addressed by another, one Évariste Galois. As well as being a brilliant mathematician, Galois was also a committed revolutionary who did some of his best mathematics while imprisoned in Paris for political agitating. Galois' work set the foundations for what would become group theory, which would change the face of algebra over the 20th century.

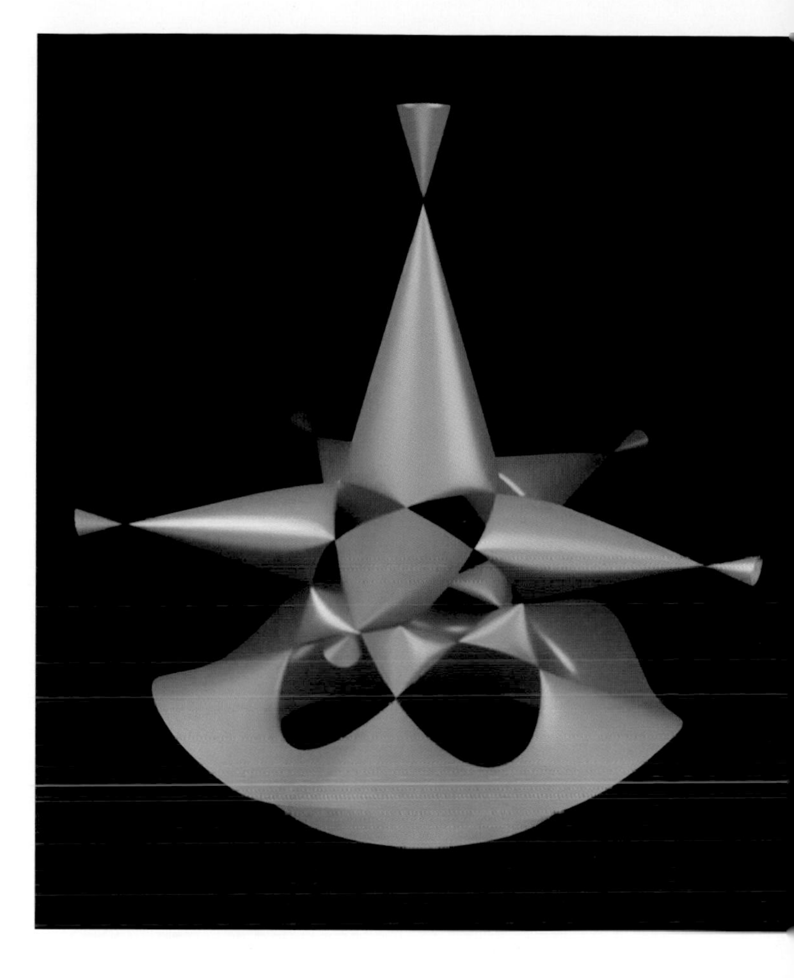

ABOVE This dervish is an example of a quintic surface, a 2-dimensional surface described by an algebraic equation of degree 5. The dervish cuts through itself in 31 different places, the most of any quintic surface.

Galois recognized an extraordinary thing: equations have a notion of symmetry, much like that possessed by a geometrical shape such as a square. Just as rotating a square by 90 degrees swaps around its edges but leaves its overall appearance the same, so the solutions of an equation could be interchanged in such a way that its overall appearance remained the same. Galois' insight was that truly understanding an equation entailed understanding these symmetries. His greatest theorem stated that an equation is solvable if and only if its group of symmetries is "solvable," whereas the group for the quintic is "unsolvable." Solvable and unsolvable groups would be central to algebra in the decades that followed. But, like Abel, Galois would not live to witness the mathematical revolution he had begun. He was killed in a duel in mysterious circumstances in 1832, at the age of 20.

44 The Navier–Stokes equations

BREAKTHROUGH NAVIER AND STOKES DERIVED THE FUNDAMENTAL EQUATIONS TO DESCRIBE FLUID FLOW.

DISCOVERER CLAUDE-LOUIS NAVIER (1785–1836), GEORGE STOKES (1819–1903).

LEGACY THE NAVIER–STOKES EQUATIONS ARE CENTRAL TO MODERN FLUID DYNAMICS. THE BIGGEST QUESTION IN THE SUBJECT TODAY IS WHETHER OR NOT THESE EQUATIONS HAVE ANY SOLUTIONS.

During the scientific revolution of the 17th century, thinkers such as Galileo Galilei, Isaac Newton and Johannes Kepler had set in place the basic principles of mechanics—the laws governing how solid objects interact. What, though, of liquids and gases? An apple dropping from a tree may fall directly downward toward the Earth, but anyone who has pulled the plug out of a bath will have noticed that fluids behave in much more complex ways. These scenarios are described by some of the most important, but trickiest, equations in science: the Navier–Stokes equations.

Fluids cannot be ignored: they surround us, from the winds which blow around the globe to the blood coursing through our veins. But they are extremely hard to describe.

The birth of fluid dynamics

The first person to launch a serious attack on the mathematics of fluid flow was Leonhard Euler, around 1757. A fundamental tenet of Newtonian solid mechanics is known as the conservation of momentum. It says that when two or more objects or particles collide, the net momentum of the system always remains unchanged (this is calculated by multiplying each object's mass by its velocity and adding up all the results). Euler's approach was to watch what happened to this law as the individual particles become smaller and smaller and their number becomes ever greater. The limit of this process is a notional perfect fluid.

OPPOSITE Lava flowing at the Volcanoes National Park in Hawaii. Lava is extremely viscous, up to 100,000 times more so than water, but the movement of both can be described completely by the Navier–Stokes equations.

ABOVE Schlieren photograph showing the flow of air around an early design of the space shuttle during a transonic wind tunnel test. The science of aerodynamics relies on the mathematics of fluid flow.

To describe such a fluid, Euler had to lift the science of calculus, as discovered by Newton and Leibniz, to a new level of sophistication. He needed to describe a system that varied across all three spatial dimensions, as well as altering over time. The resulting system of equations was a milestone in applied mathematics—a true showcase of the power of calculus. Nevertheless, there was one critical respect in which Euler's equations fell short of describing the movement of any fluid in the real world.

Thickness and viscosity

Liquids like oil and syrup are viscous. This means that there are internal frictional forces that significantly reduce the speed at which the fluid flows. In common terms, a viscous liquid is thick. For all their technical brilliance, Leonhard Euler's equations failed to take this viscosity into account. Even water has some viscosity (which is why we can swim through it), so this omission left Euler's findings all but useless for a practical analysis of the movement of real fluids.

The challenge of adapting Euler's equations to model viscous fluids was taken up independently by two 19th-century theorists. Claude-Louis Navier began his career as an engineer in France, specializing in roads, railways and above all else bridges. He achieved fame as a master bridge builder and was awarded a contract to build a suspension bridge over the River Seine; but the authorities were unhappy with the cost of the project and dismantled the bridge before it was complete. Navier was interested in the physical underpinnings of engineering, and in 1822 he produced an augmented version of Euler's equations, incorporating a new expression for the viscosity of the fluid. Unfortunately, his mathematical justification for this step was not quite right.

Twenty years later, George Stokes would provide a correct derivation of the new equations as a limit of Newton's laws. Born in Ireland, Stokes was also an expert on railways and bridges, and had been a technical consultant to the British government in the wake of several rail disasters. Stokes was also an eminent scientist who made notable discoveries in optics and the polarization of light. But his greatest work was in fluid dynamics, where his conclusions about the flow of viscous fluids exactly matched Navier's.

As the subject of fluid dynamics developed, it became entwined with countless areas of science and technology over the 20th century, from the flow of air around the wing of an airplane to the propagation of whale song across the oceans.

The Navier–Stokes question

Navier's and Stokes's work was a milestone in the history of fluid dynamics. As the subject developed, it became entwined with countless areas of science and technology over the 20th century, from the flow of air around the wing of an airplane to the propagation of whale song across the oceans. At the theoretical heart of this subject lie the Navier–Stokes equations, which offer a complete description of the conditions which any flowing fluid must satisfy. Yet, astonishingly, in all the intervening time, no one has managed to find a single satisfactory mathematical solution to the Navier–Stokes equations. Even though every fluid flow should satisfy them, all the flows that mathematicians have been able to describe suffer from the same failing: at some point, the flow breaks down. The mathematics reaches a point where it ceases to describe a physically realistic scenario. Either speeds explode out of control, infinitely tight whirlpools form, or other impossible phenomena occur.

So for over a century, the biggest question in the subject of fluid dynamics has been to solve the Navier–Stokes equations. This means exhibiting a flow that is perfectly smooth, free of singularities, and otherwise physically plausible. Of course, it must be viscous, and obey the Navier–Stokes equations. Alternatively, it may be that no solutions exist. This would be a very surprising revelation, given the evidence of numerous computer simulations suggesting that Navier–Stokes flows are excellent at modeling reality. Nevertheless, the possibility cannot be discounted. As of the year 2000, anyone successfully accomplishing either feat will earn themselves $1,000,000 courtesy of the Clay Institute, which included the Navier–Stokes equations as one of their Millennium Prize Problems.

45 Curvature

BREAKTHROUGH THERE ARE SEVERAL METHODS TO QUANTIFY HOW A SURFACE IS CURVED. THE GAUSS–BONNET THEOREM ESTABLISHED THAT THE OVERALL CURVATURE OF A SURFACE IS A FIXED NUMBER, EVEN WHEN THE SHAPE CHANGES FORM.

DISCOVERER CARL FRIEDRICH GAUSS (1777–1855), PIERRE BONNET (1819–92).

LEGACY ANALYSIS OF CURVATURE IS CENTRAL TO MODERN PHYSICS, FOR INSTANCE IN EINSTEIN'S THEORY OF GENERAL RELATIVITY.

Since the time of Euclid, geometers have studied the geometry of a flat plane, beginning with the straight lines and circles that can live on it. The geometry of a curved surface is a far trickier thing to describe. Yet, in order to be useful to physicists, mathematicians needed to get to grips with curved surfaces. The most significant breakthrough in this direction came via the work of Carl Friedrich Gauss, in the early 19th century.

Standing on a sphere, the surface curves away from the observer in exactly the same manner, no matter in which direction you look. On the other hand, a saddle slides away from you in some directions, but curls up toward you in others. Different again is a cylinder, which is curved around its center, but perfectly flat along its length. The challenge to geometers was to make sense of this variety.

Gaussian curvature

What was needed was a way to quantify the type and extent of a surface's curvature. The tools were provided by the discovery of calculus (see page 133). But, as originally conceived, calculus only dealt with curved lines, rather than surfaces. It could assign a number to a line in order to describe precisely how steep the line is at that point, but could it also provide some way to understand the curvature of a 2-dimensional surface? The answer was supplied by one of the great figures in the history of mathematics: Carl Friedrich Gauss.

OPPOSITE The curved tubes of the large helical device in Toki, Japan, where research on nuclear fusion is performed with high-temperature plasma contained by a very powerful magnetic field.

ABOVE Henry Moore's *Hill Arches,* at the National Gallery in Canberra, Australia. Surfaces of varying curvatures have long been of interest to sculptors and architects as well as mathematicians.

Gauss was a child prodigy who made several important discoveries before his 20th birthday, including a deep observation about the prime numbers (see page 49) and a major development in the ancient field of ruler-and-compass constructions (see page 194). His interest in curved surfaces, however, flourished at a later stage in his life. Gauss thought of a surface as being composed of curves. By picking a point on the surface and analyzing the various curves that pass through it, Gauss was able to produce a number quantifying the curvature at that point. This Gaussian curvature is positive on a shape like a sphere, where all the curves bend away in a similar fashion, but negative at a point on a saddle. On a cylinder, the Gaussian curvature is zero, just as it is on a flat plane. In fact, surfaces of zero curvature are exactly those which can be unrolled to lie flat.

The Gauss-Bonnet theorem

A sphere's curvature is uniform: each point is curved in exactly the same way as every other. Then again, the sphere is an exceptionally symmetrical shape in many ways. On most surfaces, the curvature varies from place to place. There is seemingly no limit to this possible variety: it may be extremely steep at some points, while at others it may be totally flat. The world is full of curved surfaces, from teapots and light bulbs to the sculptures of Henry Moore, the curvature of which varies dramatically from place to place. In technical terms, mathematicians consider curvature a local phenomenon: it describes a small region of a shape, but says nothing about its overall structure.

At the far opposite end of geometry, the subject of topology (see page 213) is concerned only with the global properties of a shape. To a topologist, it is the overall structure of a shape that is important, not the details of each tiny region. Topologists consider two shapes to be fundamentally the same if one can morph into the form of the other without cutting or gluing. Of course, this morphing process may distort the curvature at every single point. It would seem, then, that curvature and topology are completely independent notions, and simply irrelevant to each other. Yet Gauss discovered a very profound connection between these two contrasting approaches to shape analysis.

Although curvature is defined locally, Gauss found a way to average out the curvature across the entire surface. Again, the critical tool came from calculus, this time the subject known as integral calculus (see page 54). Integrating the curvature across the entire surface produces a single number. This number must describe the shape in some way; but what does it actually represent?

Integrating the curvature across the shape means somehow finding the total curvature. Critically, Gauss found that this number was far more robust than the curvature at any single point. In particular, it was resistant to topological morphing. Even as the surface changed, and the curvature fluctuated around, the Gauss number for the overall curvature remained static.

The world is full of curved surfaces, from teapots and light bulbs to the sculptures of Henry Moore, the curvature of which varies dramatically from place to place.

What was more, Gauss recognized this number as a familiar object from topology, known as the Euler characteristic. Coming out of Euler's polyhedral formula (see page 158), this is a single number which depends only on the number and type of holes within a surface. Gauss realized that integrating the curvature across the whole surface produces exactly 2π times the Euler characteristic of the shape.

In fact, Gauss never published this important result; it was later rediscovered and generalized by Pierre Bonnet. The Gauss–Bonnet theorem, as it became known, is of foundational importance in modern geometry and indeed in physics. Understanding curvature has become hugely important since the development of general relativity theory, in which gravity is interpreted as the curvature of spacetime. The Gauss–Bonnet theorem and its higher-dimensional analogs play central roles in that understanding.

46 Hyperbolic geometry

BREAKTHROUGH AN ENTIRELY NEW FORM OF GEOMETRY, MARKING THE CULMINATION OF CENTURIES OF INVESTIGATION INTO EUCLID'S PARALLEL POSTULATE.

DISCOVERER CARL FRIEDRICH GAUSS (1777–1855), NIKOLAI IVANOVICH LOBACHEVSKY (1792–1856), JÁNOS BOLYAI (1802–60).

LEGACY THIS DISCOVERY SPARKED AN OVERHAUL IN THE FUNDAMENTAL CONCEPTS OF GEOMETRY. HYPERBOLIC GEOMETRY NOW PLAYS A CENTRAL ROLE IN MATHEMATICS AND PHYSICS, NOTABLY IN THE THEORY OF RELATIVITY.

Euclid's *Elements* was a milestone in geometry, remaining the standard textbook on the subject for over 2000 years. But for all that, it did not contain a completely definitive account of the elementary geometry of a flat plane; there were still one or two loose ends. One conundrum at the very heart of *The Elements* went on to baffle numerous thinkers over the ages, before eventually being solved almost simultaneously by three 19th-century mathematicians. This was the problem of the parallel postulate. Its solution was a transformative moment in the history of geometry, just as significant as Euclid's own work.

The question involved the fundamental laws underlying the system of Euclidean geometry. What made *The Elements* such a mold-breaking work was that Euclid began by laying out explicitly his starting assumptions. He then built up his theorems step by step from there.

At the base of this intellectual edifice were his five fundamental laws, known as Euclid's postulates. None of them was complicated: they all formalized well-known characteristics of straight lines, circles and angles as they might be drawn on a plane (that is to say a very large, flat piece of paper). The first axiom says that if you take any two points, you can always connect them with a straight line. The second asserts that any segment of straight line can be extended indefinitely in both directions.

OPPOSITE Brain coral, photographed off the Andros Islands, Bahamas. The surface of many corals is essentially hyperbolic rather than Euclidean. The negative curvature of the surface gives rise to the coral's characteristic wrinkles and ridges.

Euclid's parallel postulate

Of Euclid's five laws, it was the fifth, dubbed the parallel postulate, that was the most troublesome. The difficulty began when Euclid first wrote it down. It was noticeably wordier and less elementary than the other four. Later, a somewhat neater statement was found, which was logically equivalent. It goes like this: whenever you have a straight line, and a point away from it, the line must have exactly one parallel which runs through the point. (Two lines are parallel if they can be extended indefinitely without ever crossing.) Expressed this way, the parallel postulate seems reasonable enough. Yet this law remained deeply controversial for over two millennia.

The Euclidean plane is flat. The hyperbolic plane, on the other hand, is fundamentally curved. In fact, hyperbolic space has what is known as negative curvature.

The question mark lay over its relationship to the other four simpler axioms. Was the parallel postulate really strictly necessary? Or was it actually redundant, being a logical consequence of the first four? Writing 400 years after Euclid, the astrologer Ptolemy believed that he had found a proof of the parallel postulate, meaning a logical derivation of it from the other four laws. Had he been correct, the parallel postulate could have been excised from Euclid's work as an assumption, and would instead become a theorem of Euclidean geometry. But Ptolemy was not correct, and nor were all the subsequent attempts to prove it down the ages by diverse thinkers, including the 11th-century Persian poet Omar Khayyam, the 13th-century Polish philosopher Witelo, and the 19th-century number theorist Adrien-Marie Legendre. All these attempts at proof failed. Usually these arguments accidentally smuggled in another hidden assumption, which on closer inspection could be seen to be logically equivalent to the parallel postulate. One such was the assertion that parallel lines must always be the same distance apart.

A watershed moment

It was not until the mid-19th century that the status of the parallel postulate was finally resolved, with the discovery of an entirely new form of geometry. In hyperbolic geometry, as it later became known, Euclid's first four laws hold, but the parallel postulate does not. This definitively established its logical independence from the first four laws. This watershed was arrived at independently by Carl Friedrich Gauss, Nikolai Ivanovich Lobachevsky and János Bolyai. In the hyperbolic plane, the parallel postulate is not true: given a line, and a point away from it, there are many possible lines through the point that could be drawn, each of them parallel to the original.

LEFT The discovery of hyperbolic geometry has provided inspiration to numerous artists over the last 150 years.

Curved spaces

It was Bernhard Riemann who understood how to make sense of this new geometric landscape. Riemann was one of Gauss' students, and he suggested that the time was ripe for a re-evaluation of the foundations of geometry. Riemann realized that the difference between Euclidean and hyperbolic geometry could best be understood in terms of curvature. The Euclidean plane is flat. The hyperbolic plane, on the other hand, is fundamentally curved. In fact, hyperbolic space has what is known as negative curvature. One consequence is that the three angles of a hyperbolic triangle add up to less than 180 degrees. It is a famous fact, proved in *The Elements* and memorized by generations of school students ever since, that in the Euclidean plane the three angles of any triangle always add up to exactly 180 degrees.

There are also spaces with positive curvature. The most important example is the surface of a sphere. On a sphere, the three angles of a triangle add up to more than 180 degrees. However, the sphere is a finite shape, and so does not satisfy Euclid's second postulate. There is a limit to the possible length of any straight line (meaning in this context an equator of the sphere), which is given by the sphere's circumference. Euclid's postulates were then too restrictive to deal with the new geometric possibilities. So the age of Euclidean geometry finally ended, to be superseded by that of Riemannian geometry. In the 20th century, Riemannian and hyperbolic geometry would go on to play central roles in our understanding of the physical Universe, not least in Albert Einstein's theory of relativity.

47 Constructible numbers

BREAKTHROUGH PIERRE WANTZEL TRANSLATED THE ANCIENT GEOMETRICAL QUESTION OF RULER-AND-COMPASS CONSTRUCTIONS INTO A PURELY ALGEBRAIC ONE.

DISCOVERER PIERRE WANTZEL (1814–48).

LEGACY WANTZEL'S WORK ESSENTIALLY CLOSED THE BOOK ON RULER-AND-COMPASS CONSTRUCTIONS, INCLUDING THE GREAT ENIGMA OF SQUARING THE CIRCLE.

A sure route to mathematical fame is to resolve a problem that has stood open for centuries, defying the greatest minds of previous generations. In 1837, Pierre Wantzel's seminal analysis of constructible numbers was enough to settle not just one, but an entire slew of the most famous problems in the subject, namely those relating to ruler-and-compass constructions.

As with so much in the history of mathematics, the topic had its origins in the empire of ancient Greece. The geometers of that period were interested not only in contemplating shapes in the abstract, but also in creating them physically. Initially, this was for artistic and architectural purposes, but later for the sheer challenge it posed. In time, mathematicians came to understand that the obstacles they encountered in these ruler-and-compass constructions brought with them a great deal of mathematical insight. Nowhere was this more true than in the ancient enigma of squaring the circle, and what that revealed about the number π.

OPPOSITE The Japanese art of origami has hidden mathematical depths. Just as classical ruler-and-compass problems yield the constructible numbers, so numbers producible by folding are known as origami numbers. It turns out that every constructible number is origami, but not vice versa.

Classical problems

Greek geometers decided on a set of simple rules for building shapes, using only the simplest possible tools: a ruler and pair of compasses. The ruler is unmarked, so it can only be used for drawing straight lines, not for measuring length (therefore these are sometimes called straight-edge-and-compass constructions). The compass is used to draw circles, but it may only be set to a length that has already been constructed.

RIGHT *Newton* by William Blake (1795). Newton can be seen working on a ruler-and-compass construction, considered some of the greatest mathematical problems of the era.

Today's schoolchildren still learn how to use these devices to divide a segment of straight line into two equal halves and to bisect a given angle. These were two of the very first ruler-and-compass constructions. A more sophisticated technique allows a line to be trisected, that is, divided into three equal parts. What of trisecting an angle, though? Various approximate methods were discovered, which were accurate enough for most practical purposes, but no one could find a method which worked exactly. This proved a mystery, and gave the first hint that there was real depth beneath this question. But what does it mean if one task can be carried out by ruler and compass and another cannot?

The most famous of the ruler-and-compass problems, and indeed one of the most celebrated questions in mathematics, is that of squaring the circle. The question is this: given a circle, is it possible to create, by ruler and compass, a square which has exactly the same area? At the heart of this question lies the number π (see page 54). The problem ultimately reduces to this: given a line 1 unit long, is it possible to construct by ruler and compass another line exactly π units long?

Another classical problem was that of doubling the cube. This problem had its origins in a legend from around 430 BC. To overcome a terrible plague, the citizens of the island of Delos sought help from the Oracle of Apollo. They were instructed to build a new altar exactly twice the size as the original. At first they thought it should be easy: it could be done by doubling the length of each side. But that process leads to the volume of the altar increasing by a factor of 8 (since that is the number of smaller cubes that can fit inside the new one). To produce a cube whose volume is double that of the original, the sides need to be increased by a factor of $\sqrt[3]{2}$ (that is the cube root of 2, just as 2 is itself the cube root of 8).

The question of doubling the cube therefore reduces to this: given a line segment 1 unit long, is it possible to construct another exactly $\sqrt[3]{2}$ units long?

Wantzel's deconstruction

Working in the turbulent setting of France in the early 19th century, Pierre Wantzel turned these ancient questions over in his mind. He recognized that the form question of many ruler-and-compass constructions is the same. The key to them was this: given a line 1 unit long, which other lengths can be constructed? And which cannot? If a line of length x can be constructed, then Wantzel deemed x a *constructible number*. Setting aside the geometrical origins of these problems, he devoted himself to studying the algebra of constructible numbers. Some things were obvious: for example, if a and b are constructible, then so must be $a + b$, $a - b$, $a \times b$, and $a \div b$. But these operations do not exhaust the range of constructible numbers; Wantzel realized that it is also possible to construct square roots, such as $\sqrt{a}$.

Wantzel's great triumph came in 1837, when he showed that everything constructible by ruler and compass must boil down to some combination of addition, subtraction, multiplication, division and square roots.

His great triumph came in 1837, when he showed that everything constructible by ruler and compass must boil down to some combination of addition, subtraction, multiplication, division and square roots. Since $\sqrt[3]{2}$ is a cube root, and cannot be obtained via these algebraic operations, it followed immediately that the Delians' ambition to double the cube was unattainable. A similar line of thought revealed the impossibility of trisecting an angle.

As for the greatest problem of all, squaring the circle, the final piece didn't fall into place until 1882, when Ferdinand von Lindemann proved that π is a transcendental number (see page 197). Then Wantzel's work immediately implied the non-constructibility of π, and the impossibility of squaring the circle was finally established.

48 Transcendental numbers

BREAKTHROUGH A TRANSCENDENTAL NUMBER IS ONE WHICH IS OUT OF REACH OF THE WHOLE NUMBERS USING ADDITION, SUBTRACTION, MULTIPLICATION AND DIVISION. THE FIRST EXAMPLES WERE FOUND BY LIOUVILLE IN 1844.

DISCOVERER JOSEPH LIOUVILLE (1809–82), CHARLES HERMITE (1822–1901), GEORG CANTOR (1845–1918).

LEGACY TRANSCENDENCE IS AN IMPORTANT TOPIC WHICH REMAINS POORLY UNDERSTOOD BY TODAY'S NUMBER THEORISTS.

Since the time of the Pythagoreans, and perhaps before, mathematicians have known about irrational numbers. These are numbers which can never be written as a fraction of whole numbers. But in 1844, Joseph Liouville discovered a brand-new type of number, which was even harder to describe, and further removed from the domain of the whole numbers. Liouville's number was not merely irrational, it was transcendental. Over time, the study of transcendence would revolutionize our attitude to the mathematical landscape.

The first recognized irrational number was $\sqrt{2}$. As Hippasus of Metapontum had realized to his cost, there is no possible way to write $\sqrt{2}$ as a fraction $\frac{a}{b}$, where a and b are both whole numbers. All the same, $\sqrt{2}$ can easily be described in terms of whole numbers. By the very definition of a square root, it is the number that produces 2 when multiplied by itself. So, although irrational, $\sqrt{2}$ is actually fairly close to the whole numbers—only one multiplicative step away.

Since the Pythagoreans, the numerical domain has expanded several times. The principal modern setting for mathematics is the system of complex numbers, with its imaginary unit i. Although i seems a far cry from the familiar terrain of the whole numbers, in fact it too is only separated from them by one arithmetical procedure. After all, i is defined to be $\sqrt{-1}$. So, as for $\sqrt{2}$, just multiplying it by itself returns us to the world of the whole numbers: $i \times i = -1$.

OPPOSITE A colored radar image of Tibetan mountains. With intricate patterns at all scales, from the largest to the smallest, much natural terrain is inherently transcendental, and cannot easily be modeled by simple algebraic equations.

Liouville's transcendental numbers

In 1844, the French mathematician Joseph Liouville found a brand-new type of number which defies any such simple description. A few years later, Liouville constructed one such number explicitly, decimal place by decimal place: 0.11000100000000000000000100... (The pattern is that there are 1s in the decimal places 1, $2 \times 1 = 2$, $3 \times 2 \times 1 = 6$, $4 \times 3 \times 2 \times 1 = 24$, and so on, and 0s everywhere else.) This may seem an elaborate construction, but Liouville proved something very disconcerting about this number, and others of a similar type. Liouville's numbers could not be described in terms of whole numbers, using only the ordinary arithmetical operations: addition, subtraction, multiplication and division. While $\sqrt{2}$ and i are each just one step away from the whole numbers, Liouville's numbers sat infinitely far away by the same measure. No amount of addition, multiplying, subtracting or dividing would ever return them to the whole numbers. Liouville had discovered the first transcendental numbers.

Transcendence of e and π

Liouville's numbers were a tantalizing discovery. What did they mean? Perhaps it was no more than an ingeniously constructed curiosity. After all, it hardly seemed the kind of object you would expect to run into in the usual course of scientific enquiry. The question was, are there any naturally occurring examples of transcendence? Or were Liouville's discoveries no more than numerical freaks?

After Liouville's quirky discoveries, suddenly transcendence theory was turning up at the very heart of mathematics.

This question received an unequivocal answer in 1873, when Charles Hermite proved that Euler's number e (see page 190) is transcendental. After Liouville's quirky discoveries, suddenly transcendence theory was turning up at the very heart of mathematics. In 1882, Ferdinand von Lindemann followed this up with the revelation that an even more famous number, π, is also transcendental. This breakthrough was enough to settle one of the oldest riddles in mathematics, that of squaring the circle (see page 194). Von Lindemann's theorem finally established the impossibility of the construction.

Cantor and counting transcendence

The biggest shock came in 1874, with the work of Georg Cantor. His notorious theorem established that there are different levels of infinity, some bigger than others (see page 217). Astoundingly, however, he showed that the infinity of transcendental numbers is actually larger than that of the more familiar non-transcendental or algebraic numbers.

The conclusion was that, in a very precise sense, almost all numbers are transcendental.

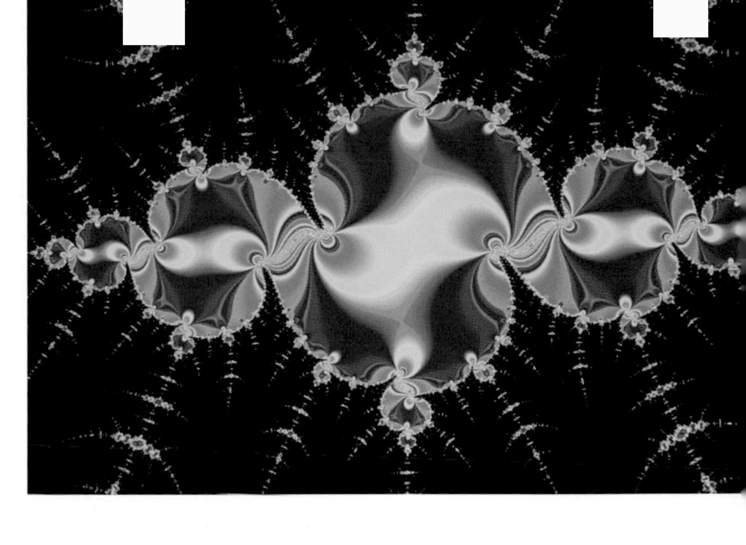

ABOVE A curve is called transcendental if it crosses some straight line infinitely many times. In the case of this Julia set, a diagonal line would cut through the outline of the bulbs at infinitely many places. There is rich interplay between the geometry of transcendental curves and the theory of transcendental numbers.

This was disconcerting to many mathematicians. The algebraic numbers include all the whole and rational numbers, and indeed almost all the numbers that mathematicians had ever contemplated, notwithstanding Hermite and von Lindemann's breakthroughs. Yet Cantor showed these familiar numbers are infinitely outnumbered by a morass of transcendental numbers, which are far harder to describe. Cantor's work was a stark reminder just how much of the numerical realm remains mysterious.

Transcendence and exponentiation

In the early 20th century, it later became clear that the number e is not just transcendental, but is the key to the entire phenomenon. Transcendence is defined by the very simplest arithmetical procedures: addition, subtraction, multiplication and division. A transcendental number is one which cannot reach the whole numbers using only these procedures. So the interaction of transcendence with ordinary arithmetic is straightforward. The missing ingredient is how transcendental numbers would behave under exponentiation, or raising to powers.

In 1900, David Hilbert considered this question in his famous address to the International Congress of Mathematicians. In the seventh of his 23 problems, Hilbert asked when one number raised to the power of another is transcendental. An initial answer came in 1934, courtesy of Aleksandr Gelfond and Theodore Schneider. This was significantly extended by Alan Baker in the late 1960s. Thanks to their work, we now know that numbers such as $3^{\sqrt{2}}$ and $3^{\sqrt{2}} \times 2^{\sqrt{3}}$ are transcendental. Yet even after all this investigation, today's mathematicians are still uncertain of the status of e^e or even $e + \pi$. All that is known is that at least one of $e + \pi$ and $e \times \pi$ must be transcendental. It seems highly likely that all such numbers should be transcendental. But transcendence remains an extremely slippery matter, even today.

49 Polytopes

BREAKTHROUGH SCHLÄFLI LIFTED THE THEORY OF PLATONIC SOLIDS INTO HIGHER DIMENSIONS WITH A COMPLETE CLASSIFICATION OF REGULAR POLYTOPES.

DISCOVERER LUDWIG SCHLÄFLI (1814–95), ALICIA BOOLE STOTT (1860–1940).

LEGACY DESPITE INITIALLY SEEMING ONLY OF THEORETICAL INTEREST, MULTI-DIMENSIONAL GEOMETRY IS NOW USED THROUGHOUT MATHEMATICS AND PHYSICS.

The classification of the Platonic solids was one of the mathematical wonders of the ancient world. In the 19th century, Ludwig Schläfli was able to emulate it in the strange new setting of higher-dimensional spaces.

The classification of the Platonic solids says that only five perfectly symmetrical 3-dimensional shapes can be built from flat faces and straight edges (see page 37). This great theorem is itself a higher-dimensional analog of the well-known fact that there are infinitely many regular polygons possible in two dimensions: an equilateral triangle, a square, a regular pentagon, a regular hexagon, and so on. In the late 19th century, the Swiss geometer Ludwig Schläfli considered a new question: what would happen if he pursued the same line of thought into four dimensions?

Probing four dimensions

Of course, Schläfli was hindered by the limitations of his human mind. Although it was a brilliant mind, like those of all dwellers in 3-dimensional space, it was unable to visualize higher-dimensional spaces. Nevertheless, it is perfectly reasonable for mathematicians to probe these sorts of questions, and Schläfli assembled the tools to do so. First among them is the notion of coordinates. Since Descartes introduced his famous coordinate system for geometry (see page 129), the science of shape has become inextricably bound up with that of number. Two-dimensional geometry, that of the Euclidean plane, is represented by pairs of numbers such as (1, 2). These two numbers represent the

OPPOSITE La Grande Arche de la Défense, in Paris, France, designed by Johann Otto von Spreckelsen. The form of the arch is that of a four-dimensional hypercube, projected into three dimensions.

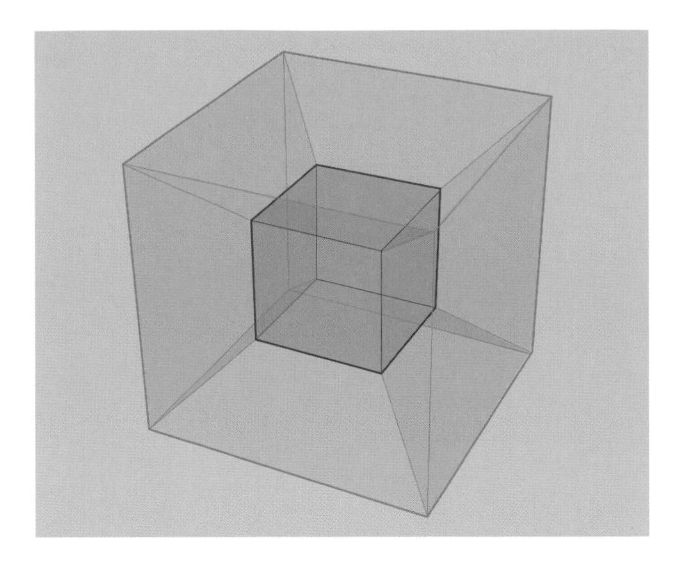

position of a point on the page. Similarly, triples of numbers such as (1, 2, 3) describe the position of a point in 3-dimensional space. So, in exactly the same way, 4-dimensional geometry can be considered as the study of quadruples of numbers such as (1, 2, 3, 4). All usual geometric notions, such as angle, length and volume, easily generalize into this new realm. The only thing that does not carry straight over is the human ability to visualize what is going on.

ABOVE A four-dimensional hypercube projected into two dimensions. Just as an ordinary cube is built from six squares folded together, so a hypercube comprises eight cubes folded together. They are the one in the center, the six around it, and the one outside.

The Platonic solids are built from 2-dimensional polygons. The most familiar of them is the cube, which is constructed from six squares folded together. Similarly the 4-dimensional analogs of the Platonic solids, the regular polychora, must be built from Platonic solids. The hypercube, for example, is constructed from eight cubes folded together. But what other examples are there?

Three of the Platonic solids are comparatively easy to generalize into higher dimensions. Just as the hypercube is a big sister to the ordinary cube, so the tetrahedron has a 4-dimensional equivalent known as the hypertetrahedron (or 4-simplex). It is built from five tetrahedra folded together in 4-dimensional space. Another way to arrive at the same shape is by considering points equal distances apart. Three points, each the same distance from both of the others, produce an equilateral triangle. Similarly, four points equal distances apart produce a tetrahedron. The same thing with five points produces a 4-simplex.

The octahedron can also be lifted into the 4-dimensional realm comparatively easily, just by looking at the coordinates of its corners. By doing this, Schläfli revealed the hyperoctahedron (also known as the 4-orthoplex).

The remaining two Platonic solids are trickier customers, the dodecahedron and the icosahedron. Nevertheless, with considerable ingenuity, Schläfli did discover 4-dimensional analogs of these too. The hyperdodecahedron is a much larger beast, being constructed by folding together 120 ordinary dodecahedra, while the hypericosahedron is built from 600 tetrahedra.

To discover these 4-dimensional versions of Theaetetus' ancient solids was a tremendous achievement. But the huge question which then confronted Schläfli was whether there are any other 4-dimensional regular shapes beyond these—shapes that have no equivalent in three dimensions. Schläfli did indeed find one. The octaplex is built by folding 24 octahedra together.

Platonic polytopes

Schläfli's greatest triumph was to prove that there are no others beyond these: these six Platonic polychora are the only regular shapes in four dimensions. By 1852, Schläfli had accomplished for four dimensions what Theaetetus had achieved for three 2,000 years earlier. But he did not stop there. Schläfli then turned his attention to the geometry of even higher-dimensional spaces: five, six, seven, eight dimensions, and so on. In each of these settings, it was easy to concoct analogs of the tetrahedron, cube and octahedron; the answers jumped straight out from looking at their coordinates. So every space has its own simplex, hypercube and orthoplex. But what of the other shapes that don't fit the pattern so easily: the dodecahedron, icosahedron (and their hyperversions) and the octaplex?

All usual geometric notions, such as angle, length and volume, easily generalize into this new realm. The only thing that does not carry straight over is the human ability to visualize them.

Schläfli might have expected that, as he probed ever higher dimensions, he would find ever more of these anomalous shapes, and the Platonic classification would become ever lengthier to state and more difficult to prove. But this was where Schläfli had his greatest insight of all. He realized that the story in five dimensions and onward is actually simpler than in three or four. There are no anomalous shapes there, no analogs of the icosahedron or octaplex. From five dimensions upward there are only ever three Platonic polytopes: the simplex, hypercube and orthoplex.

Schläfli's analysis of higher-dimensional geometry was one of the great accomplishments of human reason. Yet his work did not receive the acclaim it deserved. In the early 20th century, the self-trained prodigy Alicia Boole Stott rediscovered for herself much of Schläfli's work, including the classification of the six Platonic polychora in four dimensions. Her research attracted the interest of several European mathematicians, and it was only then that Schläfli's pioneering work in multi-dimensional geometry began to receive its rightful recognition.

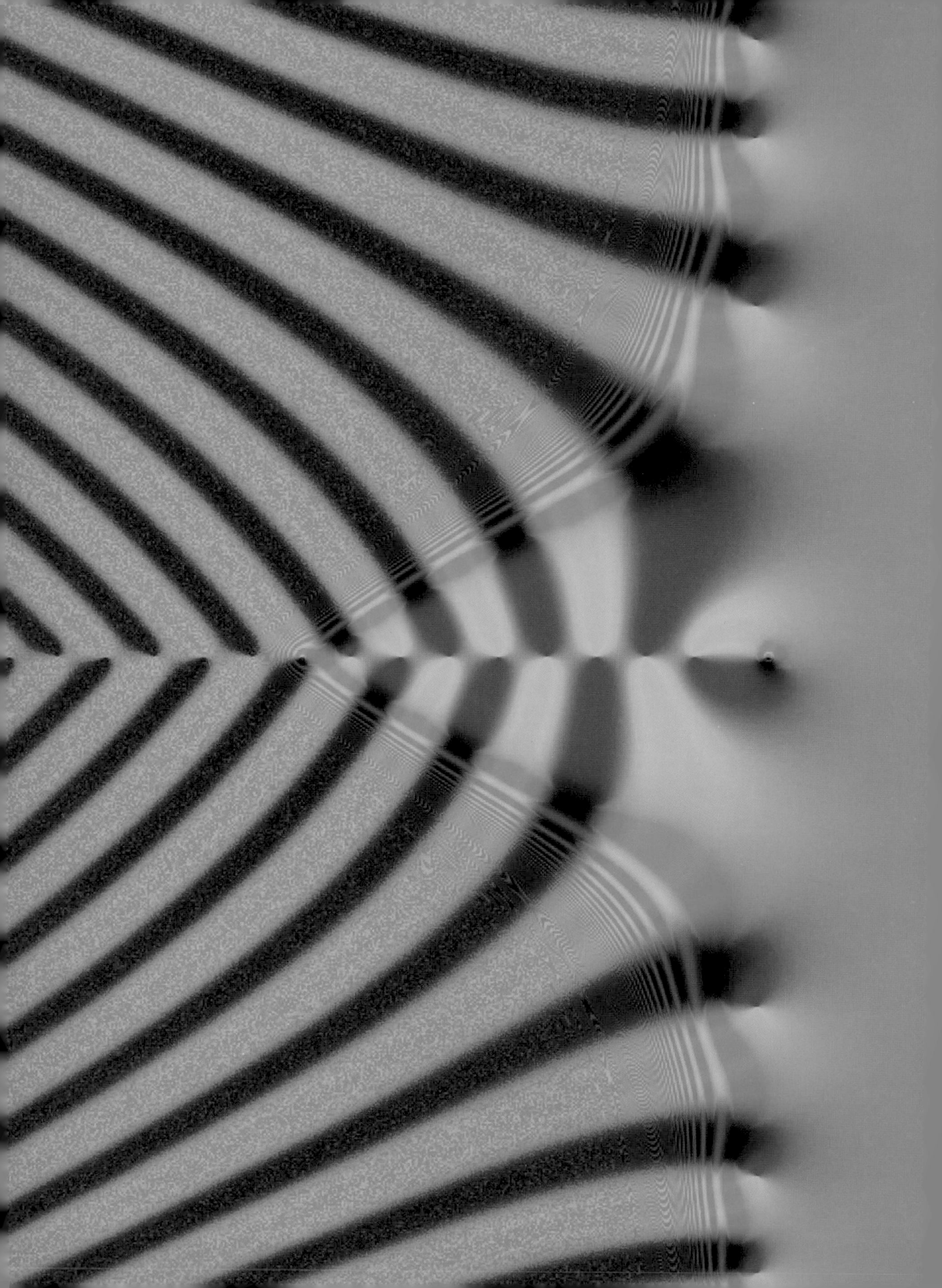

50 Riemann's zeta function

BREAKTHROUGH RIEMANN FOUND A NEW WAY TO ANALYZE PRIME NUMBERS BY INCORPORATING TECHNIQUES FROM COMPLEX ANALYSIS. HIS MAJOR DISCOVERY WAS AN OBJECT CALLED THE RIEMANN ZETA FUNCTION.

DISCOVERER BERNHARD RIEMANN (1826–66).

LEGACY THE RIEMANN HYPOTHESIS WOULD GIVE US THE BEST POSSIBLE PICTURE OF THE PRIME NUMBERS. PROVING IT REMAINS THE BIGGEST CHALLENGE IN MATHEMATICS.

The aim of number theory is to understand the whole numbers: 1, 2, 3, 4, 5, 6, and so on. But, as the ancient Greeks knew, all of these numbers are built from a special subcollection: the prime numbers. For years mathematicians have sought to understand the primes, and in 1859 there was a gigantic leap forward.

Every whole number is constructed from primes in a straightforward way. The non-prime 6 is equal to 2×3, for example. As the building blocks of all other numbers, an understanding of the prime numbers has been one of the longest-running mathematical quests. Many of history's greatest thinkers have used their talents to try to uncover the secrets of the primes, but progress has been painfully slow. Numerous fundamental questions remain unanswered even today. Of all of them, the most important is the one posed by Bernhard Riemann in 1859.

Counting the primes

There are many possible lines of enquiry about the primes and the possible sizes of the gaps between them. But one crucial question was formulated by Carl Friedrich Gauss in 1792: if you choose a number, such as 15, 302 or 1,000,000, is there a quick way to tell how many primes there are beneath it?

OPPOSITE The Riemann zeta function, as seen through Chris King's RZViewer program. The setting is the plane of complex numbers, and the color of each spot denotes the value of zeta there.

A complete answer to this question would give a huge amount of insight into how the primes are distributed through the whole numbers. The difficulty is that they seem to contain no predictable pattern. But if

Gauss's question could be answered in the affirmative, it would bring sense to this apparent randomness.

At just 15 years old, Gauss already possessed astonishing mathematical insight. His approach to the prime-counting problem was not too ambitious. Instead of aiming for a precise method to calculate the number of primes, he just wanted an approximate answer. If you pick a number at random between 1 and 1,000,000, what is the rough likelihood of hitting a prime? Is it reasonably likely, near impossible, or what? Gauss said that if you pick a number at random from the collection {1, 2, 3, ..., n}, the chance of hitting a prime should be around $\frac{1}{\ln n}$ (where "ln n" represents the natural logarithm of the number n; see page 109). In fact, there are 78,498 primes below 1,000,000, while Gauss's estimate predicted just 72,382.

In his more mature years, Gauss was able to refine this estimate further. He claimed that the number of primes below n should be around Li n, where "Li n" stands for the subtler logarithmic integral of n. This new formula was still not intended as an exact method for counting primes. All the evidence suggested that it was a good estimate. For example, Li 1,000,000 = 78,628 – a fairly small error. Yet Gauss was not able to provide any compelling argument as to why this was the case.

The Riemann hypothesis

In the 19th century, the geometer Bernhard Riemann, Gauss's former student, also contemplated this question. To the astonishment of his contemporaries, Riemann published a paper entitled "On the Number of Primes Less Than a Given Magnitude," which contained not an estimate, but an exact answer to his teacher's question. At the center of Riemann's paper was a mysterious object, known as the zeta function (because he chose to represent it with the Greek letter zeta, or ζ). This function was a fantastic discovery. It encapsulated a huge amount of information about the spread of the prime numbers. By exploiting the power of the zeta function, Riemann did indeed produce an explicit formula for the number of primes below any limit.

Riemann's breakthrough was to relocate the zeta function to the complex numbers, where he developed sophisticated new techniques that allowed him to express its full power.

In fact, the zeta function was not absolutely new. Leonhard Euler had already partially realized its potential. But Riemann's breakthrough was to relocate the zeta function to the complex numbers (see page 105), where he developed sophisticated new techniques that allowed him to express its full power.

A function is a mathematical gadget which takes in inputs and produces outputs. Once Riemann had weaved his magic, the inputs and outputs of the zeta function were both complex numbers. In his formula for the primes, the critical information consisted of those inputs whose corresponding output was zero. These special numbers appeared at the very heart of his formula. But what exactly are they? This question was easier asked than answered. Some of the zeroes were easy to identify: they appeared at -2, -4, -6, and so on. Feed any of these into the zeta function and 0 would pop out. But there were others, too, that were harder to find. Riemann believed that the remaining zeroes all appeared on a certain vertical line in the complex plane (which comprises complex numbers of the form $\frac{1}{2} + iy$, where y is any real number). This later became to be known as the critical line.

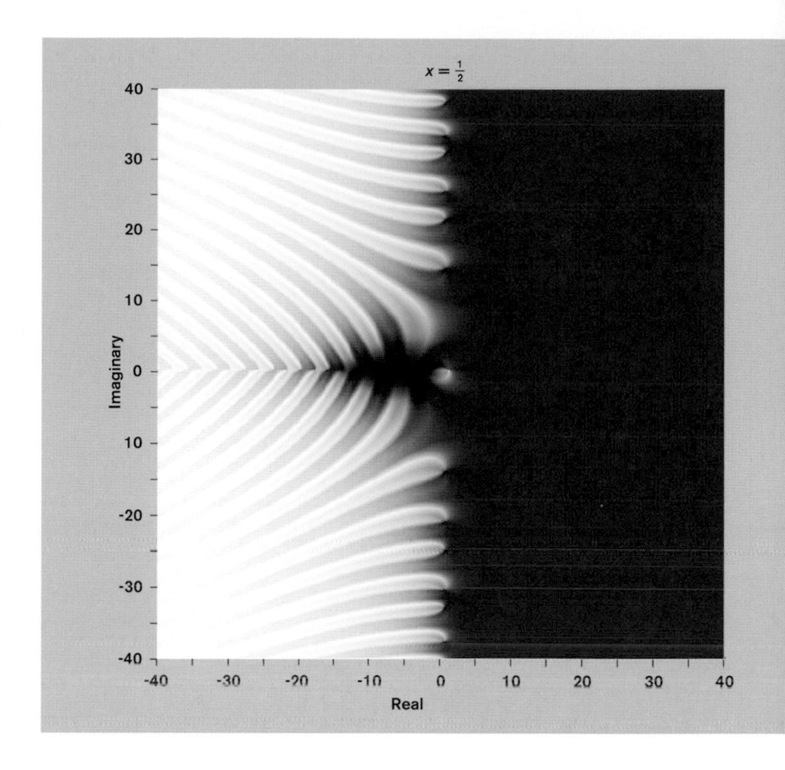

ABOVE The Riemann zeta function on the complex plane. Red indicates where the output is a real number, and black spots indicate an output of zero. The non-trivial zeroes can be seen along the critical line at $x = \frac{1}{2}$.

Riemann gave up on his attempts to prove this, but remarked that he thought it "very probable." As other mathematicians pored over his work, many attempted to resolve the matter of the location of the zeta function's zeroes, which became known as the Riemann hypothesis. It resisted all such efforts, and continues to do so today.

The prime number theorem

However, some progress was made. In 1896, Jacques Hadamard and Charles de la Valée Poussin independently proved that Riemann's zeroes must all lie inside a certain critical strip, surrounding the critical line. This breakthrough enabled them to prove the validity of Gauss's estimates for the number of primes, a celebrated fact now known as the prime number theorem. However, the Riemann hypothesis itself has eluded mathematicians for over 150 years. Because of the light it would shine on the prime numbers, it is widely considered the most important open problem in mathematics.

51 Jordan curve theorem

BREAKTHROUGH ANY LOOP MUST DIVIDE A PAGE INTO AN INSIDE AND AN OUTSIDE. DESPITE THE SEEMING SIMPLICITY OF THIS STATEMENT, IT WAS TRICKY TO PROVE.

DISCOVERER CAMILLE JORDAN (1838–1922), HENRI LEBESGUE (1875–1941), L.E.J. BROUWER (1881–1966), JAMES WADDELL ALEXANDER (1888–1971).

LEGACY LIFTING JORDAN'S CURVE THEOREM TO HIGHER DIMENSIONS WAS NOT STRAIGHTFORWARD, LEADING TO THE DISCOVERY OF STRANGE SHAPES INCLUDING ALEXANDER'S HORNED SPHERE.

The theorem proved by Camille Jordan in 1887 seems so obvious as hardly to be worth stating. It says that when a loop is drawn on a piece of paper, it will divide the page into two regions: an "inside" and an "outside," each consisting of just one piece. But in mathematics, nothing is known until it is established with precision and rigor. Despite first appearances, this result was surprisingly difficult to prove. What is more, when mathematicians lifted this result to higher dimensions, they were in for a shock.

Although Camille Jordan was principally an engineer, he performed several important pieces of research in the purer parts of mathematics. He played a significant role in the early theory of groups (see page 389), but is best remembered for his theorem about loops on a 2-dimensional plane. It is a testament to his mathematical maturity that Jordan realized such a seemingly obvious fact should even require a proof.

OPPOSITE A plan of the labyrinth on the floor of Chartres Cathedral in France. Unlike other types of maze, the labyrinth comprises a single path from the start to the center, making it impossible to get lost. The Jordan curve theorem guarantees that, despite appearances, this path is topologically equivalent to a simple straight path or circular disc.

Continuity and topology

In order to formalize the notion of a loop, Jordan needed to know what distinguishes a curved line from a sequence of dots or dashes. In mathematical terms, the critical property is that of continuity, meaning that there can be no jumps or breaks. On top of this, a loop must never touch or cross over itself, but should eventually cycle round to end where it began. This sounds a reasonable definition. But today we realize that loops may be far stranger objects than this immediately suggests.

Since the discovery of fractals (see page 253), mathematicians have identified many exotic and unnatural shapes that nevertheless satisfy this property of continuity. It is easily possible for a loop to be infinitely long, or infinitely wiggly, meaning that, no matter how far we zoom in, we will never find a section that is perfectly smooth. (Smoothness is a far stricter condition than mere continuity.)

In more recent times, this idea has been expressed in terms of topology: the perspective by which two shapes are the same if one can be pulled into the shape of the other, without cutting or gluing. In this language, every loop is simply a circle. Examples include shapes with sharp corners, such as triangles, or intricate fractal designs such as the Koch snowflake.

Just as a loop is a circle which has been subjected to topological morphing, so a bubble is nothing more than a sphere which may similarly be deformed.

But just how strange can loops, or topological circles, be? Might there even be one so unnatural and unexpected that somehow its wiggles manage to carve up the page into many regions, instead of just two? Or indeed, might there be a loop which is somehow unexpectedly trivial, and fails to divide up the page at all? That was the question Jordan addressed in 1887, and after a considerable amount of work he was able to provide a reassuring answer.

Jordan–Brouwer separation

Jordan's curve theorem established that any loop crisply divides a 2-dimensional plane into two regions. It is easy to see that the same thing does not apply in 3-dimensional space. A loop floating freely in space cannot divide a room into two regions. But replacing the 1-dimensional loop with a 2-dimensional bubble will divide the ambient space into an "inside" and an "outside." Just as a loop is a circle which has been subjected to topological morphing, so a bubble is nothing more than a sphere which may similarly be deformed.

This, then, was the appropriate way to lift Jordan's curve theorem into higher dimensions. In 1911, Henri Lebesgue and L.E.J. Brouwer did exactly that, proving what became known as the Jordan–Brouwer separation theorem: every 2-dimensional bubble will automatically divide 3-dimensional space into exactly two parts. Indeed, the same thing holds in all higher dimensions too: any $(n-1)$-dimensional bubble will split the ambient in n-dimensional space into an "inside" and an "outside."

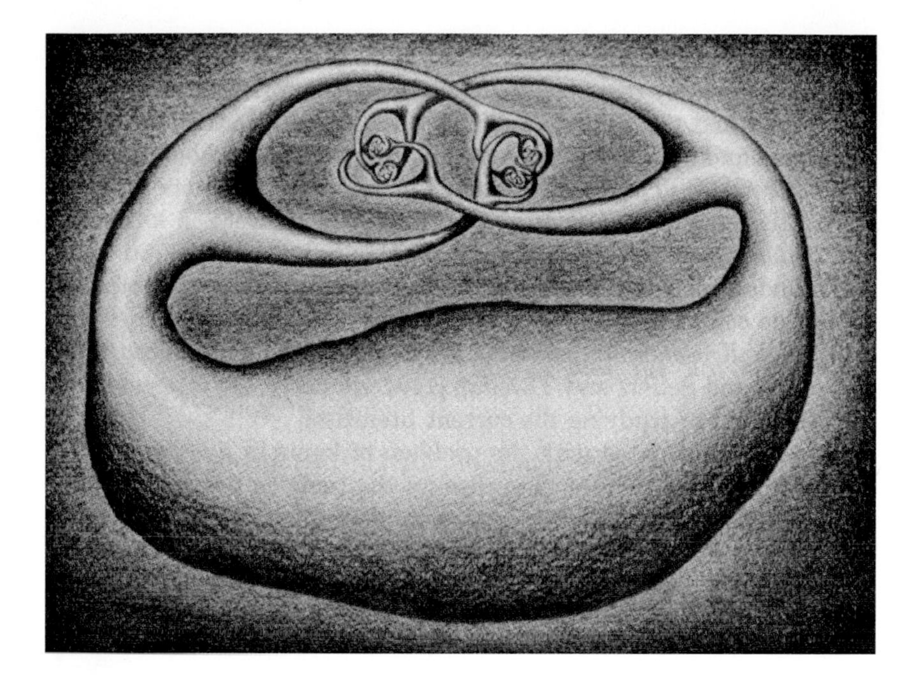

LEFT Alexander's horned sphere. Like a sphere, it divides the ambient space into an inside and an outside, but the infinite dividing and intertwining of the horns ensures that the outside of this shape is topologically unlike the outside of an ordinary sphere.

Alexander's horned sphere

The Jordan–Brouwer separation theorem suggested that geometry in all dimensions behaved in accordance with common sense. But a closer analysis revealed a dramatic twist. In 1924, James Waddell Alexander produced an extraordinary shape, the likes of which no one had seen. In topological terms, his shape was a sphere. But he had deformed it to have strange horns, which kept dividing and subdividing into ever smaller horns. What is more, these infinitely many horns were also inextricably intertwined.

The result was shocking. Alexander's horned sphere was topologically spherical. So, as it must, it divided the ambient space into an inside and an outside. Topologically, the inside resembled the inside of any other sphere. But the outside of the shape was not topologically similar to the outside of an ordinary sphere. Because of the ways the sphere's horns divided and interlinked, the external region was an infinitely more complex place than the outside of an ordinary sphere.

In topology, the complexity of a space can be measured with loops. On these terms, the outside of an ordinary sphere is the simplest sort of space, where any loop can freely morph into the form of any other. But around the horned sphere, loops could become entangled in the horns in infinitely many different ways showing that the outside of Alexander's sphere was a supremely complicated and inhospitable space.

52 Classification of surfaces

BREAKTHROUGH WORKING IN THE NEW SUBJECT OF TOPOLOGY, WALTHER VON DYCK WAS ABLE TO PRODUCE A COMPLETE CATALOGUE OF EVERY POSSIBLE 2-DIMENSIONAL SURFACE.

DISCOVERER JOHANN LISTING (1808–82), AUGUST MÖBIUS (1790–1868), FELIX KLEIN (1849–1925), WALTHER VON DYCK (1856–1934).

LEGACY THIS CLASSIFICATION WAS ONE OF THE FIRST MILESTONES IN TOPOLOGY, WHICH WOULD TRANSFORM MATHEMATICS OVER THE 20TH CENTURY.

Just as a 1-dimensional shape is a curve, so a 2-dimensional shape is called a surface. The most common example is a sphere. But there are many others, such as the bagel-shaped torus with its hole. When Felix Klein discovered an extraordinary new surface, the Klein bottle, in 1882, this paved the way for the first great theorem in the new mathematical discipline of topology: a complete listing of all possible surfaces.

The idea of naming every possible surface seems ridiculous. Indeed, from the standpoint of traditional geometry, it is. From teacups to pretzels, the world contains a huge variety of shapes (and geometry textbooks offer even more). But in the new subject of topology, which began to emerge in the mid-19th century, many of these shapes were deemed to be essentially identical. If one shape can be distorted into another, then to a topologist, they are one and the same. This morphing may involve extreme stretching and twisting, so a strand of spaghetti is judged to be the same as a sphere. But it may not include cutting or gluing. So a teacup is not the same as a sphere, because the hole through its handle can never be eliminated.

OPPOSITE A glass Klein bottle, which has no inside or outside, so water cannot be trapped within it, unlike a sphere. In three dimensions any Klein bottle must have a flaw where the surface cuts through itself; this happens at the very top in this case.

Spheres with handles

A teacup, however, is topologically the same as a torus. The main section of the cup can be deformed to a sphere, as small as required, but, critically, the handle will remain. Adding handles to a sphere is a way to generate genuinely new shapes: a sphere with two handles (or double torus) is different from an ordinary torus. A teapot is topologically a double torus; a pretzel, meanwhile, is topologically a sphere with three handles added, or a triple torus.

All of these surfaces are closed, meaning they have no edges, unlike a sheet of paper. They come in just one piece, and what is more they have a finite area, unlike an infinite plane. In 1863, August Möbius, then aged 71, was able to show that every surface that satisfies these criteria and could possibly exist in our 3-dimensional Universe must be, in topological terms, a sphere with some number of handles added.

The Möbius strip

A great achievement though this was, there were already suggestions that this understanding of surfaces might be somehow incomplete, stemming from a curious discovery. In 1847, Johann Listing contemplated a rather simple object: a rectangular strip of paper. If he glued one end to the other, Listing produced a cylinder. This was not a closed surface, of course, since it had an edge, and nor was it particularly insightful; after all, cylinders had been well understood since the time of Archimedes. But a slight change to the construction produced something altogether more exciting. When Listing put a half-twist in the strip before attaching the ends, the result was not a cylinder, but something new. Fascinatingly, this object only had one side. When Listing ran his finger along the inside of the paper, it would travel across both the inside and the outside before returning to the starting point.

When Listing put a half-twist in a strip of paper before attaching the ends, the result was not a cylinder, but something new. Fascinatingly, this object only had one side.

In 1858, Möbius would independently make the same discovery, and it was his name that would become forever associated with this Möbius strip. This delightful object has many fascinating properties which continue to bewilder today. If you cut it in two along its center, for instance, the result is not two new Möbius strips, but a single strip containing a double twist.

The Klein bottle

As well as having some wonderfully unexpected properties, the Möbius strip also presented geometers with a more serious challenge. How did this one-sided object fit into the broader world of surfaces? Of course, the strip itself was not a closed surface, since it had an edge, so it did not contradict Möbius' theorem. All the same, it raised a question: all the previously known surfaces had two sides. Might there be a closed surface with no edges and only one side?

The answer was a resounding "yes," as Felix Klein discovered in 1882. But his answer came with a qualification. The beautiful Klein bottle, as it became known, was the result of taking two Möbius strips and fusing them along their edges. However, if you try this at home with two paper rectangles and a pot of glue, all you are likely to get is a mess. Klein found a way to do it, but he had to allow the surface to pass through itself. Today people do make Klein bottles from glass, but they will always have a flaw: a place where the surface cuts through itself. Yet on its own terms, a Klein bottle is a perfectly coherent surface. The difficulty only arises when one tries to construct it within 3-dimensional space. In a 4-dimensional world, a flawless Klein bottle could easily be formed.

Von Dyck's theorem

The Klein bottle was not the only new surface to be discovered. Just as Möbius had begun with a sphere and then attached handles, so it was now possible to cut holes in a shape and sew in Möbius strips. This led to a brand-new family of non-orientable surfaces which cannot exist perfectly in 3-dimensional space.

In 1888, Klein's student Walther von Dyck produced a wonderful theorem, upgrading Möbius' original analysis and flexing the muscles of the new subject of topology. Every 2-dimensional closed surface, he showed, must be topologically equivalent either to a plain sphere, or to a sphere with handles added, or to a sphere with some number of Möbius strips sewn in.

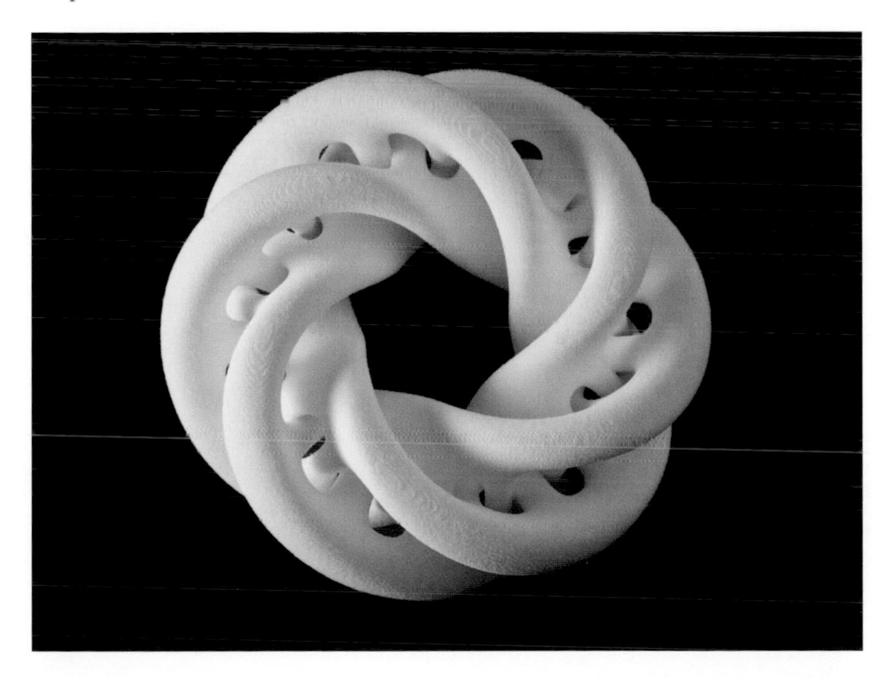

LEFT A sculpture consisting of two interlinked skeletal Möbius strips, produced by 3D printing. Möbius strips are the keys to understanding the topology of unusual shapes like the Klein bottle, and have long fascinated artists too.

53 Cardinal numbers

BREAKTHROUGH STARTING WITH ONE INFINITE SET, CANTOR WAS ABLE TO MANUFACTURE A BIGGER ONE. HENCE CANTOR'S DIFFERENT LEVELS OF INFINITY ARE CALLED CARDINAL NUMBERS.

DISCOVERER GEORG CANTOR (1845–1918).

LEGACY SET THEORY BEGAN WITH CANTOR'S WORK AND NOW FORMS THE MAIN LOGICAL FOUNDATION FOR MATHEMATICS. QUESTIONS ABOUT INFINITE SETS CONTINUE TO FASCINATE TODAY'S LOGICIANS.

Of all the ideas contemplated over the centuries by mathematicians, scientists and philosophers (not to mention mystics, eccentrics and cranks), few have the enduring allure of infinity. Scientists have long posed the question of whether time has a beginning or an end, and whether space is infinite in extent or merely unimaginably large. But it was the German logician Georg Cantor who finally got to grips with the mathematics of infinity.

Infinity has long been a source of wonder and even fear—as if excessive contemplation of the infinite could be hazardous for the human mind. Meanwhile theologians from all the world's religions have pondered the presumed infinitude of God and compared it to our own finite nature. Within mathematics, for a long time infinity occupied an uncomfortable middle ground. On the one hand, it is self-evident that there is no biggest whole number: you can always add one to any number you have in mind. So it must be that the entire collection of whole numbers is infinite. On the other hand, mathematicians had held back from investigating the consequences of this fact, perhaps fearing that the treasured precision of mathematics could dissolve into meaninglessness.

When Georg Cantor turned his mind to these matters in the late 19th century, the theorems he proved were astounding on their own terms. But perhaps his greatest triumph was in recognizing that infinity is not something magical or beyond human comprehension; it is amenable to intellectual analysis in exactly the same way as any other idea. It is a

OPPOSITE The Tarantula Nebula is situated 160,000 light-years away and is one of the most densely packed regions of space we know of. Whether space and matter are finite or infinite is one of the oldest and biggest questions in science.

sad irony that Cantor, who accomplished this triumphant feat of reason, suffered terribly from mental illness in his later years.

The beginnings of set theory

The subject that Cantor created is known as set theory. Sets, to Cantor, were nothing more than collections of objects. He was particularly interested in comparing the different sizes of sets. If one begins with a collection of 8 objects, and then removes half of them, the set that remains is half as big. This sounds so obvious as to be trivial. But what had alarmed earlier thinkers is that the same reasoning spectacularly fails when applied to infinite sets. If one begins with the set of all whole numbers {1, 2, 3, 4, 5, ...}, and then removes the odd numbers, what remains is the set of even numbers {2, 4, 6, 8, 10, ...}. One might expect that this set should be half as big. The surprise is that it is not: the set of even numbers is exactly the same size as the full set of whole numbers. How so? Well, the two collections can be paired off exactly: 1 with 2, 2 with 4, 3 with 6, 4 with 8, 5 with 10, and so on.

Prior to this, there had been an unspoken assumption that all infinite sets must be the same size. This was the myth that Cantor exploded.

This idea of two sets being the size if one can be paired off with the other was at the heart of Cantor's work. It is a very natural idea; after all, this is how we learn to count things. To calculate the number of biscuits on a plate, a child would assign each biscuit a number: biscuit 1, biscuit 2, biscuit 3. In other words, she pairs off the set of biscuits with a set of numbers, such as {1, 2, 3}. A set theorist would say that the two sets are measured by the same cardinal number.

Prior to Cantor, insofar as people had considered it at all, there had been an unspoken assumption that all infinite sets must be the same size. This was the myth that Cantor exploded. Another famous infinite set is the collection of all decimals, or, as mathematicians call them, real numbers. To represent a single real number requires infinitely many digits. A famous example is π, whose digits continue forever without termination or repetition. But π is just one number. How many others like this can there be? Infinitely many, of course. But Cantor demonstrated, with a startlingly clear and concise argument, that this infinity is larger than the infinity of the whole numbers. Any attempt to marry up the two sets was bound to fail. Some real numbers would always be left out.

This was a shocking discovery; but Cantor didn't stop there. He showed that the sets of whole and real numbers are represented by two of the smallest cardinal numbers in an infinite hierarchy; above them come

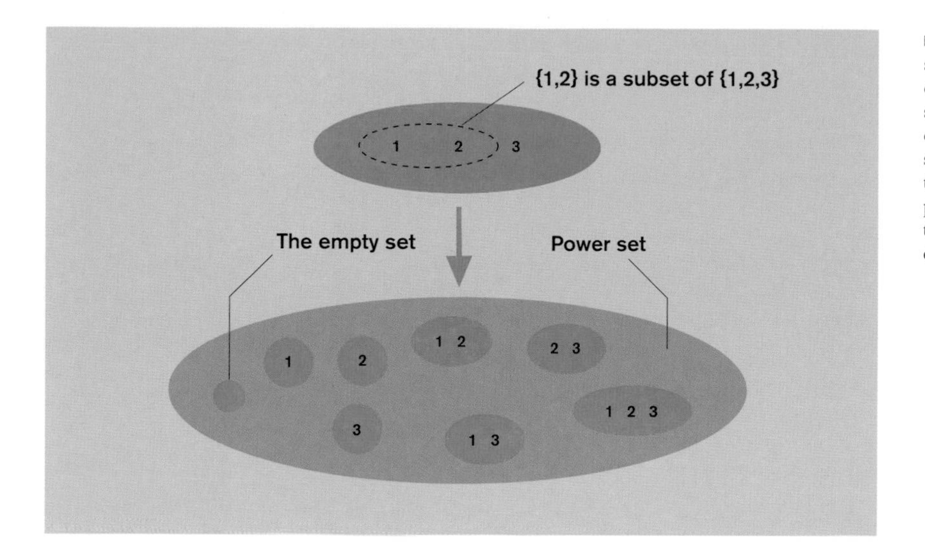

LEFT The three-element set {1,2,3} has eight elements in its power set. The surprise comes when this simple idea is applied to infinite sets, and produces infinite sets that are of genuinely different sizes.

still bigger levels of infinity. This seems an astonishing claim, but again, the argument that Cantor provided was short, slick and very difficult for his critics to gainsay.

Power sets

The idea behind Cantor's greatest theorem was that of the power set. Every set, such as {1, 2, 3}, has a range of subsets, such as {1} and {1, 2}. (Here {1, 2} is the same as {2, 1} and {1, 1, 2}. Only the contents of a set are important; reordering or repetition makes no difference.) How many subsets are there? In this example eight:

$$\varnothing, \{1\}, \{2\}, \{3\}, \{1,2\}, \{1,3\}, \{2,3\}, \{1,2,3\}.$$

(The symbol ∅ represents the empty set: a set with nothing in it.)

Taking all of these together, the set of all subsets is called the power set of the original. Again, it was when this simple idea was applied to infinite sets that the surprise came. Cantor's theorem states that there can never be any matching between a set and its power set. The power set will always be bigger.

This had seismic repercussions. Starting with the set of whole numbers, the power set is bigger; in fact, it is the same size as the collection of real numbers. But then one can take this particular power set to produce something still bigger. This can be repeated as many times as desired, thus deconstructing the single notion of infinity into an infinite hierarchy of cardinal numbers.

54 Wallpaper groups

BREAKTHROUGH FEDOROV AND SCHÖNFLIES PROVIDED A COMPLETE LIST OF THE POSSIBLE WAYS IN WHICH PATTERNS CAN PROPAGATE THROUGH 2- AND 3-DIMENSIONAL SPACE.

DISCOVERER EVGRAF FEDOROV (1853–1919), ARTUR SCHÖNFLIES (1853–1928).

LEGACY FEDOROV AND SCHÖNFLIES' THEOREMS ARE CORNERSTONES OF THE SCIENCE OF CRYSTALLOGRAPHY, AS WELL AS FUNDAMENTAL FACTS FOR TODAY'S GRAPHIC DESIGNERS.

From the intricate tilings on the walls of an ancient temple to the computer-generated wallpaper on your laptop, abstract patterns have always had a fascination for humans. But behind these alluring designs lies a piece of mathematics even more elegant than the most intricate pattern: the classification of wallpaper groups. Though artists had hinted at it for centuries, it was fully formalized in the 19th century independently by two mathematicians: Evgraf Fedorov in Russia in 1890 and Artur Schönflies in Germany in 1891.

The difference between a single picture and the types of pattern that appear on modern wallpaper and on the mosaic floors of ancient buildings is that patterns are repetitive. If you focus on one patch of the pattern, you should be able to find identical regions elsewhere. Repeating patterns like this pose a fundamental mathematical conundrum: what are the different ways that a pattern can repeat itself?

Repeating patterns are not merely aesthetically beautiful; they are of central importance in the study of crystals. The molecules which make up solids such as diamonds are arranged in a repetitive lattice. Of course, crystals are 3-dimensional, so these patterns repeat in three different directions: left–right, up–down and forward–backward. A wallpaper pattern, meanwhile, is only 2-dimensional, repeating left–right and up–down. Nevertheless, Fedorov and Schönflies saw that the same mathematical analysis could be brought to bear in both cases.

OPPOSITE An image from a kaleidoscope. Symmetry has long been recognized as a source of beauty. In this example, the different sections are exact rotational copies of each other, and also exhibit mirror symmetry.

LEFT A tiling in Damascus, Syria. Decorative tiling has a long history within Islamic culture. This example illustrates the wallpaper group with rotational centers of order 2, 3, and 6, and no mirror symmetries.

Possible and impossible symmetry

Starting with a patch of a wallpaper pattern, you can slide it to the right (or left) until it lies exactly over another identical patch. This is what mathematicians call translational symmetry, meaning that the whole pattern appears unchanged when it is shunted one unit rightward. The same thing is true moving the pattern upward or downward. So double translational symmetry was the starting point for the analysis of patterns. The complicating factor is that the patch might have other internal symmetries of its own. It is these extra symmetries which lead to more intricate compositions. For instance, if the design depicts a butterfly, it may have a mirror symmetry, meaning that the left wing is an exact reflection of the right. Alternatively, if the pattern includes a flower with six petals, the pattern may have rotational symmetry, whereby rotating the entire pattern around the center of the flower leaves the whole design looking the same.

But what additional symmetries are possible? The first deep insight was that rotational symmetry is possible only with order 2, 3, 4 or 6, that is, patterns of flowers with these numbers of petals. Surprisingly, rotations of order 5, 7, 8 or larger are all incompatible with double translational symmetry, and thus impossible to include in any repeating pattern. (Of course, you can include a five-petaled flower in a pattern, but the rotation can only ever be a symmetry of the flower, not of the whole pattern.) The same thing is true inside a 3-dimensional crystal: again, rotations of order 2, 3, 4 and 6 are the only ones permitted. This fundamental fact goes by the name of the crystallographic restriction theorem (see page 319).

The 17 wallpaper groups

Crystallographic restriction is a profoundly important limitation. But in fact its proof is surprisingly short and simple. Fedorov and Schönflies' real achievement was to see that this opened the door to something even greater: a full classification of all patterns—a complete listing of all the possible ways in which rotational, mirror and other symmetries can coexist with a pattern's basic double translational symmetry. The result is a total of 17 possible configurations, known as the 17 wallpaper groups.

These sorts of patterns were particularly popular in medieval Islam, where a religious injunction against figurative art encouraged the creation of imaginative abstract designs.

Over the centuries, innumerable artists and craftsmen have experimented with symmetrical patterns. We now know that every such example is an instance of one of the 17 wallpaper groups. These sorts of patterns were particularly popular in medieval Islam, where a religious injunction against figurative art encouraged the creation of imaginative abstract designs. Perhaps these artists had already arrived at 17 as the number of fundamentally different patterns? Indeed, it is sometimes claimed that all 17 groups are represented in the Alhambra palace in Spain, although some skepticism is warranted here. All the same, the work of Fedorov and Schönflies is a triumphant example of science illuminating art.

Space groups

The classification of wallpaper groups was a magnificent achievement, but Fedorov and Schönflies did not stop there. By extending their analysis into three dimensions, Fedorov and Schönflies were also able to show that there are exactly 230 space groups, meaning 230 distinct 3-dimensional patterns, each representing one way by which a crystalline structure can be arranged.

In 1900, the German mathematician David Hilbert challenged the mathematical community to pursue the study of patterns into even higher dimensions. Might it even be possible, he asked, that in a 4- or 5-dimensional world, free from the limitations of the crystallographic restriction theorem, there could be infinitely many fundamentally different patterns? In 1911, Ludwig Bieberbach provided the answer. Although the patterns in higher-dimensional spaces differ from the 2- and 3-dimensional varieties, nevertheless the number of space groups does remain finite in every dimension. Today we know that in four dimensions there are exactly 4895 possible hypercrystals, in five dimensions there are 222,097, while in six dimensions the total stands at 28,934,974.

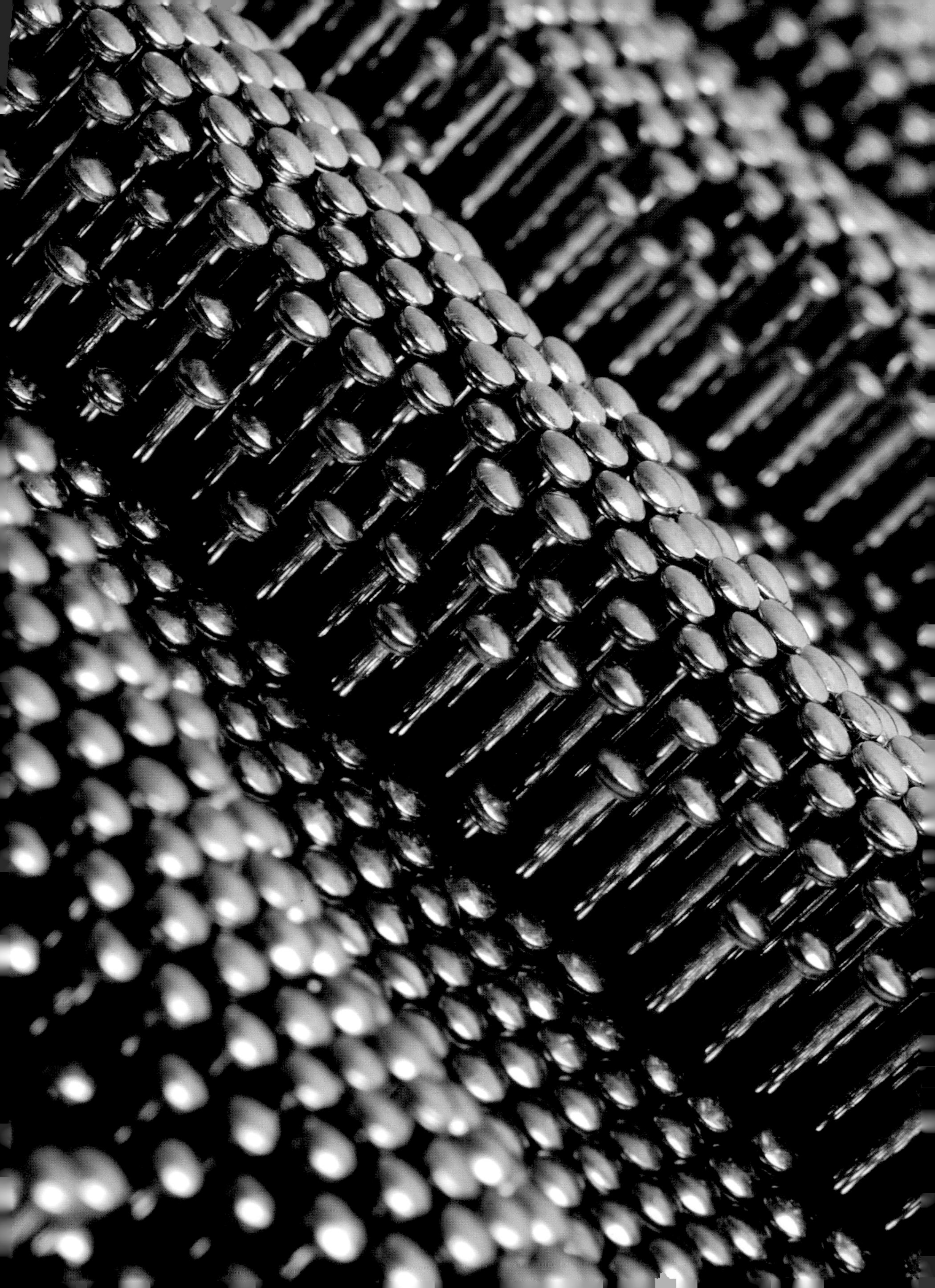

55 Digital geometry

BREAKTHROUGH PICK DISCOVERED A SIMPLE FORMULA FOR CALCULATING THE AREA OF ANY SHAPE, SO LONG AS ITS CORNERS SIT ON A SQUARE GRID.

DISCOVERER GEORG PICK (1859–1942).

LEGACY PICK'S THEOREM AND ITS HIGHER-DIMENSIONAL ANALOGS ARE OF FOUNDATIONAL IMPORTANCE IN DIGITAL GEOMETRY, A SUBJECT WHICH IS BECOMING INCREASINGLY IMPORTANT IN THE COMPUTERIZED WORLD.

Since the time of the ancient Egyptians, and probably even before that, geometers have devoted time and effort to quantifying the sizes of certain shapes, in terms of their areas and volumes. Many types of shape come with their own formulae for doing this, the most famous being Archimedes' expression for the area of a circle (see page 53). In 1899, Georg Pick proved an extraordinary result that is valid for all 2-dimensional shapes, so long as they are built from straight edges and have their corners lying on a square grid. Lifting Pick's theorem into higher dimensions has produced wonderful mathematical results with important applications in computer science.

The flat Euclidean plane is host to many and varied shapes, from the smooth curves of ellipses and circles (see page 47) to the sharp corners of triangles and trapezia. There is a sense in which shapes built from straight edges are easier to understand than curvy ones. After all, any shape built from straight edges can be chopped up into triangles. If we want to know the area of the entire shape, one approach is to calculate the area of every triangle and then add up the results. However spiky and irregular the shape may be, this will always work—although it may be a time-consuming process.

OPPOSITE Detail from a pin matrix bearing the impression of a hand. Pin matrices produce images which are necessarily approximate; the only points represented are those which coincide with pins, and smaller details get missed out. George Pick's theorem guarantees that such approximations are far easier to compute with than precise representations.

Pick's theorem

In 1899, Georg Pick made an extraordinary discovery. He found a single, simple formula to calculate the area of any straight-edged shape.

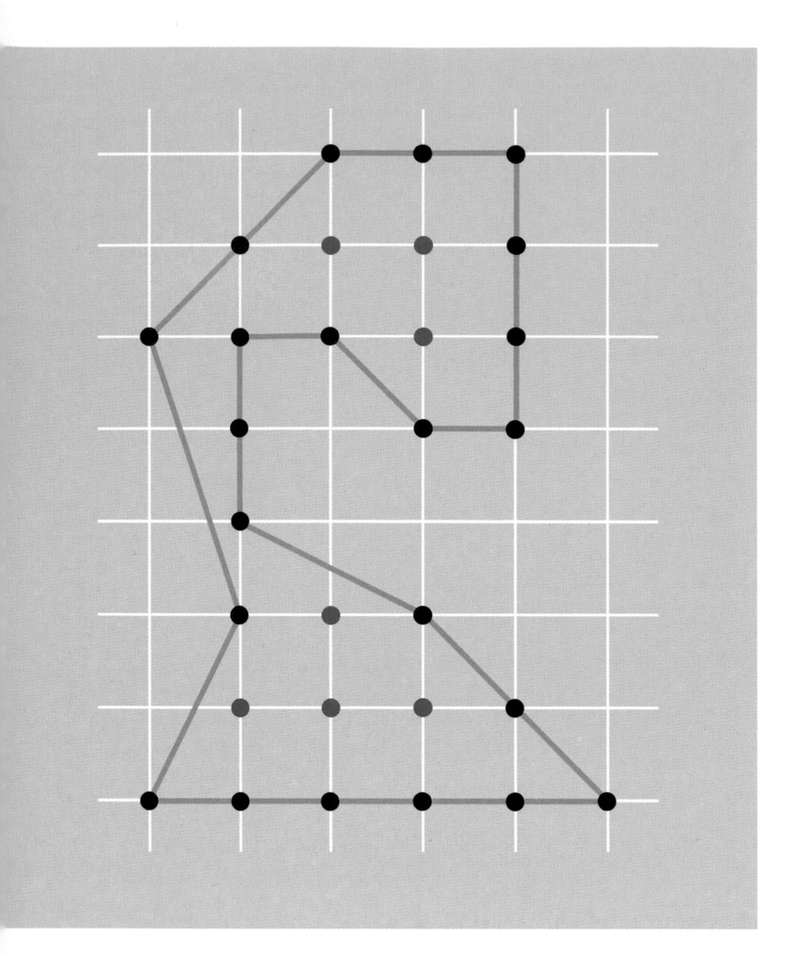

The only criterion was that the corners of the shape had to lie on a square grid. Apart from that, the shape could be as irregular and asymmetrical as desired.

Pick's formula entails counting the number of grid points that lie inside the shape. That number is called C. For a very large polygon, one might expect that the area of the shape should be approximately C, since this gives a rough estimate of the number of whole squares the shape contains. This guess might be improved by considering the grid points that lie exactly on the edge of shape. Suppose there are B of these. These might be considered half in, half out of the shape, so one might hope that adding an extra $\frac{B}{2}$ to C might produce a number a little nearer to the exact value.

ABOVE This irregular shape contains $C = 7$ lattice points inside, and $B = 22$ on its boundary. So Pick's theorem immediately tells us that its area is $7 + \frac{22}{2} - 1 = 17$.

This line of thought is not very scientific, so one should certainly not expect it to work perfectly. But here Georg Pick produced a surprise. In 1899, he showed that the formula $C + \frac{B}{2}$ always overestimates the area, and by exactly one unit. Pick proved that the exact value of the area of any shape will always be exactly $C + \frac{B}{2} - 1$.

Pick's theorem is a wondrous thing, and particularly useful in the digital era. On a computer screen, despite appearances, all shapes have straight edges, and their corners lie on a square grid (albeit a very fine one). So Pick's theorem will automatically apply in this context, and can be used to calculate the number of pixels within any shape.

Reeve tetrahedra

When presented with a beautiful theorem such as Pick's, a geometer's immediate reaction is to search for a similar result in higher dimensions. One can easily impose a cubic lattice on 3-dimensional space. So what are the volumes of shapes whose corners lie on lattice points? By counting the points inside and those on the boundary, would some variant of Pick's theorem allow the volume to be calculated quickly?

Unfortunately, in 1957, John Reeve showed that this would never work. He achieved this by constructing a family of irregular tetrahedra: four-sided shapes built from triangles. All his tetrahedra shared three of the same corners, but the fourth roamed around, producing tetrahedra of different shapes and volumes. None of his shapes contained any lattice points inside the shape. That is to say, $C = 0$ for all of them. What was more, they each contained only four lattice points on their boundaries, namely their four corners. So the values of B and C were the same for all Reeve's shapes, yet their volumes were different. This was enough to rule out the possibility of any simple Pick-like formula for deducing the volume.

On a computer screen, all shapes have straight edges, and their corners lie on a square grid. So Pick's theorem will automatically apply in this context, and can be used to calculate the number of pixels within any shape.

Ehrhart's analysis

Reeve had shown that a simple approach could not work for lifting Pick's theorem into three dimensions. So, in the 1960s, Eugène Ehrhart attempted something a little more sophisticated. His work would come to be of great importance in mathematics and computer science. Starting with a 3-dimensional shape built from straight edges, with its corners on the lattice, Ehrhart looked at what happened when the whole shape was expanded by a certain amount. Of course, this would be likely to bring new lattice points inside the shape. But how many? In answer to this question, Ehrhart formulated a precise and elegant law.

The resulting *Ehrhart polynomial* encodes a large amount of information about the shape. Indeed, the volume of the original unexpanded shape can simply be read off it. But other important data also emerges from Ehrhart's analysis, such as the Euler characteristic of the shape (see page 157). In recent years, Ehrhart polynomials have found surprising applications in computer science in ways that are not explicitly geometric, particularly in analyzing and improving the use of data caches.

The analysis of Pick and Ehrhart has also found new applications in the theory of optimization, when faced with complicated practical problems, such as how to divide up a workload between factories with different capacities, timing and prices. Surprisingly, finding the optimal solution can often be reduced to geometrical questions about multidimensional polytopes. Ehrhart polynomials can produce quick and easy information, that would otherwise be very difficult to extract.

56 Russell's paradox

BREAKTHROUGH BERTRAND RUSSELL'S FAMOUS PARADOX ILLUSTRATED THE DANGERS OF SELF-REFERENCE IN MATHEMATICS.

DISCOVERER BERTRAND RUSSELL (1872–1970).

LEGACY THE PARADOX PRECIPITATED A CRISIS IN MATHEMATICS. EFFORTS TO NAVIGATE AROUND IT WOULD PRODUCE DEEP INSIGHTS INTO THE LOGICAL FOUNDATIONS OF NUMBERS.

For as long as people have studied logic, they have been fascinated by paradoxes. The earliest and most famous is the liar paradox, which simply says: "This statement is false." In 1900, Bertrand Russell imitated this paradox in the new branch of logic known as set theory. His work would trigger a crisis in the foundations of mathematics, and it continues to trouble philosophers and logicians today.

Eubulides of Miletus was a philosopher around the fifth century BC. Like Zeno of Elea before him (see page 33), Eubulides' enduring contribution to the subject was in assembling a list of paradoxes.

Eubulides' paradoxes

One famous example from Eubulides' list is the paradox of the masked man. A woman is asked, "Do you know this masked man?" She answers, truthfully, that she does not. Yet the masked man is her father. Does it follow that she does not know her own father?

Ultimately, the masked-man paradox stems from ambiguous language, in the interpretation of the word "know." The paradox can be avoided if the woman replies that she is not sure whether she truly knows the masked man—all she can say is that she does not recognize him at that moment.

Eubulides' most famous paradox is altogether harder to get around than the masked man. The liar paradox—the statement "this statement

OPPOSITE A mirrored cube gives an illusion of infinite regress. The mirrored wall contains a reflection of the entire cube, including a reflection of itself. Objects which contain themselves can produce startling consequences, as Bertrand Russell observed.

is false"—presents us with a problem, as this sentence seems to resist ordinary classification in terms of truth and falsity. If it is true, then according to what it asserts, it must be false. But if it is false, then what it asserts is indeed the case, so it is true. Neither classification works. At the core of Eubulides' paradox is its self-reference—the fact that the sentence makes some assertion about itself. This phenomenon has fascinated and perplexed philosophers ever since.

Russell's paradox

In the late 19th century, a brand-new branch of mathematics developed. It was set theory, and it sprang from the work of Georg Cantor (see page 217) and Gottlob Frege. Many objects in mathematics have difficult, technical definitions. A set, however, was nothing complex, but simply a collection of objects. Cantor's work was remarkable in many ways, not least in seeming to illustrate that such a simple idea can produce startling consequences.

Yet the simple notion of a set was not as robust as it seemed. In fact, Bertrand Russell discovered in 1901 that the idea was essentially contradictory. He produced his famous paradox by creating a set which places impossible demands on itself, just as Eubulides' sentence had done. Russell reasoned that if a set is any collection of objects, then it is reasonable to speak of sets of sets. In fact, one might consider the set of all sets. At this point, the initial tame definition of a set began to produce some rather wild consequences. Unlike sets of numbers, or sets of people, the set of sets must contain itself. The possibility of sets containing themselves was the means by which self-reference sneaked into the theory.

BELOW Visual illusions share several traits with logical paradoxes. No visual interpretation of this picture is stable, in the same way that the liar paradox cannot consistently be said to be either true or false.

It was from this observation that Russell was able to craft his paradox. Eubulides' paradox stems from the division between sentences which are true and those which are false. Similarly, Russell divided sets into two types: those which contain themselves and those which do not. Just as Eubulides created a paradox with a sentence asserting its own falsity, so Russell contemplated a paradoxical object: the set of all sets which do not contain themselves. Calling this *X*, the question is: does *X* contain itself?

Like Eubulides' sentence, this question cannot be resolved either way. If *X* contains itself, then since all the members of *X* do not contain themselves, *X* cannot either. On the other hand, if *X* does not contain itself, then it fulfills all the requirements to be a member of *X*, and so must contain itself after all.

To illustrate his paradox, Bertrand Russell used an analogy away from set theory. In a small village, a barber shaves all the men who do not shave themselves, and only those men. The question is: does the barber shave himself? If he does, then as one of the men shaved by the barber, he must not shave himself. But if he doesn't, then he fulfills all the requirements to be shaved by the barber, and so must shave himself. Neither assumption is tenable.

Axiomatic set theory

With regard to the barber, the conclusion was clear: there can be no such person and no such village; the entire scenario is a logical impossibility. As Russell put it, "the whole form of words is just a noise without meaning." The conclusion for the theory of sets was equally stark: set theory, as it had been hitherto been studied, was simply a logical impossibility.

Russell reasoned that if a set is any collection of objects, then it is reasonable to speak of sets of sets. At this point, the initial tame definition of a set began to produce some rather wild consequences.

This was bad news for Georg Cantor's theory of infinities, and worse still for Gottlob Frege. Frege was in the process of publishing a book using set theory as a foundation for the whole of mathematics when Russell wrote to him with news of the paradox.

In the decades that followed, mathematicians sought to extricate the subject from the crisis at its foundations and formulate a concept of set which avoided Russell's paradox. The result was a far more abstract approach to mathematical logic, which would culminate in a new understanding of the subject via Gödel's incompleteness theorems (see page 277).

57 Special relativity

BREAKTHROUGH SPECIAL RELATIVITY GAVE US A NEW PICTURE OF OUR UNIVERSE IN WHICH THE SPEED OF LIGHT APPEARS CONSTANT, NO MATTER THE MOTION OF THE OBSERVER. STUDYING THIS SCENARIO REQUIRED NOVEL MATHEMATICAL METHODS.

DISCOVERER HENRI POINCARÉ (1854–1912), HERMANN MINKOWSKI (1864–1909), HENDRIK LORENTZ (1853–1928), ALBERT EINSTEIN (1879–1955).

LEGACY SPECIAL RELATIVITY WAS THE FOUNDATION FOR THE LATER DEVELOPMENT OF GENERAL RELATIVITY, AND REMAINS A CRITICAL PART OF OUR UNDERSTANDING OF THE UNIVERSE.

Albert Einstein's famous theory would forever change the way that we view our Universe. Yet his reimagining of space and time was underpinned by earlier approaches to geometry, coming from the work of Henri Poincaré and others. To understand the new physics it was necessary first to get to grips with these subtle mathematical matters.

Relativity first became an important topic in science through the work of Galileo. It was in his 1632 work *Dialogue Concerning the Two Chief World Systems* that Galileo publicly threw his weight behind the heliocentric cosmology of Copernicus, and against the Earth-centered model preferred by the religious authorities of the time.

Galilean relativity

Galileo's *Dialogue* was persuasively written, intellectually radical and hugely influential. He needed to address the obvious counterargument to the suggestion that the Earth is constantly spinning on its own axis and orbiting around the Sun: if this is so, why when water drips from a tap does it not feel the drag from the moving Earth and fall back toward the west?

OPPOSITE Light has been a major topic of scientific enquiry for hundreds of years. Evidence that the speed of light, uniquely, is absolute and does not depend on the speed of the observer was the spur for the development of special relativity.

To counter this, Galileo performed an important thought experiment. He imagined a man inside a ship that is moving perfectly smoothly and silently, without rocking or splashing. (Today, we might place the imaginary man in an airplane, train or spacecraft.) The man is in a sealed room from which he cannot see out. But he has various items

RIGHT Bode's galaxy. The light which reaches us now left the galaxy 12 million years ago. A consequence of special relativity is that there is no way for Bode's galaxy to affect us (or vice versa) on causal time scales shorter than this.

with him: a fish tank, some butterflies, a bottle of water, and so on. Sometimes the ship is moving, but at other times it weighs anchor. The question is: can the man know whether at a given moment the ship is moving smoothly or is docked? If he releases the butterflies, or dribbles water out of the bottle, is there some experiment he can perform which will tell the difference?

Galileo's answer was no. The physics within a room moving at a constant speed in a fixed direction is literally indistinguishable from that which is stationary. Copernican astronomy became the scientific standard, but Galilean relativity was no less important a development: it stated that there can be no way to distinguish stationarity from movement at a fixed speed in a fixed direction. This principle subsequently became a law of Newtonian physics.

The speed of light

Galilean relativity remained an important tenet of science for hundreds of years. But it was finally challenged by the Michelson–Morley experiment of 1887, which was designed to discover the nature of light. The purpose of the experiment was to test for ether, the substance which supposedly carried light. As well as killing off the ether theory, it uncovered the first evidence of something rather unsettling. Light travels at a fixed speed, which we now know to be around 299,792,458 meters per second (often abbreviated as c). However, the difficulty is that this speed defies Galilean relativity. While Galileo had postulated that all speeds are fundamentally equivalent, light seemed to behave uniquely. For instance, if a woman is traveling at high speed, in the same direction as a beam of light, one might expect that the beam should appear slower to her than to a stationary observer. But all experiments suggested that this didn't happen. Astonishingly, the speed of light appears exactly the same, irrespective of the speed of the observer.

Galilean relativity had to be overthrown. Galileo's principle was replaced with two new laws: firstly, that all speeds below c are equivalent, and relative to the frame of reference of the observer. But secondly, c itself is constant and absolute. This theory of special relativity was partially investigated by the mathematicians Henri Poincaré and Hendrik Lorentz, before being proposed in its most elegant and complete form by Albert Einstein. Einstein took these two rules as his axioms and deduced everything else from them.

Lorentz transformations

Lorentz and Einstein each recognized some of the strange consequences of the constancy of light. Even when two observers are traveling at high speed relative to each other, every beam of light must travel at the same speed in relation to each of them. So, when these two looked at each other, they would notice some very unusual effects. Events seem to happen at a different rate in a fast-moving frame than in a stationary one. The observers' wristwatches will each appear to be running slower to the other than to the wearer. But it is not only time that dilates. Each observer will appear both heavier and thinner to the other.

Astonishingly, the speed of light appears exactly the same, irrespective of the speed of the observer.

Lorentz introduced a simple mathematical device called a Lorentz transformation, which could convert information about a stationary frame of reference into a description of how it appears as a fast-moving frame. Lorentz transformations are fundamental tools of relativity theory. Indeed, it was through thinking about Lorentz transformations that Einstein arrived at the most famous consequence of special relativity, the fact that an object's mass is a measure of its energy, or to put it another way: $E = mc^2$.

Minkowski space

How can one mathematically model a universe in which special relativity holds? Hermann Minkowski provided the answer in 1907. Minkowski spacetime is a 4-dimensional structure (three of space and one of time), which drew on recent work in both the complex numbers (see page 105) and hyperbolic geometry (see page 189). Critically, Lorentz transformations represent symmetries of the structure, just as a rotation of a square is a symmetry of that shape.

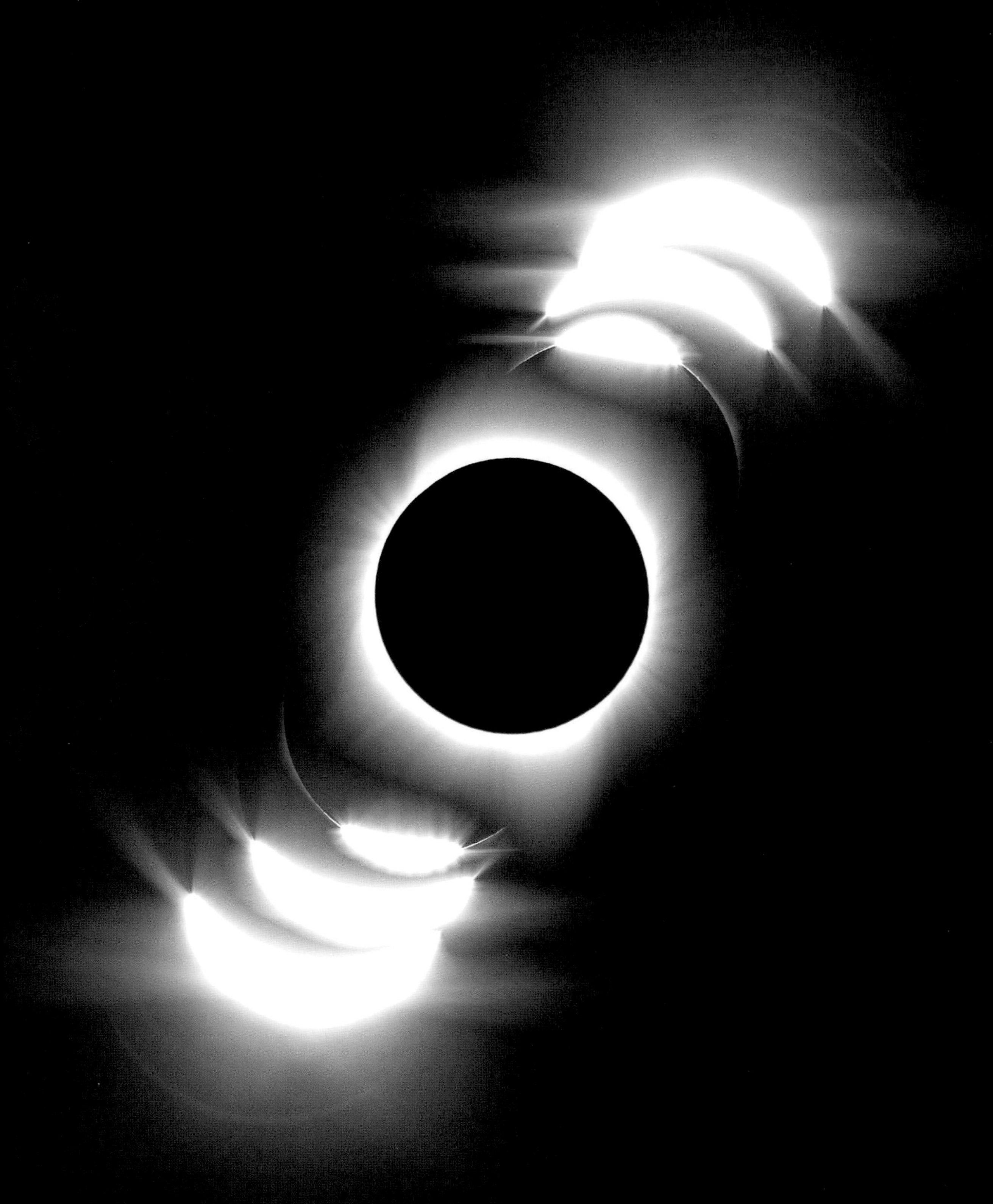

58 The three-body problem

BREAKTHROUGH WHAT PATHS WILL THREE OBJECTS FOLLOW IF THEY ARE LEFT TO TRAVEL FREELY UNDER GRAVITY? THIS QUESTION LAUNCHED THE SUBJECT OF CHAOS THEORY. A SOLUTION WAS PROVIDED BY KARL SUNDMAN IN 1909.

DISCOVERER HENRI POINCARÉ (1854–1912), KARL SUNDMAN (1873–1949).

LEGACY SUNDMAN'S SOLUTION WAS A BREAKTHROUGH IN THE THEORY OF CHAOTIC SYSTEMS, BUT WAS OF LIMITED PRACTICAL USE. MANY MYSTERIES IN THE AREA REMAIN.

Our Earth orbits our Sun in a simple, repetitive, elliptical orbit. But what happens in solar systems containing more than two stars? This problem launched the subject of chaos theory, and continues to challenge mathematicians and physicists today.

The best-known constellation in the night sky is Ursa Major, or the Great Bear (also known as the Plough and the Big Dipper). It is usually described as consisting of seven stars. But one of the seven actually consists of two separate points of light. These are Mizar and Alcor, sometimes known as the Horse and Rider.

The Horse and Rider

It was a matter of debate for medieval astronomers whether the two stars were connected in any manner, or simply appear that way from Earth. The modern telescope revealed a surprising answer: the two are indeed close enough to influence each other. But more excitingly still, each star is itself a composite system. Mizar comprises four stars locked together by gravity, while Alcor is a binary system, consisting of two stars.

In fact, the Universe is full of star systems very unlike our own, where stars, black holes and other heavy bodies dance around each other, held together by mutual gravitational attraction. Yet the mathematics of these situations is astonishingly complex. Indeed, many such systems are technically chaotic, which makes any form of prediction very hard. Many mathematicians have worked on understanding the gravitation of

OPPOSITE A sequence of images showing a solar eclipse, with the Moon blocking out the Sun. The Earth, Moon and Sun are humanity's most important astronomical bodies, and their interaction lies at the heart of the three-body problem.

multi-element systems. But the major breakthrough came in 1909, when Karl Sundman produced an expression for the development of a three-body system. Nevertheless, multi-body systems remain a topic of research today, representing the archetypal example of chaos in our Universe.

Two- and three-body systems

A binary system, such as Alcor, comprises just two stars orbiting around each other. The mathematics of this situation is simple and elegant, as Johannes Kepler showed in 1609. Kepler modeled our Solar System as a single planet orbiting a single Sun, assuming that the gravitational influence of the other planets, and far-off stars, was negligible. The resulting laws of planetary motion were exceptionally clear and beautiful. The path of the planet would be an ellipse, or in a double-star system the two stars would trace out interlocking elliptical paths. There are other possibilities for two-body systems. When a fast-moving comet travels close to a star, it may be unable to escape the tug of gravity, consequently smashing straight into it. Alternatively, the comet may boomerang around the star, tracing out the path of a parabola (see page 57).

The resulting laws of planetary motion were exceptionally clear and beautiful. The path of the planet would be an ellipse, or in a double-star system the two stars would trace out interlocking elliptical paths.

But what happens when a third object is added? Of course, we live in such a three-body system: that of the Sun, Earth and Moon (ignoring the occasional influence of Jupiter and the other planets). So we know what happens: the Earth traces out an elliptical orbit around the Sun, while the Moon revolves around the Earth in a similar fashion. By ignoring the direct influence of the Sun on the Moon, the mathematics becomes tractable. For most purposes, we can think of this as two two-body systems.

A more complicated three-body system might involve three stars of comparable masses, each of them falling under the gravitational influence of the other two. In this situation, the elegant geometry of elliptical and parabolic orbits gives way to something incomparably more complex. If you trace the path of one of the three stars as it is pulled this way and that by the other two, the curve it traces out is far from a simple shape like an ellipse. In fact, unlike our Earth with its repetitive orbit, such a system may never repeat itself at all.

Chaos

It was Isaac Newton who demonstrated that Kepler's laws of planetary motion were the consequence of a single law of gravity. Newton then attempted to mimic the analysis for a three-body system, but soon gave

LEFT The constellation of Ursa Major. At the top left of the picture, the double point of Mizar and Alcor is visible. Mizar, the brighter of the two, is a four-star system, and Alcor is a binary system. They are around 1 light-year apart and 83 light-years from Earth.

up, declaring that the problem "exceeds, if I am not mistaken, the force of any human mind." Thus the challenge was laid down, and in 1887, it acquired a price tag. King Oscar II of Sweden, a notable patron of the sciences, announced a prize of 2500 crowns for the solution of the three-body problem.

No less a figure than Henri Poincaré turned his mind to the challenge. He made considerable progress, reducing the problem to a set of ten equations to be solved. In the course of this work, Poincaré became the first mathematician to seriously analyze the phenomenon of chaos. The signature of chaos is that if you restart the system, having made just a tiny adjustment to the stars' initial positions or speeds, the result will evolve in an entirely different way. Although he did not completely solve the problem, Poincare's advances were deemed significant enough for him to be awarded King Oscar's prize.

Sundman's series

In 1912, the astronomer Karl Sundman finally produced a mathematical expression for the three-body problem. It is given by a series (see page 97), meaning that infinitely many numbers have to be added to describe the state of the system at a specific time. Theoretically, Sundman's series is a perfect solution to the problem. The trouble is that it converges exceptionally slowly. This means that extracting any useful information from it requires summing incomparably more numbers than any person or computer could ever practically manage. Similar solutions were found for the four-, five- and more body problems by Qiu-Dong Wang in 1991. These were wonderful mathematical accomplishments. Yet, because of their slow rates of convergence, multi-body problems continue to challenge today's chaos theorists.

59 Waring's problem

BREAKTHROUGH EDWARD WARING ASKED WHETHER EVERY WHOLE NUMBER CAN BE BROKEN DOWN INTO A LIMITED NUMBER OF POWERS. DAVID HILBERT PROVIDED A POSITIVE ANSWER, FOR EVERY POSSIBLE TYPE OF POWER.

DISCOVERER EDWARD WARING (1736–98), DAVID HILBERT (1862–1943).

LEGACY DESPITE HILBERT'S THEOREM, THE EXACT NUMBERS THAT SOLVE WARING'S PROBLEM REMAIN A SOURCE OF MYSTERY TODAY.

In 1621, the Renaissance mathematician and linguist Claude Bachet was working on a translation of the greatest classical work of number theory, Diophantus' *Arithmetica*. He realized that within the work Diophantus was implicitly assuming something rather extraordinary: a universal law of numbers which was far from obviously true. It revolved around the strange arithmetic of powers. In time, this chance observation would grow into one of the deepest topics in number theory.

A "power" is the name given to a number multiplied by itself. So 3^2 (which is short for 3×3) is a second power (second powers are also known as squares), while 6^5 (or $6 \times 6 \times 6 \times 6 \times 6$) is a fifth power. The essential idea is simple enough, but the arithmetic of powers contains many dark mysteries. One of the oldest ideas (its exact origins are lost in the mists of time) involves breaking whole numbers down into squares. For example, 5 is not a square number. But it can be written as two squares added together: $2^2 + 1^2$. On the other hand, 7 cannot be written that way, and nor can it be written as three squares added together. Four squares, however, are enough: $2^2 + 1^2 + 1^2 + 1^2$. This observation leads to several questions of real mathematical profundity.

OPPOSITE
Since Pythagoras, mathematicians have been fascinated by the unexpected patterns produced by squares, cubes and higher powers. Edward Waring asked whether any whole number could be described by small collections of such powers.

Lagrange's four-square theorem

Two very long-standing problems are: which numbers can be written as two squares added together? And secondly, is there some number of squares that is sufficient to describe any number? The first of these

RIGHT Lagrange's four-square theorem guarantees that every whole number can be expressed as four squares added together. For example, 15 = 9 + 4 + 1 + 1.

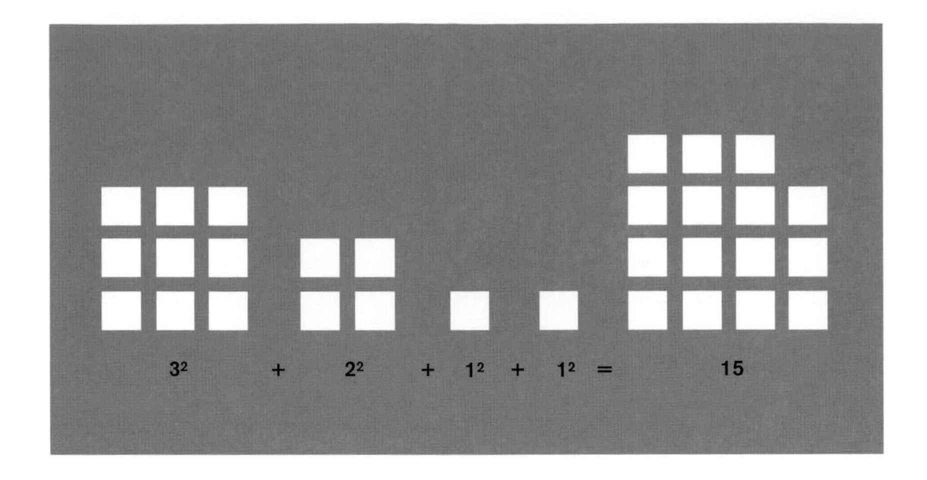

questions was addressed by the first great number theorist of the modern era, Pierre de Fermat, on Christmas Day in 1640. He provided a neat characterization of those numbers which can be expressed as two squares added together and those which cannot. For prime numbers, the condition is particularly simple: primes which are one greater than a multiple of four (such as 5 and 13) can, but those one fewer than a multiple of four (such as 7 and 11) cannot.

Fermat's theorem was a beautiful piece of mathematics which laid to rest an ancient number-theoretical problem. The second question, however, took even longer to resolve, although Claude Bachet had caught a glimpse of the answer within the *Arithmetica*. Diophantus seemed to believe that any whole number can be expressed as four squares added together. Many centuries later, Bachet came to share this certainty. But he too was unable to provide a proof. Certainly this rule is true for any number you can think of: $100 = 9^2 + 3^2 + 3^2 + 1^2$, for example. But could it really be true for all whole numbers? In 1770, Joseph-Louis Lagrange finally proved Diophantus and Bachet right: four squares are always enough.

Waring poses his problem

Lagrange's theorem settled the question of breaking numbers down into squares, but it didn't conclude the whole line of enquiry. Just a few months earlier, the Oxford University researcher Edward Waring had figured out how to extend this investigation to dig even deeper into the mathematics of powers. He wanted to lift the question from squares to higher powers. If every number can be written as the sum of four squares, then what of cubes (meaning third powers)? The number 11 can be written as four cubes: $2^3 + 1^3 + 1^3 + 1^3$, but 23 cannot be described by fewer than nine cubes:

$2^3 + 2^3 + 1^3 + 1^3 + 1^3 + 1^3 + 1^3 + 1^3 + 1^3$? But must it even be true that there is an answer—that there is some number of cubes which is adequate to describe any number?

The same questions could then be posed for fourth powers, fifth, sixth, and so on. Edward Waring conjectured that nine cubes should be enough to describe any number. For fourth powers he said the limit should be 19. But his boldest claim was that all of these questions must have answers: for every type of power, there would be some limit to how many are needed to describe every possible number. This was by no means obvious. It might have been that for fifth powers, say, no figure is adequate; as you look at ever larger and larger numbers, it could be that the required quantity of fifth powers simply goes up and up. Waring's problem went to the core of the arithmetic of powers and their role in number theory.

The Hilbert–Waring theorem

The breakthrough came in 1909, when David Hilbert managed to prove that Waring's optimism was justified. At every level, say that of nth powers, there is some other number g, so that every whole number can be written as g many nth powers added together. This Hilbert–Waring theorem was one of the great number-theoretical breakthroughs of the 20th century. But it did not completely settle the matter. In particular, it did not provide any clue as to what the answer (the number g) should actually be.

Waring wanted to lift the question from squares to higher powers. If every number can be written as the sum of four squares, then what of cubes?

Further progress came later in 1909, when Arthur Wieferich solved the problem for third powers (or cubes). Again, Waring had been correct: every whole number can be written as nine cubes added together.

Then, in 1936, three researchers found an explicit recipe which they claimed gave the number g exactly, for powers of every level. Dickson, Pillai and Niven's formula has subsequently been verified for all powers up to $n = 471,600,000$. Once it is established for all possible values of n, Edward Waring's theory of powers, first hinted at in Diophantus' *Arithmetica* so many years ago, will finally be laid to rest.

Markov processes

BREAKTHROUGH FIRST FORMULATED BY ANDREI MARKOV IN AROUND 1910, MARKOV PROCESSES MODEL SITUATIONS WHICH EVOLVE RANDOMLY. THEY ALLOW SHORT-TERM UNCERTAINTY WHILE PERMITTING LONG-TERM PREDICTION.

DISCOVERER ANDREI MARKOV (1856–1922), ANDREY KOLMOGOROV (1903–87), GEORGE PÓLYA (1887–1985).

LEGACY MARKOV PROCESSES ARE NOW APPLIED WIDELY ACROSS THE PHYSICAL AND SOCIAL SCIENCES TO HELP UNDERSTAND A LARGE VARIETY OF UNCERTAIN SITUATIONS.

Many processes in the Universe are—or seem—random. This is particularly true in the social sciences, which seek to understand the messy and unpredictable world of human affairs. But how can one analyze random processes, or make predictions about them? To answer this question, scientists and economists turn to mathematics. The most fruitful idea here is that of a Markov process, named after Andrei Markov, who formalized it in the early 20th century.

The word "random" suggests anything can happen. Yet even in situations where the short-term outcome is completely unpredictable, it is an important insight that one may still be able to make accurate forecasts over the longer term. If you toss a fair coin 100 times, it is possible that you will get 100 heads—possible, but vanishingly unlikely. More significantly, if you keep tossing the coin, it is a certainty that sooner or later you will get a tail. Coin tosses and dice rolls are beloved of probability theorists, but situations in other scientific disciplines tend to be more complex. In the early 20th century, Andrei Markov was able to extract the key principles from a random process, retaining enough structure for analysis and prediction to be possible.

OPPOSITE It is extremely hard to predict the short-term movement of smoke particles; there are too many factors involved. Modeling it as a Markov process allows long-term predictions to be made, such as the inevitability of the smoke eventually dissipating.

The drunkard's walk

The purest form of Markov process is a random walk, in which someone or something navigates a grid or a maze, deciding at random which direction to take at each junction. The simplest example is the drunkard's walk. A drunkard is staggering home along a pavement. With each step forward, he may move leftwards toward the wall, rightward toward the gutter, or he may maintain his current line. Each of these options comes with a certain probability (perhaps depending on how drunk he is). Unfortunately, though, the pavement is only four paces wide. Once he reaches the wall, he can only go directly forward or lurch rightward toward the gutter again.

If the road is long enough, the drunkard is certain to collapse into the gutter sooner or later. But how long is it likely to take him to do so? If the road is 100 paces long, what are his odds of making it to the end? These are the types of questions which the theory of Markov processes can answer.

Snakes and Ladders

Needless to say, the theory has applications beyond dipsomaniacs' traveling arrangements. In fact, playing roulette against a casino can be modeled as a drunkard's walk, where "falling in the gutter" amounts to going broke. Because the casino stacks the odds in its favor, anyone who plays for long enough is guaranteed to go bust eventually.

Like the drunkard teetering on the edge of the gutter, all that matters is where you are now, not how you got there. This crucial criterion of forgetfulness is the defining characteristic of a Markov process.

Another famous Markov process is the board game Snakes and Ladders. This is a game of pure luck, involving no skill, in which the first player to reach the final square is the winner. Progress is assisted by ladders which carry you nearer your goal, and hindered by snakes which drag you backward. The critical point is that the future development of the game depends only on the state of board now, not the past. If you are on the 25th square while your opponent is on the 3rd, your prospects of winning are identical, irrespective of whether you have been playing for 30 seconds or three hours. Like the drunkard teetering on the edge of the gutter, all that matters is where you are now, not how you got there. This crucial criterion of forgetfulness is the defining characteristic of a Markov process.

Random walks

The drunkard's walk is essentially a 1-dimensional problem, since his position vacillates only between the wall and the gutter. In higher-dimensional random walks, a wanderer may navigate an entire city or more complex network. These more sophisticated scenarios were studied by George Pólya in 1921. He contemplated a traveler in a 2-dimensional square grid (such as the streets of Manhattan). This mystery tourist

LEFT Markov theorists often contemplate Manhattan, since its roads are mostly laid out in a rigid grid system, where a random walker can choose between four routes at each junction. The walker has a 100% chance of eventually reaching any specified point, even on an infinite grid.

navigates each crossroads at random, taking a 1 in 4 chance of turning left, right, heading straight on, or turning back. A natural question is whether she is likely to find herself outside her own house at some stage.

Pólya was able to deduce that this, too, is a certainty. Given enough time, the random traveler will make it home. But the same is not true if the network is 3-dimensional. In a 3D city, in which elevators are installed at every junction allowing a choice of left, right, forward, backward, up or down, the random roamer has no guarantee of eventually arriving home. In fact, the odds are around 34%. In higher-dimensional spaces still, the chance drops even further.

Randomness and non-randomness

Today, random walks and other Markov processes appear throughout science, modeling random phenomena such as chemical reactions, radioactive decay, or the motion of clouds of particles. In the second half of the 20th century, it became clear that their great expressive power makes them useful in situations that are not truly random, but whose behavior depends on so many factors as to make accurate predictions impossible. Examples include fluctuations in financial markets, or among populations of organisms, or errors in computer software. In developing information theory (see page 289), Claude Shannon even modeled human language as a Markov process, in which one symbol follows another with a given probability.

61 General relativity

BREAKTHROUGH EINSTEIN'S GREATEST ACHIEVEMENT WAS TO DESCRIBE THE WAY IN WHICH THE FABRIC OF THE UNIVERSE IS DEFORMED BY GRAVITY. TO MANAGE THIS, HE NEEDED THE SOPHISTICATED GEOMETRY OF TENSOR CALCULUS.

DISCOVERER ALBERT EINSTEIN (1879–1955).

LEGACY GENERAL RELATIVITY HAS RECEIVED CONSIDERABLE SUPPORTING EVIDENCE AND FORMS THE FOUNDATION OF OUR UNDERSTANDING OF THE UNIVERSE ON THE VERY LARGEST SCALES.

The theory of special relativity (see page 233) was a wonderfully accurate account of the ways in which matter and light move. But it did not address the very reason that stars form and celestial objects move in the first place: gravity. To model a gravitational universe required the updating of the old Newtonian account of gravity in a new relativistic setting.

Albert Einstein's theory of general relativity is best understood in terms of the curvature of spacetime itself. His great insight was that space and time are bent by the presence of matter (or, equivalently, energy). The idea is striking in its simplicity, but to quantify and describe the curvature of a 4-dimensional object is technically extremely demanding. Indeed, Einstein was himself troubled by the geometrical depths he needed to explore in order to formulate general relativity as a mathematically rigorous theory.

Tensor calculus

The theory of curvature had been initiated in earnest by Carl Friedrich Gauss (see page 185). Gauss's work led to an analysis of the ways in which a 2-dimensional surface may be curved; this was followed by Bernhard Riemann's analysis of curved spaces, coming out of the discovery of hyperbolic space (see page 189). With this, curvature took center stage in a new era of geometry. However, higher-dimensional spaces, such as the 4-dimensional spacetime of our Universe, are much harder to analyze. In the first place, there is the simple biological problem that humans are not able to visualize objects in four or more dimensions. The second obstacle

OPPOSITE Black holes are powerful evidence for the theory of general relativity. In the two-galaxy system 3C321, a jet of energy from a supermassive black hole at the center of one galaxy (top left) smashes into the second (center).

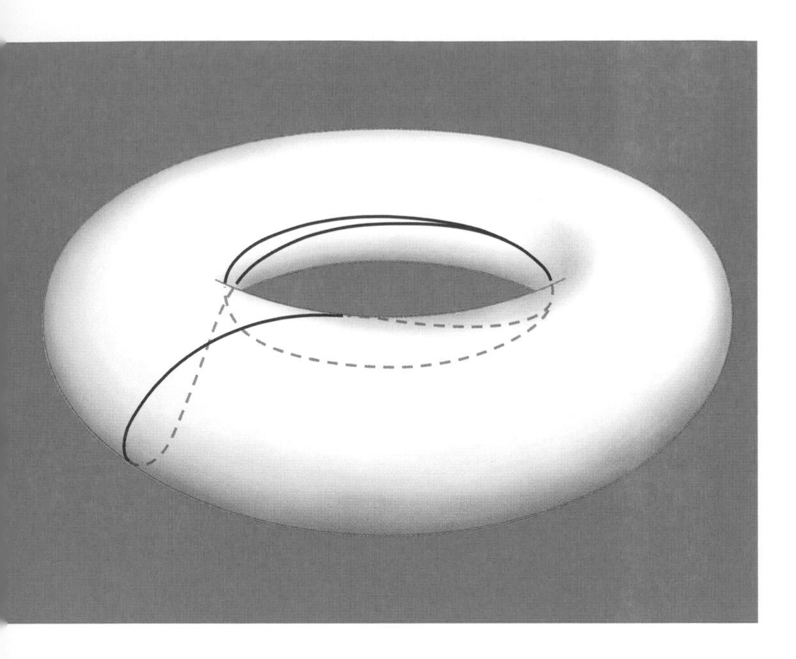

is that such spaces can be curved in a great variety of ways. To analyze these possibilities requires some heavy conceptual machinery.

It was in the late 19th century that the appropriate tools began to be assembled, by Gregorio Ricci-Curbastro and his student Tullio Levi-Civita. While Gauss had been able to describe the curvature of a 2-dimensional surface using a single number, Ricci and Levi-Civita deployed higher-dimensional arrays known as tensors.

ABOVE A geodesic on a torus. General relativity relies on the theory of geodesics (or shortest paths), but even for 2-dimensional surfaces these can be subtle. For 4-dimensional spacetime, the heavy mathematical machinery of tensor calculus is needed.

Einstein's field equations

In Ricci and Levi-Civita's theory of tensor calculus, Einstein found the tools he needed to describe the shape of a universe curved by gravity. His work boiled down to a single fundamental equation, known as Einstein's field equation. It involves the Ricci curvature tensor, which is a fundamental measure of the curvature of a space. In the context of spacetime, this is modified to give the Einstein tensor, denoted G. Einstein's insight was that the curvature of spacetime is completely determined by the amount of matter or energy present. This in turn is expressed by an object called the stress–energy tensor, denoted T. In suitable units, then, the field equation says: $G = T$.

In order to find a geometrical description for the shape of spacetime, it is necessary to solve this equation, which is a challenging geometrical problem. In fact, Einstein's original account of the theory was far from perfect. To correct the errors, he entered a correspondence with Levi-Civita, remarking to him that "it must be nice to ride through these fields upon the horse of true mathematics while the likes of us have to make our way laboriously on foot."

Geodesics and freefall

The first modern theory of physics came out of the work of Galileo, Kepler and Newton (see page 181). A fundamental insight of that work had been that objects traveling in a fixed direction at a given speed were indistinguishable from objects at rest. Such motion would continue perpetually unless some external force intervened.

The same principle would play a central role in Einstein's new theory. However, the concept of a straight line no longer made sense in an intrinsically curved universe. Straight lines, as analyzed by Euclid (see page 45), can only exist against a background of a perfectly flat space. On a sphere, for example, it is impossible to draw straight lines. Instead, their role is played by geodesics.

In general relativity, the geodesics of spacetime describe the paths of freefalling objects. So it is gravitational freefall, rather than motion with constant velocity, that is fundamental to general relativity.

Geodesics are the answers to the question: "What is the shortest path between two points?" The results often have pleasing geometric interpretations. For instance, on the surface of a sphere, the geodesics are the segments of great circles (that is to say, the result of slicing through the sphere with a flat plane passing through its center, resulting in the largest circles the sphere can contain). On other surfaces, geodesics are subtler. Even on the torus (or donut), some careful analysis is required to reveal the various possibilities. In general relativity, the geodesics of spacetime describe the paths of freefalling objects. So it is gravitational freefall, rather than motion with constant velocity, that is fundamental to general relativity.

Black holes

General relativity does not immediately seem a natural idea, because we humans live in a relatively light gravitational field. Nevertheless, over the 20th century, Einstein's theory received extensive experimental support. As starlight reaches us from distant points in the galaxy, it is often possible to detect its deflection as it passes by a very massive object. In 1919, Arthur Eddington observed this gravitational lensing directly. During a solar eclipse, Eddington noticed that the stars behind the Sun appeared to slightly alter their position.

The most spectacular evidence for general relativity comes from places in the Universe where spacetime is warped to such an extreme degree that light itself cannot escape. The existence of such black holes was predicted in 1916 as a geometrically unusual solution to Einstein's field equation. It was not until 1971 that the first black hole, Cygnus X-1, was identified, through the motion of another star orbiting a dark patch of sky.

62 Fractals

BREAKTHROUGH FRACTALS ARE PATTERNS THAT LOOK SIMILAR AT ANY SCALE. GASTON JULIA WAS THE FIRST TO ILLUSTRATE HOW THEY MAY EMERGE FROM ITERATING SIMPLE PROCEDURES.

DISCOVERER GASTON JULIA (1893–1978), BENOÎT MANDELBROT (1924–2010).

LEGACY BEAUTIFUL FRACTAL PATTERNS ARE NOW UBIQUITOUS IN MODERN DESIGN, WHILE THE MATHEMATICS BEHIND THEM FINDS EVER BROADER APPLICATIONS.

In 1919, Gaston Julia performed a mathematical experiment. He took a simple mathematical rule and then repeated it over and over again, watching what happened as the repetitions increased. The result was fascinating, producing an exceptionally intricate structure, the likes of which no mathematician had seen before. Although Julia's work brought him fame among his contemporaries, it would be decades before it could be appreciated by the world at large. In the days before computer-generated imagery, it was impossible for anyone to appreciate the astounding beauty of the structures he had discovered, which would later become known as fractals.

Gaston Julia was born in French Algeria and studied mathematics in Paris in the run-up to the First World War. In the first year of the war, he was called up to fight in the east of France, and almost immediately suffered a terrible injury when a bullet hit him full in the face. Showing astonishing courage, he refused evacuation until the attack had been repelled. When he was finally transported to hospital, unfortunately the doctors were unable to save his nose, and for the rest of his life he wore a leather patch across his face in public. It was during this period in hospital that Julia worked on his doctoral thesis, which would in time lead to some of mathematics' most beautiful images.

OPPOSITE A Julia set, typical of the infinite intricacy of fractal patterns. These sets arise by starting with a number and then repeatedly applying a simple algebraic rule. The number is marked white if the sequence remains bounded, rather than exploding beyond all limits.

Julia sets

The idea behind a Julia set was wonderfully simple. Begin with a rule, such as "multiply a number by itself and then subtract one." In algebraic terms, this takes the starting number z to the number $z^2 - 1$. What happens, Julia asked, as this rule is repeated over and over again? When he began with 0 and then applied the rule, the next number was -1. Applying the rule again took it back to 0, then back to -1, and so on. So this sequence flickers forever between 0 and -1. But with a different starting point, the sequence behaves differently. Starting at 2, the rule first produces 3, and then 8, 63, 3968 and so on. This time, the sequence grows with great speed, never settling down.

Although some attempts were made to represent the Julia set graphically, it was not until the dawn of the computer that its beautiful filigree structure could be fully revealed.

Julia realized that any rule divides numbers into two types: those such as 0, for which repeated applications remain within a limited range (whether settling down on a single value, or flickering back and forth between two or more), and numbers like 2, whose sequence explodes beyond all bounds. The collection of all values for which the rule remains bounded is called the rule's Julia set.

But what does it actually look like? Julia sets are best appreciated in the context of the complex numbers (see page 105). For the rule $z \mapsto z^2$, the Julia set is nothing complicated: it is a circle with radius 1, centered on 0. Under repeated application of the rule, points inside the circle gradually close in on zero, while those outside disappear off to infinity. But with a rule only slightly more complicated, $z \mapsto z^2 - 1$, the corresponding Julia set becomes infinitely more complex. Although some attempts were made to represent the resulting set graphically, it was not until the dawn of the computer that its beautiful filigree structure could be fully revealed.

Mandelbrot's fractal revolution

Julia sets were among the first fractals, though Gaston Julia would not have recognized the term. The word was coined decades later by Benoît Mandelbrot, who sparked major interest in the subject in the 1970s and 1980s, both within the scientific community and in the public at large with mesmerizing pictures in books such as *The Fractal Geometry of Nature*.

The defining property of fractals is that they continue to look similar no matter how far one zooms in. Another early pioneer of fractals was Wacław Sierpiński, who, around the same time that Julia began his

LEFT The edible flower of Romanesco broccoli has a natural fractal structure. The entire flower closely resembles a single bud, a pattern which continues down several levels.

work, discovered how to construct fractals from simple geometrical ingredients such as squares or triangles. One of his most famous creations is the Sierpiński gasket. Beginning with an ordinary equilateral triangle, he divided it into four triangles of equal sizes and then threw away the central one. Then he repeated this move with each of the three remaining triangles: each was divided into four, with the central one thrown away. Continuing this process ad infinitum produced a beautiful design, with the defining property of a fractal: self-similarity. Each sub-triangle is identical in shape to the whole, and this remains true, no matter how tiny a segment one focuses on.

The fractal world

As well as wowing the world with his fractal designs, Mandelbrot also made a strong case for their broader scientific importance. Geometers have long observed, perhaps with some disappointment, that the physical world is a far messier place than traditional geometry, with its straight lines and clean curves. Yet many aspects of our world demonstrate fractal patterns, from snowflakes and fern leaves to river deltas. This occurs when physical processes are scale-invariant: the physics describing how a river splits into smaller streams as it hits a bank of sediment does not depend in any essential way on the size of the waterway involved. Hence a river delta will look approximately the same whether viewed on a large scale or a small one, giving the network of capillaries, streams and rivulets a strikingly fractal appearance. Similarly, the mechanism by which a fern grows is similar to the process by which each of its individual fronds grows, leading to a plant with a beautiful fractal look.

63 Abstract algebra

BREAKTHROUGH THE RULES GOVERNING THE USUAL SYSTEMS OF ARITHMETIC ALSO APPLY IN MORE ABSTRACT SETTINGS, FIRST INVESTIGATED BY DAVID HILBERT (1862–1943) AND EMMY NOETHER IN THE 1920S.

DISCOVERER EMMY NOETHER (1882–1935).

LEGACY TODAY'S MATHEMATICS IS FUNDAMENTALLY ABSTRACT. THE NEW APPROACH TO ALGEBRA WENT ON TO TRANSFORM GEOMETRY AND NUMBER THEORY.

One of the most notable trends in the history of mathematics is toward abstraction. During the 20th century, mathematicians embraced this philosophy with gusto, and were richly rewarded across many branches of the subject. A foundation was laid in Emmy Noether's seminal algebraic work of the 1920s.

The earliest mathematicians studied plain whole numbers, along with familiar geometrical shapes such as straight lines and circles. As a result their work remains accessible and comprehensible to all. Today's researchers, on the other hand, study systems of far greater obscurity, making their work unfathomable to the uninitiated. This high level of abstraction is not the product of intellectual narcissism, nor even a case of mathematicians deliberately challenging themselves. Rather this approach can bring considerable power and an unexpected clarity.

OPPOSITE The Guggenheim museum in Bilbao, Spain. The project pioneered the use of 3-dimensional computer aided design in modern architecture. This technology is reliant on an algebraic approach to geometry, and has opened up many tools and techniques unavailable to designers of earlier generations.

A running theme in the history of mathematics is of successive generations realizing that the ordinary whole numbers are far deeper and less straightforward than they at first seemed. Enduring questions such as the Riemann hypothesis (see page 206) and abc conjecture (see page 349) are testament to this fact, as are Gödel's incompleteness theorems (see page 277). To address the difficulties among the whole numbers, it is worthwhile mentally separating out the factors which arise. This leads to structures which are more abstract, but in many ways logically simpler than the system of whole numbers.

In geometry, meanwhile, people have come to recognize new settings such as the non-Euclidean spaces discovered in the 19th century (see page 189), and the higher-dimensional realms of Ludwig Schläfli (see page 201). At the same time, the shapes that people have sought to study in these settings have become increasingly complex, both for purely mathematical reasons, and due to the demands of an increasingly sophisticated understanding of the physical Universe.

Noether's rings

A turning point in abstraction came in the 1920s, in the work of the German mathematician Emmy Noether. In algebraic terms, the fundamental properties of whole numbers are that they can be added, subtracted and multiplied. A little earlier, David Hilbert and others had observed that other systems share these properties. Such arithmetical systems came to be known as *rings*. This prompted Noether to undertake a completely abstract analysis of rings. Noether did not assume that the objects she was working with need be numbers at all, merely that there were some operations on them known as "addition" and "multiplication." These might bear minimal resemblance to the familiar processes we know by these names, except in so far as they obey the rules which define a ring, such as $a + b = b + a$.

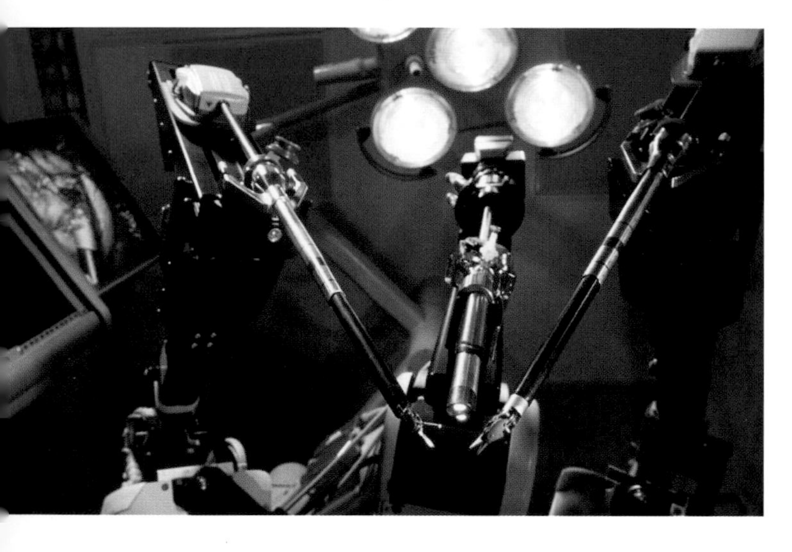

ABOVE The Da Vinci surgical robot system in the operating theatre. Precision robotics requires complex equations to be solved to determine how each joint should be bent to maneuver the tool into exactly the right position. Ingenious algebraic techniques are used in this process.

Noether's fresh approach provided a completely new perspective on the ancient processes of addition and multiplication. She saw how rings were constructed from special subsystems known as ideals, which threw considerable light on major mathematical topics such as modular arithmetic (see page 77), which emerge when the ring of whole numbers is divided by a certain ideal. Arithmetic modulo 5, for instance, can be thought of as the usual ring of whole numbers, divided by the ideal consisting of multiples of 5.

Over time, many other kinds of rings have been discovered, which behave subtly differently from rings of numbers. Matrices, for example, are not single numbers, but arrays of numbers, such as $\begin{pmatrix} 1 & 0 \\ 0 & 1 \end{pmatrix}$. This example is a 2×2 matrix. Matrices of the same size can be added, subtracted and multiplied and satisfy all the conditions for a ring. Yet they fail to satisfy another basic law, which might unthinkingly be taken for granted. If you take two matrices, A and B, it will usually be the case that $A \times B \neq B \times A$. So, matrices form what is known as non-commutative rings. What is more, it is not always possible to divide one matrix by another. Because Noether's analysis of rings did not assume commutativity or divisibility, it remained equally valid for these unexpected structures.

Noether's work on rings was an investigation into pure algebra, but the abstract revolution spread throughout mathematics. A particularly dramatic case was in the field of geometry.

Algebraic geometry

Noether's work was an investigation into pure algebra, but the abstract revolution spread throughout mathematics. A particularly dramatic case was in the field of geometry. When Descartes first unveiled his Cartesian coordinates in the 17th century (see page 129), this allowed a fruitful cross-pollination of algebraic and geometric techniques. It became possible to speak about the equation of a shape, at which stage the study of equations (algebra) and shapes (geometry) became inextricably linked. This algebraic approach to geometry began to flourish in the mid-20th century through the work of Oscar Zariski. As a result it became possible to make meaningful geometrical statements, even in situations where there was no obvious physical interpretation—no "shapes" as they are conventionally understood.

Emmy Noether's study of rings came to occupy a central position in this new abstract type of geometry. In the 1950s, Zariski had focused on special shapes called varieties, which are defined by equations. He knew that every variety gives rise to a ring. In order to study these special shapes, it was necessary to study the corresponding ring, which immediately brought Noether's abstract algebraic techniques to bear. The trouble was that the correspondence was only one way: although every variety corresponded to a ring, this did not mean that every ring corresponded to a variety.

This asymmetry was the spur to one of the 20th century's most profound leaps into abstraction. In the 1960s, Alexander Grothendieck once again overhauled geometry by finding a way to do it using arbitrary rings. The resulting highly abstract structures are known as schemes and are the central objects of study to modern algebraic geometers. Despite their highly rarefied nature, the varieties and schemes of today's algebraic geometry are the direct descendants of the straight lines and circles of Euclid and Descartes.

64 Knot polynomials

BREAKTHROUGH IN 1923, ALEXANDER APPLIED ALGEBRAIC TECHNIQUES TO THE STUDY OF KNOTS, PROVIDING A METHOD FOR TELLING TWO KNOTS APART.

DISCOVERER JAMES ALEXANDER (1881–1971), VAUGHAN JONES (1952–).

LEGACY KNOTS APPEAR IN A WIDE RANGE OF SCIENCES, FROM BIOCHEMISTRY TO ELECTRONICS. ALEXANDER'S AND JONES' ALGEBRAIC DESCRIPTIONS ARE CENTRAL TO THESE APPLICATIONS.

In the late 19th century, a small group of scientists cultivated an interest in knots. The work they did would lay the foundations of the subject of knot theory, which gradually gathered momentum over the 20th century, eventually becoming a vital tool in several areas of science, including biochemistry and quantum physics. A crucial moment came in 1923, with the discovery of the first knot invariant by James Waddell Alexander.

The impetus behind knot theory originally came from two Scottish physicists in the 1870s: William Thomson (who later became Lord Kelvin) and Peter Guthrie Tait. They were soon joined by a geometrically minded vicar, Thomas Kirkman, and Charles Little of the University of Nebraska in the United States.

These were not the first thinkers to be intrigued by knots. Their use in technology and design is as old as civilization, but it was only in the early 19th century that Carl Friedrich Gauss had begun to employ calculus to analyze the ways that two loops of string might become entangled. But, with no particular motivation to dwell on the subject, Gauss moved on to other matters. Thomson and Tait, on the other hand, had a powerful purpose behind their work. They had come to believe that knots formed the very fabric of matter.

OPPOSITE Knots have been used for practical purposes for thousands of years. But it is only recently that they have been taken seriously in scientific circles, posing deep mathematical conundrums and arising in various unexpected contexts.

The vortex theory of atoms

Like many of his contemporaries, Kelvin postulated the existence of a substance called ether, which permeated the Universe. He believed that physical matter formed where ether became excited, rather as an airplane whips up air into vapor trails. This animated ether took the form of miniature vortices, or whirlpools, which curved around themselves into loops—Kelvin likened these to smoke rings. These notional vortices were held to be the fundamental units of matter: atoms. But what makes an atom of carbon different from an atom of silver? Thomson argued that tubes of ether could become knotted in different ways, and the geometry of the knot supposedly determined the chemistry of the atom.

The vortex theory did not survive long. It was an imaginative idea, but—not to put too fine a point on it—completely wrong.

The vortex theory did not survive long. It was an imaginative idea, but—not to put too fine a point on it—completely wrong. Nevertheless, the theory of knots which Kelvin had initiated would in time yield genuine insights into the nature of reality.

The Alexander polynomial

To a mathematician, a knot is a tangled piece of string, much like a sailor's or scout's knot. The difference is that the ends are then fused, forming a loop. The simplest knot is a plain circular hoop, known as the unknot. Of course an unknot is not much use for tying anything up; the first truly knotted knot is the overhand knot, or, as it is known to mathematicians, the trefoil, which crosses over itself exactly three times.

The early knot theorists sought to understand the knots that emerge as the number of crossings increases. They found that there are eight knots with up to six crossings, and seven more with exactly seven. As they raised the number of crossings, however, they encountered the fundamental difficulty of knot theory. Two knots that appear different on first sight might, after a suitable manipulation, turn out to be the same.

This began as an inconvenience, but as investigations continued, it became clear that the question of determining whether two knots are actually the same is one of huge geometric depth—and seemingly intractable difficulty. Even the unknot can appear tangled if the loop is twisted around enough. The first theoretical breakthrough came in 1923, when James Waddell Alexander found a way to describe knots using algebraic techniques. To every knot, he assigned an object called a polynomial. The unknot, for example, has a polynomial "1," while the trefoil has "$x - 1 + x^{-1}$." Because these algebraic expressions are different, it follows that the two knots must be essentially different too. No manipulation of the trefoil would ever reduce it to being an unknot.

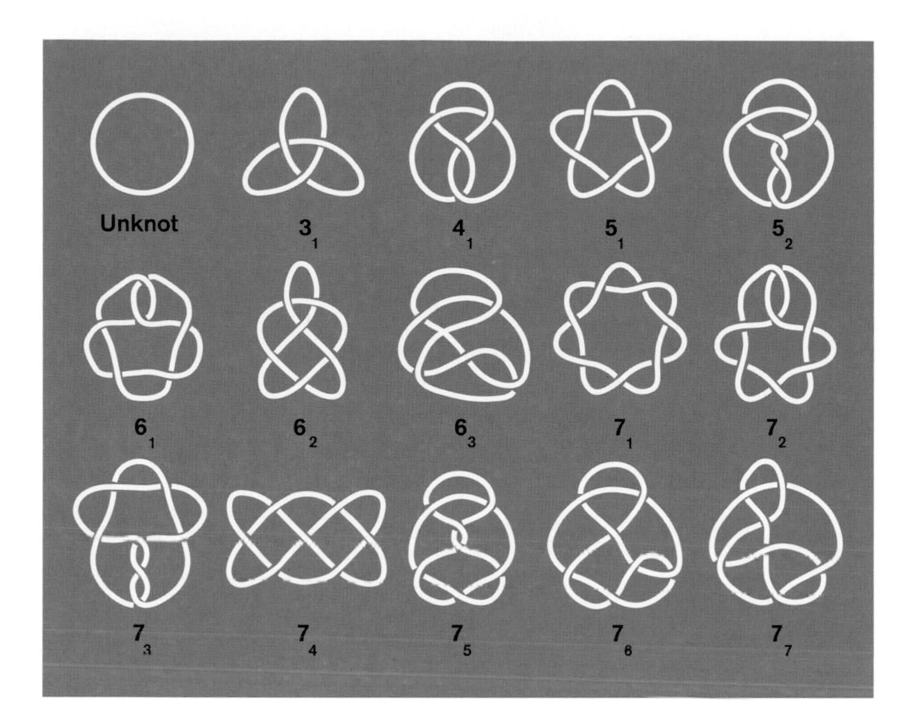

LEFT The 15 knots with up to 7 crossings. These knots are all genuinely different; no amount of manipulation will turn one into the other. Up to 16 crossings, there are 1,701,936 different knots.

For the first time, Alexander's work presented mathematicians with a reliable way to determine that two knots are truly different.

Knot invariants

A milestone though it was, Alexander's method was not perfect. There remained pairs of knots which shared the same algebraic description, despite being genuinely different. For much of the 20th century, the search continued for an improvement to Alexander's work. Major progress arrived in 1984, with Vaughan Jones's discovery of a different algebraic description of knots. The Jones polynomial had a power that Alexander's did not: it could distinguish most knots from their mirror images.

Jones's breakthrough soon found scientific applications. Ultimately, knot theory is about the interaction between geometries of different dimensions. The knot itself is a 1-dimensional object sitting inside 3-dimensional space. Similar considerations arise in quantum field theory, making knot-theoretic tools such as Jones's indispensable in that subject. On a more practical level, the Jones polynomial is used by biochemists to understand the way that enzymes with cells manipulate strands of DNA. Yet not even the combination of Alexander's and Jones's work is enough to distinguish any two knots. So the humble knot continues to be a source of frustration and fascination to today's mathematicians.

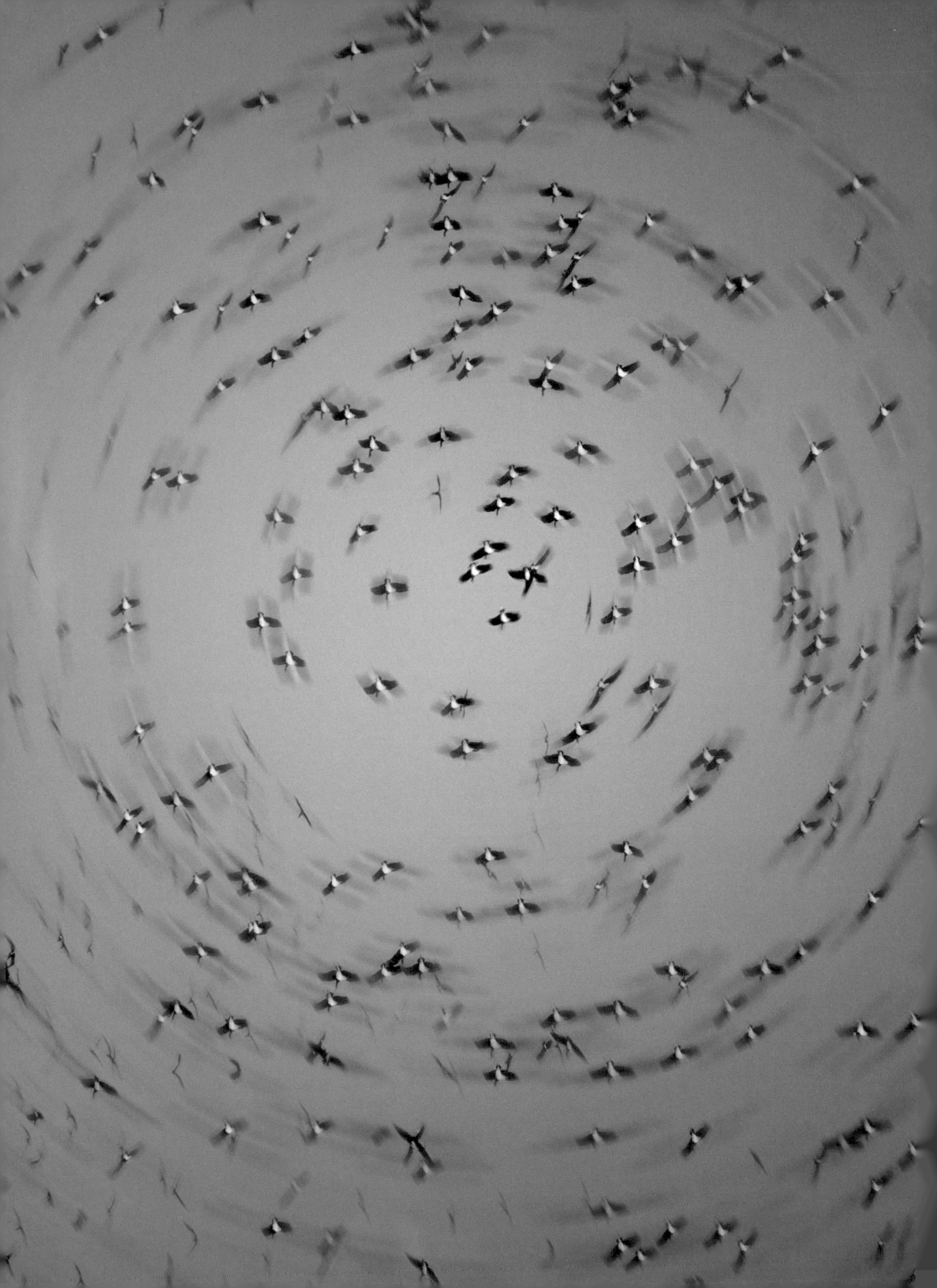

65 Quantum mechanics

BREAKTHROUGH IN THE EARLY 20TH CENTURY, EXPERIMENTS BEGAN TO SUGGEST THAT LIGHT CAN BEHAVE AS BOTH A PARTICLE AND A WAVE. THE CHALLENGE WAS TO FIND A MATHEMATICAL MODEL TO DESCRIBE THIS NEW PHYSICS.

DISCOVERER MAX PLANCK (1858–1947), ALBERT EINSTEIN (1879–1955), ERWIN SCHRÖDINGER (1887–1961), WERNER HEISENBERG (1901–1976).

LEGACY QUANTUM MECHANICS, AND THE SCHRÖDINGER EQUATION, ARE RIGHTLY HAILED AMONG THE GREATEST DISCOVERIES IN THE HISTORY OF SCIENCE.

Throughout history, physicists have turned to mathematics to model and analyze all manner of physical phenomena. This was never truer than in the early 20th century, when investigations into the nature of matter produced some shocking results. To make sense of this new physics of quantum mechanics, scientist urgently needed to develop some new mathematics.

The experiment which turned physics on its head involved shining a light source at a wall, with a screen between the two acting as an obstacle. By opening one or two slits in the screen, light was allowed through. But what would the pattern look like on the wall? The answer would reveal the nature of light.

The double-slit experiment

When the double-slit experiment was first performed in the early 19th century by Thomas Young, it was intended to resolve the longstanding question of whether light comes in particles or waves. Young reasoned that if light was composed of particles, then when he shone light through two slits, all he would see was two bright concentrations of light—one behind each slit. But if light was really a wave, then he would see something else. Unlike particles, waves interfere. The light which passed through the two slits should interfere, resulting in a stripy pattern of dark patches where the two waves cancel each other out, and bright patches where they reinforce each other. When he performed the experiment, the wall lit up with interference patterns, unmistakably indicating the presence of

OPPOSITE Many species of bird migrate long distances each year. How they navigate has always been a mystery, but some scientists are now investigating whether quantum effects may allow them to sense the direction of the Earth's magnetic field, as they head south.

waves. So the particle theory of light was abandoned, and later in the 19th century the mathematics of waves was studied in great detail through the work of Joseph Fourier (see page 169).

Yet the 20th century would bring a shock. Firstly, the particle theory of light would make an astonishing comeback, beginning with the unexpected observation that light shone at a piece of metal causes electrons to be emitted. This photoelectric effect seemed to defy all explanation, since the energy of the electrons did not seem to depend on the intensity of the light in the expected way. Albert Einstein explained this anomaly by conceiving of light as coming in quanta of a fixed size. These are now called photons.

Particles had always been thought of as having a definite position at a given moment. But with the arrival of quantum mechanics, this all changed.

The second important piece of evidence was a rerun of the double-slit experiment by Geoffrey Taylor. This time, more advanced technology allowed light to fire toward the screen at the much-reduced rate of one photon at a time. The result was extraordinary: gradually, the familiar picture of an interference pattern appeared. Clearly, each individual photon was more likely to arrive at one of the bright patches than a dark one. But how could a single photon interfere with itself? The answer had to be that light was mysteriously exhibiting properties of both a wave and a particle. Wave-like, it passed through both slits, and then interfered on the screen. However, the resulting interference pattern could be interpreted as the probability of a single photon arriving at each particular location on the wall.

Wave functions

To begin to make sense of quantum mechanics, some new mathematics was required. An essential ingredient was probability theory. In the past, particles had always been thought of as having a definite position at a given moment. But with the arrival of quantum mechanics, this all changed. A better model assigned the particle a certain probability of being found at a given location. The mathematics of probability theory was well developed by this time, so an obvious starting point was to attempt to model the particle's behavior with a standard tool such as the normal distribution (see page 145).

However, the same experiment that ushered in the new mathematics also demonstrated that classical probability theory would never be enough; probability distributions cannot interfere as waves do. Probabilities are always zero or above, never negative. So while two distributions may be

LEFT Colored image showing a collection of tracks left by subatomic particles in a bubble chamber. Any charged particle leaves a trail of tiny bubbles as the ambient liquid boils. The paths are curved due to an intense magnetic field. The tightly wound spiral tracks are due to electrons and positrons.

able to reinforce each other, they could never cancel each other out. But that is exactly what the dark patches on the wall showed to be happening.

Probability distributions on their own could not capture the wave-like side of light. A fuller story required the development of what are now known as quantum wave functions. Unlike classical probability distributions, a wave function assumes values in the complex numbers (see page 105). Light should be viewed as a wave function, which takes different (complex) values at different positions in space. Such objects can interfere with each other as waves do; they also encode the many properties of the particle, including its probability of being found at each position.

Schrödinger's equation

A critical question in so many areas of science is how an object changes over time. The Navier–Stokes equations (see page 181) describe how a fluid's flow progresses, while Kepler's laws (see page 121) predict the path of a planet locked in orbit around a star. So, while the major philosophical question posed by quantum mechanics was the physical interpretation of a complex-valued wave function, the principal scientific question was to predict how such an object should evolve over time. The answer was provided by Erwin Schrödinger in 1926. Schrödinger's equation elegantly described the way in which the propagation of a wave function through space depends on its energy. This was a wonderful achievement, as it set aside the philosophical difficulties presented by the new theory and presented scientists with a definite way to analyze a given situation: simply solve the appropriate Schrödinger equation.

66 Quantum field theory

BREAKTHROUGH THE INCORPORATION OF QUANTUM MECHANICS IN A BROADER FRAMEWORK, TAKING INTO ACCOUNT FIELDS SUCH AS ELECTROMAGNETISM.

DISCOVERER PAUL DIRAC (1902–84), RICHARD FEYNMAN (1918–88), SIN-ITIRO TOMONAGA (1906–79), JULIAN SCHWINGER (1918–94).

LEGACY QUANTUM FIELD THEORY IS THE BASIS OF THE STANDARD MODEL OF PARTICLE PHYSICS, WHICH DESCRIBES THE WORLD OF SUBATOMIC PARTICLES WITH CONSIDERABLE ACCURACY.

Quantum mechanics is undoubtedly one of the greatest triumphs in science (see page 265). Nevertheless, as a complete analysis of nature, it has several shortcomings. To begin with, it does not take into account the various ways that particles can influence each other. Quantum mechanics therefore needed to be embedded inside a larger theory. The result is quantum field theory, the language of modern particle physics. However, the mathematics behind it remains extremely mysterious.

The most obvious shortcoming of quantum mechanics was that it failed to take into account that other revolution in physics, relativity theory (see page 233). Paul Dirac set about bridging this gap by taking the fundamental formula of quantum mechanics, Schrödinger's equation, and developing a relativistic version. The resulting Dirac equation was a milestone, exhibiting several features of what would later become known as quantum field theory.

The Dirac equation

It is has long been assumed that particles (whether atoms, protons, electrons or photons) are the fundamental building blocks of the Universe—indestructible entities whose existence spans the lifetime of the Universe. However, special relativity, especially the celebrated equivalence of mass and energy ($E = mc^2$) warns against this perspective. It holds that particles may spring in and out of existence. (Indeed, our current scientific understanding is that they do so constantly.)

OPPOSITE An artistic conception of the Higgs field. This Higgs boson was predicted in 1964, and is a major plank in the Standard Model of particle physics. The first direct evidence of it was discovered at the CERN Large Hadron Collider, in 2012.

Instead, Dirac suggested, electrons should be viewed as excitations of something more primitive: an electron field—just as photons are viewed as excitations of the electromagnetic (or light) field. The same is true for the many other particles discovered during the 20th century: quarks, neutrinos, Higgs bosons and the rest. Each is viewed as an excitation of its own field.

Quantum electrodynamics is celebrated for its stunningly accurate predictions about the behavior of electrons, which have been validated in the laboratory with errors as low as one part in a trillion.

Dirac's proposition was arrived at through contemplation of the mathematics behind quantum mechanics and special relativity. But it quickly received spectacular experimental support. Dirac realized that his equation for the electron unexpectedly allowed a second solution, one whose electric charge and other physical attributes are the electron's opposites. Thus he predicted the existence of an antiparticle, a positron. This was discovered in 1932 by Carl Anderson. Today, scientists believe that all elementary particles have equivalent antiparticles. (One of the enduring mysteries of our Universe is why matter is so much more common than antimatter.)

Quantum electrodynamics

Although Paul Dirac's work was a great breakthrough, it did not take into account the physical processes through which particles interact. It was in the 1940s that the first true quantization of a classical piece of physics was accomplished, namely the interaction of matter and electromagnetism. Developed by Richard Feynman, Sin-Itiro Tomonaga and Julian Schwinger, the resulting field theory is quantum electrodynamics (or QED). Building on Dirac's work, QED is celebrated for its stunningly accurate predictions about the behavior of electrons, which have been validated in the laboratory with errors as low as one part in a trillion.

The Standard Model of particle physics

Electromagnetism is one of the four fundamental forces of nature. Over the coming decades, mathematical physicists built on the success of QED to develop quantum field theories for two of the remaining three. The second force is the weak nuclear force, which accounts for why nuclei are prone to decay. It is mediated by particles called W and Z bosons, discovered in 1983. The electroweak theory is the appropriate field theory for this force, which actually holds that at very high energy, the weak and electromagnetic forces are united—they are really two aspects of one single force.

The third force is the strong nuclear force, the glue that binds atomic nuclei together. This strong force is postulated to have its own field, fancifully called the color field, and corresponding particles known as gluons.

ABOVE The ATLAS detector at CERN's Large Hadron Collider (LHC). ATLAS is a broad-range detector that measures all particles and physical processes that emerge during proton collision within the LHC. The eight magnetic rings which confine the particles are visible.

Putting these together, the resulting combined framework is known as the Standard Model of particle physics, which has been the bedrock of our understanding of particle physics since the 1970s. A wonderful piece of evidence was the apparent discovery in 2012 of the Higgs boson. This had been predicted in 1964, and is a long-standing part of our understanding of electroweak field theory.

Renormalization and Yang-Mills

The Standard Model of particle physics has been immensely successful, correctly predicting the existence of many new types of particle, most notably quarks and the Higgs boson. Yet there are still several outstanding problems. In physical terms, the biggest question is how to resolve the Standard Model with the final force of nature: gravity, as described by general relativity (see page 249).

However, this is not the only problem. In the early days of quantum field theory, it was noticed that the weak nuclear force, when viewed on the tiniest scales, was expressed by a divergent expression (see page 97), which sometimes yields an infinite result. Of course this is of no practical use. This was addressed by a piece of trickery called renormalization, about which Feynman commented, "Having to resort to such hocus-pocus has prevented us from proving that the theory of quantum electrodynamics is mathematically self-consistent." This question remains unresolved, as do several related matters including the famous Yang–Mills problem, which the physicist Edward Witten said "would essentially mean making sense of the Standard Model." Despite its impressive evidence base, from a purely mathematical perspective, quantum field theory remains extremely poorly understood, posing a major challenge for the future generations of mathematical physicists.

67 Ramsey theory

BREAKTHROUGH BY ANALYZING HOW ORDERED CONFIGURATIONS MUST EMERGE EVEN IN THE MOST DISORDERED SITUATIONS, RAMSEY SPAWNED A NEW AREA OF MATHEMATICS.

DISCOVERER FRANK RAMSEY (1903–30).

LEGACY RAMSEY THEORY IS USED AS A TOOL THROUGHOUT MATHEMATICS. YET MANY QUESTIONS REMAIN UNANSWERED; IN PARTICULAR, THE EXACT NUMBERS THAT SOLVE RAMSEY-TYPE PROBLEMS ARE USUALLY VERY HARD TO FIND.

In 1930, just a few months before his death, Frank Ramsey published an important paper launching a new subject which became known as Ramsey theory, probing the boundary between order and disorder. The exact numbers which emerge from the subject remain deeply mysterious.

Ramsey had scholarly interests outside mathematics. A close friend of John Maynard Keynes, he published three papers on questions of economics, including one on optimal taxation. His ideas in this area would later be recognized as important early insights into emerging fields. Ramsey was also fascinated by questions of philosophy, and published several philosophical papers of his own as well as translating Ludwig Wittgenstein's *Tractatus Logico Philosophicus* into English from German (a Cambridge colleague later recalled that he had learnt to read the language in "almost hardly over a week"). Later, Ramsey traveled to Austria to discuss the work with Wittgenstein. By the time of his death due to liver disease at the age of 26, Ramsey had made a major impression on the 20th century's intellectual landscape.

Within the field of mathematics, Ramsey made several important contributions, notably to logic and the foundations of mathematics. Today, though, he is best known for the solution to a puzzle known as the "dinner party problem." This is no mere riddle, but a theoretical breakthrough that many mathematicians have subsequently come to rely upon.

OPPOSITE The starting point of Ramsey theory is a riddle about planning a dinner party. How many people need to be invited to guarantee that there will be either three mutual friends or three mutual strangers present?

ABOVE Graphs representing different possible relationships between guests at a five person dinner party. Strangers are marked in blue, and friends in white. The circled graph does not contain either three mutual friends, or three mutual strangers, proving that five is not a solution to the dinner party problem.

The dinner party problem

The problem Ramsey set out to solve can be expressed as a dilemma faced by the organizer of a dinner party. She wants to ensure that the mix of guests is a sociable one. As a way of striking the right balance, she wants to ensure that there are either three mutual friends present, or three mutual strangers. The question is: how many people need to be invited for this combination to be guaranteed?

Three people are not enough. It may be that one is a mutual friend of the other two, who have never met. In fact, four people are not enough either, since people may know each other in pairs, meaning every guest has one friend and two people they do not know. It's also possible to concoct a party of five without either of the required configurations. With six people, however, every possible dinner party must have either three mutual friends or three mutual strangers.

Mathematically, the best way to express one of these dinner party problems is in the language of colored graphs (see page 149). If every guest at the dinner party is represented by a dot, then two dots are connected with a white edge if the two people are friends, or with a blue edge if they are strangers. So the question becomes whether the resulting network of dots and edges somewhere contains either a white or blue triangle. The solution to the original dinner party problem is that every network with at least six dots must indeed contain a triangle of a single color.

On its own this makes a satisfying puzzle, though it hardly seems like a major mathematical breakthrough. But the question easily generalizes. Suppose that instead we want a dinner party where there are either four mutual friends or strangers. In terms of graphs, we are looking for the minimum size that guarantees the presence of four points connected with edges of the same color. This number is much harder to find, because the variety of possible graphs grows astronomically quickly. There are already around 10 million ten point graphs, so checking them all for the required configuration quickly becomes impractical. Yet, in 1955, the answer was pinned down to exactly 18. If you invite 18 guests to your home there are bound to be either four mutual friends, or four mutual strangers. The same question for six guests remains uncertain even today, but is known to lie between 43 and 49.

This is the archetype of a problem in Ramsey theory. Mathematicians identify a simple criterion and prove that in any large enough structure the chosen configuration must arise.

Infinite Ramsey theory

What use is Ramsey theory, one might wonder, if its questions are so difficult to answer? What Frank Ramsey proved was that every dinner-party-type question must have an answer. There will always be some number that guarantees the presence of a suitable configuration. This is by no means obvious. It might have been true that one could find arbitrarily large graphs that avoid six dots joined by edges of the same color. Thanks to Ramsey, we now know that, after some point, every such graph will do this. (In fact, the threshold is somewhere between 102 and 165.)

At a dinner party with infinitely many guests, there must be an infinite subset of guests who are either all mutual friends, or all mutual strangers. Expressed in terms of graphs, this turns into a very beautiful result.

Ramsey's theorem takes on a new flavor when applied in infinite situations. Expressed in terms of graphs, it turns into a very beautiful result. Imagine an infinite graph where every pair of points is connected by an edge, which can be one of a limited number of colors. Ramsey's theorem says that there will be some infinite collection of points which together form a graph of a single color.

This is truly a wonderful theorem, as the original graph may be a hugely complex or even random structure. Nevertheless, Ramsey showed that one can always find a very simple, highly ordered structure lurking inside. This ability to conjure order from confusion is what makes Ramsey theory such a powerful weapon for today's mathematicians.

68 Gödel's incompleteness theorem

BREAKTHROUGH BY UNDERSTANDING HOW TO ENCODE LOGIC WITHIN NUMBERS, AND VICE VERSA, GÖDEL SHOWED THAT NO COMPLETE SET OF RULES FOR MATHEMATICS WILL EVER BE WRITTEN DOWN.

DISCOVERER KURT GÖDEL (1906–78).

LEGACY PERHAPS THE MOST IMPORTANT MOMENT IN THE HISTORY OF LOGIC, GÖDEL'S THEOREM CAUSED MATHEMATICIANS TO LOOK AT THE FOUNDATIONS OF THEIR SUBJECT WITH FRESH EYES.

Mathematics began as a purely practical subject: a tool for calculating facts about the world. For millennia it has been of incalculable value to farmers, merchants, engineers and scientists. But around the beginning of the 20th century, some more philosophical questions came to the fore. In the renaissance of mathematical logic, no theorem had a bigger impact than the incompleteness theorem proved by Kurt Gödel in 1931.

Just as physicists have always sought the basic laws which govern the physical Universe, so in the late 19th century logicians such as Gottlob Frege and Giuseppe Peano began to seek the fundamental axioms which underlie the numerical domain. This question was considerably harder than might be expected. It is easy to write down some laws of numbers. For instance, it should always be true that when you add together two numbers, the order does not matter: $a + b = b + a$, no matter what a and b are. The same thing should hold for multiplication: $a \times b = b \times a$. These are both perfectly valid laws of numbers. The difficulty was not so much in framing individual laws, as knowing when to stop.

The logical pioneers of the 19th and early 20th centuries had an ambitious goal: a complete rule book for numbers. This would list the fundamental properties of numbers, from which all other mathematical facts should be deducible. Each individual rule should be as simple as possible. (In fact, the two laws above, $a + b = b + a$ and $a \times b = b \times a$, are not usually taken as fundamental; they are consequences derived from even more primitive rules.)

OPPOSITE Gödel designed a mirror between the logical world and that of numbers. Just as a logical system can be used to describe numbers, so numbers, he realized, can describe logic. Gödel's incompleteness theorem was the dramatic consequence of this new type of reflection.

RIGHT An extract from Bertrand Russell's and Alfred North Whitehead's *Principia Mathematica.* It was a technical tour de force, but put extreme demands on the reader. The purpose was to build up standard mathematical results from secure logical foundations.

∗54·43. $\vdash :. \alpha, \beta \in 1 \,.\, \supset : \alpha \cap \beta = \Lambda \,.\, \equiv \,.\, \alpha \cup \beta \in 2$

Dem.

$\vdash . {*}54{\cdot}26 \,.\, \supset \vdash :. \alpha = \iota{'}x \,.\, \beta = \iota{'}y \,.\, \supset : \alpha \cup \beta \in 2 \,.\, \equiv \,.\, x \neq y \,.$

[∗51·231] $\equiv \,.\, \iota{'}x \cap \iota{'}y = \Lambda \,.$

[∗13·12] $\equiv \,.\, \alpha \cap \beta = \Lambda$ (1)

$\vdash . (1) \,.\, {*}11{\cdot}11{\cdot}35 \,.\, \supset$

$\vdash :. (\exists x, y) \,.\, \alpha = \iota{'}x \,.\, \beta = \iota{'}y \,.\, \supset : \alpha \cup \beta \in 2 \,.\, \equiv \,.\, \alpha \cap \beta = \Lambda$ (2)

$\vdash . (2) \,.\, {*}11{\cdot}54 \,.\, {*}52{\cdot}1 \,.\, \supset \vdash .$ Prop

From this proposition it will follow, when arithmetical addition has been defined , that $1 + 1 = 2$.

Hilbert's program and *Principia Mathematica*

The early 20th century saw a flurry of activity around the foundations of mathematics, as advances in logic seemed to make the prospect of a complete mathematical rulebook a reality. The influential thinker David Hilbert saw this as the biggest intellectual challenge of the age, and threw his weight behind a concerted effort to bring it about. One of the most serious attempts was made by Bertrand Russell and Alfred North Whitehead. Their three-volume magnum opus was called *Principia Mathematica,* and it developed a new kind of logic known as type theory. This was robust enough, they hoped, to act as a foundation for the entirety of mathematics. Famously, it was not until the second volume that they got around to proving that $1 + 1 = 2$. The later sections derive many of the familiar properties of whole numbers before laboriously deducing several more sophisticated mathematical theorems.

Though not without its critics, *Principia Mathematica* was an extraordinary achievement—perhaps the greatest step forward in mathematical logic since Aristotle. But was it the ultimate mathematical rule book? Could every mathematical question be resolved, simply on the basis of Russell and Whitehead's laws?

These debates made the early 20th century a thrilling time for logicians. But it was not long until Kurt Gödel dealt all such hopes a devastating blow. At the time, Gödel was a little-known Austrian researcher who had just finished his PhD at the University of Vienna. Gödel took it upon himself to read and thoroughly analyze the *Principia Mathematica.* The difficulty of this task alone should not be underestimated, but the result of Gödel's investigations came as a bombshell to mathematical logic.

Gödel's theorem

In 1931, Gödel published a paper entitled "On Formally Undecidable Propositions of *Principia Mathematica* and Related Systems," which contained his famous incompleteness theorem. The upshot of this work was that all the recent efforts to draft the definitive rule book for mathematics were doomed. The shock was not that *Principia Mathematica*, or any of the other logical theories doing the rounds in mathematical circles, should be incomplete. Rather, Gödel's incompleteness theorem asserted that no complete set of rules for arithmetic could ever be written down (or indeed generated by a computer). Every such system would be guaranteed to contain gaps: assertions about numbers that could be neither proved nor disproved from the given rules. Such an assertion is called "formally undecidable."

Of course, once identified, any given gap could be plugged just by drafting a suitable new rule to incorporate into the book. The trouble was, Gödel realized, that this process of plugging gaps one at a time would never come to an end.

Gödel-numbers

At the center of Gödel's work was the phenomenon of self-reference. Other theorists had been using logic to describe numbers. Gödel's insight was that you could equally well use numbers to describe logic. This technique of Gödel-numbering axioms, proofs and theorems revealed that any logical system powerful enough to describe arithmetic must also be able to describe itself. Gödel's final flourish was to encode in numbers the statement "this statement has no proof." Interpreting this as a purely numerical statement, it had to be either true or false. But if it was false, this meant that the logical system could provide a proof for an untrue statement, which would be a disaster. Hence the statement had to be true. With that, Gödel had found an example of a true but unprovable statement: a gap in the logical system.

Principia Mathematica was to act as a foundation for the entirety of mathematics. Famously, it was not until the second volume that they got around to proving that 1 + 1 = 2.

Gödel's theorem was a transformative moment in the history of logic. It drew a line under the foundational crisis of the early 20th century and forced mathematicians to take a more considered view of the central tenets of their subject. From it several brand-new mathematical disciplines were spawned. The new subject known as proof theory compares the strengths of different known mathematical rule books. Meanwhile, reverse mathematics asks for the exact assumptions needed to prove various standard theorems of mathematics.

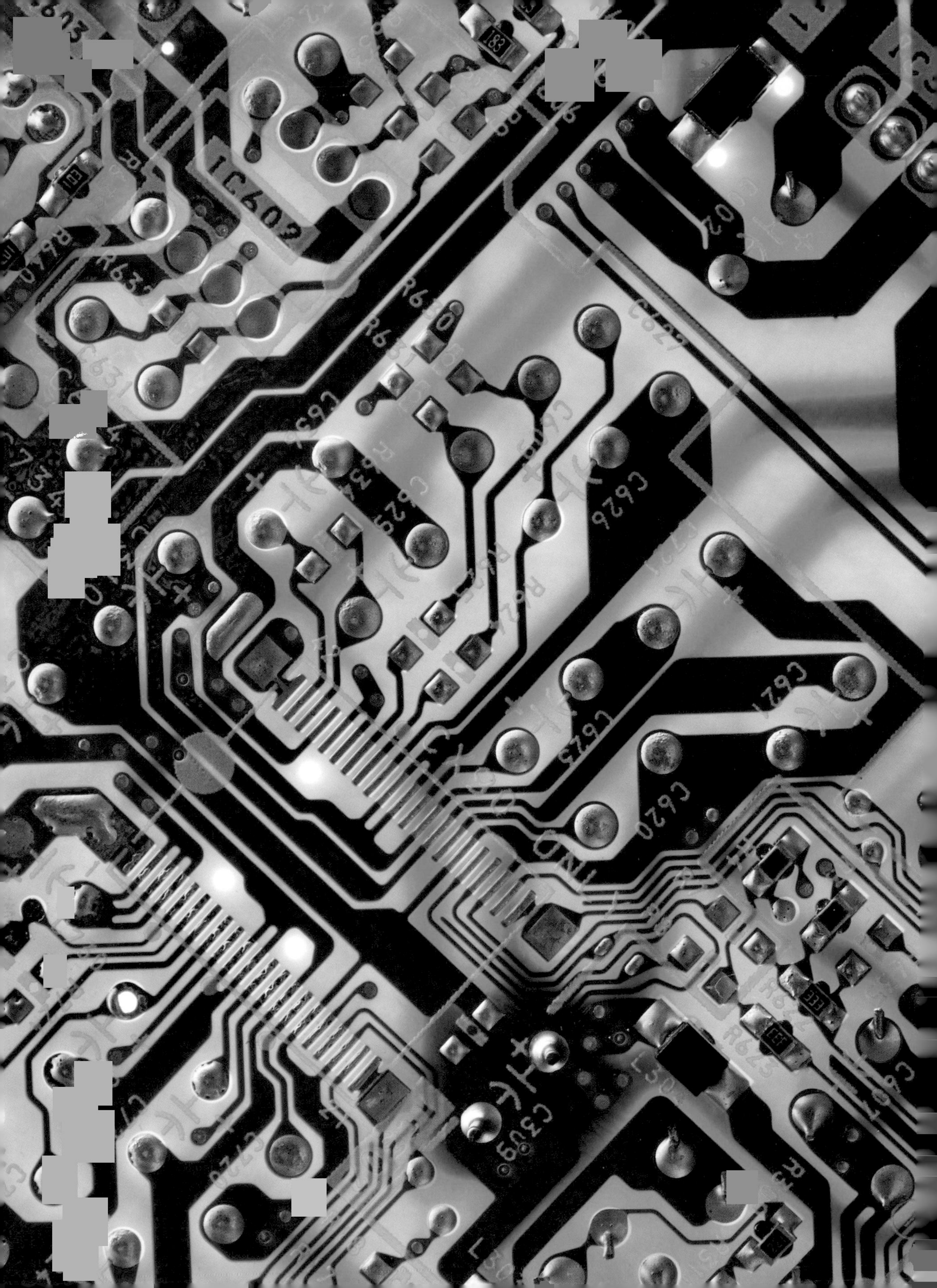

69 Turing machines

BREAKTHROUGH TURING MACHINES ARE THEORETICAL DEVICES WHICH CARRY OUT A PROCEDURE ACCORDING TO STRICT LOGICAL INSTRUCTIONS. INVESTIGATIONS INTO TURING MACHINES LED TO DEEP INSIGHTS INTO THE LOGIC UNDERLYING THE WHOLE NUMBERS.

DISCOVERER ALONZO CHURCH (1903–95), ALAN TURING (1912–54).

LEGACY AN EPOCHAL MOMENT IN THE 20TH CENTURY, TURING'S ANALYSIS WAS THE THEORETICAL WORK THAT USHERED IN THE COMPUTER AGE

The work that Alan Turing and Alonzo Church carried out in the 1930s was not intended to bring about technological innovation. Rather, it was an attempt to address a technical problem in mathematical logic, proposed by David Hilbert a decade earlier. But the repercussions of their work quickly exploded far beyond scientific circles, underpinning the single most important invention of the twentieth century: the programmable computer.

In the late 1930s, a researcher at Cambridge University by the name of Alan Turing turned his mind to a question posed by David Hilbert a decade earlier. In essence, the question was how to determine whether or not a mathematical assertion is true. Kurt Gödel had already blown an irreparable hole in Hilbert's objective by showing that, no matter what logical framework you adopt, there will be some mathematical statements which are unprovable (see page 329).

OPPOSITE A printed circuit board: the standard way to connect the components of modern computers. Computers have changed beyond recognition over the last 50 years. While there are many possible designs of computer and ways to compute, the Church–Turing thesis guarantees that all are ultimately equivalent.

Algorithms and proofs

Gödel's work was about the existence (or not) of mathematical proofs. Turing, meanwhile, was interested in procedures for determining truth. Could there be a single process capable of deciding the truth or falsity of any statement, Gödel's unprovable assertions aside?

There are many unsolved problem in mathematics: the Riemann hypothesis (see page 206) and the Birch and Swinnerton–Dyer conjecture (see page 359) for example. Hilbert had envisioned nothing less than a single process to resolve them all. It would begin with a mathematical statement, follow some predetermined rules, and then arrive at an answer: "true" or "false." If such a procedure could be invented, it would mark the end point of mathematics as it had been known. Hilbert had dared to hope that it was possible. Might he be right?

Turing machines

To answer that question, Turing needed to understand what such a "process" might consist of. The answer he landed upon was a Turing machine. This notional device came in two parts: a long tape, decorated with symbols from some alphabet (these days it usually just features the symbols 0 and 1), and a reader which would creep along the tape, interpreting each symbol as a simple instruction. According to the instructions it received, the machine could read, erase and overwrite symbols on the tape, and it could step forward and backward along it. Turing realized that by feeding in the right instructions, this simple machine could perform many mathematical tasks. Not only addition and multiplication, it could test numbers to determine whether they are prime, and it could solve equations. In fact, the more Turing experimented, the less there seemed to be that a Turing machine could not do.

Turing realized that by feeding in the right instructions, his simple machine could perform many mathematical tasks. In fact, the more he experimented, the less there seemed to be that a Turing machine could not do.

Turing machines perfectly captured the notion of "processes." Turing then went one better by devising a universal Turing machine. Extraordinarily, this single machine could mimic any other. If Hilbert's question was to have a positive answer, it surely lay in the universal Turing machine. But it was not to be. Turing was able to pose a fiendish puzzle which even a universal Turing machine could not answer. Specifically, he asked it to determine which machines will eventually complete all their instructions and stop, and which will keep running forever. Turing was able to encode this as a purely mathematical question, but the universal Turing machine was unable to provide an answer.

Church–Turing thesis

The previous year, Alonzo Church had also addressed Hilbert's question, but via a totally different means. Machines had played no role in his thinking. Instead he had set up an abstract system of semantics known as lambda calculus, and shown that within this system there is no way to tell whether or not two expressions represent fundamentally the same mathematical object.

Church and Turing collaborated to understand how their own work related to each others'. In fact, although expressed in completely different languages, the underlying mathematics was the same. Anything that could be computed by a Turing machine could also be expressed in Church's lambda calculus, and vice versa. Over time, other logicians came up with further interpretations of what it means to be computable. All these, too, were shown to be equivalent to Church's and Turing's theories.

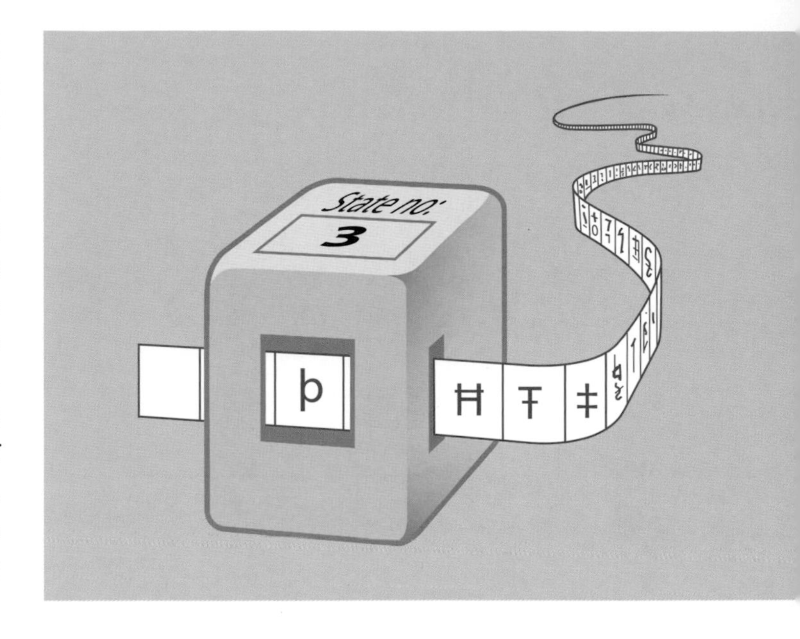

ABOVE An artistic representation of a Turing machine, reading, writing, and erasing symbols on its tape, and switching internal state according to predefined rules. Although far from a practical design, anything which any computational device can achieve can in principle be completed by a Turing machine.

This Church–Turing thesis, that computability is a robust notion, not reliant on any particular system of computation, is a hugely important one. Numerous such systems are now known, principally programming languages such as C++, Java and Python, but also some systems with radically different appearances, such as the game of Life (see page 333). Of course, today's computers can be improved, made faster, furnished with more memory, and so on, and programming languages can be made more user-friendly and better tailored to certain tasks. But the Church–Turing thesis carries a warning that no tweak in the design can ever turn an uncomputable object into a computable one. In particular, no computer will ever be built that satisfies David Hilbert by being capable of answering every possible mathematical question.

Programmable computers

After the theoretical triumph of the Turing machine, some thinkers, among them Turing's friend John von Neumann, became determined to construct a universal Turing machine in the real world. In the days before electronic engineering, these early computers such as the EDVAC were gigantic contraptions consisting of thousands of connected valves. Whole rooms were required to house their workings, not to mention large teams of people to operate them. The role of the tape was played by punch cards initially, carrying early computer programs, and later magnetic tape. Nevertheless, they brought Turing's vision to life—and they were the forebears of the smartphones and computers upon which the modern world depends.

70 Numerical analysis

BREAKTHROUGH NUMERICAL ANALYSIS IS THE ART OF SOLVING EQUATIONS APPROXIMATELY. ITS ROOTS GO BACK TO THE BEGINNING OF MATHEMATICS, BUT IT WAS ONLY IN THE 20TH CENTURY THAT IT GREW INTO A SUBJECT IN ITS OWN RIGHT.

DISCOVERER RICHARD COURANT (1888–1972), JOHN VON NEUMANN (1903–57).

LEGACY NUMERICAL ANALYSIS IS EMPLOYED EVERY DAY BY ENGINEERS AND SCIENTISTS ENGAGED IN A BROAD RANGE OF ACTIVITIES. IN RECENT DECADES, IT HAS BECOME INSEPARABLE FROM THE DEVELOPMENT OF COMPUTATIONAL MATHEMATICS.

Mathematics, it is rightly said, is an exact science. Yet the tools of mathematics are applied daily across countless areas of science, technology and engineering, where the total precision prized by mathematicians is unavailable, and in any case unnecessary. Instead, we are confronted with all the uncertainties and inaccuracies of the physical world. In order to be useful in this realm, mathematicians have developed a range of methods for solving problems by approximate means. This area is known as numerical analysis, one of the most useful branches of mathematics today.

Numerical analysis is nearly as old as mathematics itself. One of the earliest mathematical artifacts is a Babylonian tablet, on which a square is drawn and the diagonal marked (see page 31). The length of the diagonal is marked as being (in modern notation) 1.41421297..., a good approximation to the correct value of $\sqrt{2} = 1.41421356 \ldots$ The "Babylonian method" we believe they employed is still used today for approximating square roots.

Centuries later, Archimedes found a beautiful way to approximate the number π (see page 54). Archimedes sandwiched a circle between two polygons each with 96 straight sides, one inside and one outside. By calculating the lengths of the perimeters of these two shapes, Archimedes was able to place π somewhere between $3\frac{10}{71}$ and $3\frac{10}{70}$.

OPPOSITE Fluid flow is a highly complex phenomenon, and the equations governing it are currently too difficult for us to solve exactly. Instead the science of fluids demands a numerical approach.

Newton's method

Those of us who studied mathematics at school are used to questions where is a single, exactly correct answer. But, even within mathematics, this is a misleading picture, and in broader science it is very far from the truth, even when it comes to that enduring mathematical theme, solving equations. Given some process which takes in numbers as inputs, and gives out other numbers as outputs, the challenge is to find an input whose corresponding output is zero. Over the centuries, various methods have been found, beginning with Al-Khwarizmi's work on quadratic equations. But the work of Abel and Galois on quintic equations (see page 177) reveal that, even for comparatively simple algebraic processes, finding solutions is by no means trivial.

The inconvenient fact is that many equations do not necessarily have exact solutions, and even if they do, they commonly will not be expressible in any simple way. To extract useful information from such situations, numerical methods are indispensable.

For other processes, perhaps originating in astronomy or another branch of science, very often there is no immediate way to find a solution. In these situations, approximate numerical methods are essential. An early breakthrough in this area was the Newton–Raphson method. This works by beginning with a first guess at a solution to the equation. Then, Isaac Newton and Joseph Raphson described a technique which starts with one approximation, and then uses it to manufacture a better one. Repeating this produces a sequence of ever closer approximations to the solution. However, the Newton–Raphson method does not invariably work; it depends in a subtle way on the underlying geometry of the function. This illustrates an important moral: that numerical methods typically require deep theoretical justifications.

Differential equations

With the development of calculus (see page 133), mathematics presented the world of science a wonderful gift. Calculus is exactly the right tool to describe many systems that evolve over time. Many scientific questions, however, run the other way: they begin with information about how a system varies over time, and the objective is to derive a mathematical description of the system's condition at a particular moment. This often involves finding the solution to a differential equation. Famous examples are the Navier–Stokes equations (see page 181) for describing the flow of a fluid, the Schrödinger equation in quantum mechanics (see page 265), and Einstein's field equations for describing the shape of the Universe, part of the theory of general relativity (see page 249). There are innumerable others too, across the world of science and engineering.

Unfortunately, however, solving differential equations is by no means straightforward, as the Navier–Stokes problem notoriously illustrates. While novice calculus students are accustomed to equations with comparatively simple answers, the inconvenient fact is that many do not necessarily have exact solutions, and even if they do, they commonly will not be expressible in any simple way. To extract useful information from such situations, numerical methods are indispensable.

ABOVE The finite element method is commonly used for modeling the stresses and strains solid objects come under. Among many other places, this technique is valuable in the transport industry's design of safety systems.

Scientific computing

During the 20th century, numerical analysis developed into a subject in its own right, and differential equations became a particular focus. One milestone was the work of Richard Courant, who in the 1940s developed the finite element method. This works by approximating the original troublesome equation with a geometrical mesh of simpler equations, each of which is easier to solve. His approach would later grow into an important intellectual tool for scientists and engineers; it is particularly valuable for studying situations where solid materials may twist and bend under pressure, the underlying geometry of which can be extremely complex.

In the second half of the 20th century, numerical analysis has become inseparable from mathematical computing. Now numerical analysts search for computational methods for solving equations. Early work on this was performed by John von Neumann, an early pioneer of both numerical analysis and computer design (see Turing machines, page 281). Today's engineers draw heavily on von Neumann's innovations in both areas, using sophisticated modeling programs to understand the behavior of solids under pressure, for example in the design of pre-stressed concrete beams for use in modern buildings.

Numerical analysis is not only important to the engineering of solid objects such as vehicles and buildings. In any branch of science which deals with fluids, from weather forecasting to aerodynamic design, numerical methods as implemented on powerful modern computers are indispensable tools.

71 Information theory

BREAKTHROUGH SHANNON INVESTIGATED THE PROPERTIES OF INFORMATION FLOWING ALONG A CHANNEL. HE WAS ABLE TO PROBE THE LIMITS OF EFFICIENT INFORMATION TRANSFER USING THE NOTION OF ENTROPY.

DISCOVERER CLAUDE SHANNON (1916–2001).

LEGACY SHANNON'S INFORMATION THEORY FORMS THE BASIS OF ALL MODERN COMMUNICATION TECHNOLOGY, INCLUDING THE INTERNET.

The second half of the 20th century saw the arrival of the information age. Today information streams around the world non-stop, connecting homes, businesses, charities and governments via underground fiber-optic cables and satellite link-ups. But how can information flow along such channels? And how can we guard against data becoming corrupted as it does so? The mathematics of information was first described by Claude Shannon in 1948, and it would go on to become one of the defining sciences of the era.

There are many ways to send a message, from handwriting, sign language or Braille to smoke signals or semaphore. But in the modern era, most technology first translates the content into binary.

The binary system

The binary system only has two symbols: 0 and 1, known as bits (short for binary digits). Binary was first studied by Gottfried Leibniz in 1666, but it is only in the last century that it has come to play a central role in the world. Any number can be expressed in binary, just as easily as it can be written in decimal notation. When we write a number such as 312, the right-hand column stands for units: the 2 represents 2. But the 1 in the middle stands for one 10. Similarly, the 3 stands for three 100s. Working from right to left, the columns represent: 1, 10, 100, 1000. In binary, or base two, they stand for 1, 2, 4, 8, 16, and so on. So 17 in binary is 10001, while 26 is 11010.

OPPOSITE
Increasingly, modern communication takes place through fiber-optic cables. Information travels at light-speed through the fibers with little interference or noise. At the transmission end, the light flickers on and off, encoding a binary message where "1" is represented by the light being on and "0" by it being off.

It is not only numbers which need to be interpreted as binary strings. To translate a letter or punctuation symbol into binary, the standard system today is ASCII (the American Standard Code for Information Interchange). According to ASCII, "1101010" represents a lower-case letter "j," while "0100110" represents the ampersand symbol "&."

Of course, letters and numbers are not the only types of information that need to be transmitted. We live in an age of spreadsheets, smartphone apps, games based on 3D graphics, and all manner of software and data. Information comes in a much wider variety of forms than ever before. Nevertheless, there are standard protocols for describing all these various types of information as strings of binary.

Information transfer

In a modern communication system, the main task is to send a binary string from one location to another. This could be achieved, for example, by having a pulse of high-energy electricity to stand for 1, and a lower-energy pulse to stand for 0. An appropriate sequence of pulses can then carry any information along a wire.

This is indeed the basic method commonly used today. However, it suffers from two significant drawbacks. Suppose that the string to be sent reads 00000000 … 0000, making a total of a million bits all set to 0. It could perfectly well be sent with a million pulses of electricity, but this would be an extremely inefficient way to proceed. It would be much quicker, for example, to convert "The message is one million zeroes!" into ASCII code, and send that. This is an extreme example of a more general phenomenon: with limited time and bandwidth available, it is important to find ways to encode and transmit information efficiently.

What was needed was a way to encode information efficiently for quick transmission, yet robustly in the face of any errors creeping in.

The second problem with the naïve approach is that it is extremely intolerant of error. A glitch in the system that flipped just one bit from a 0 to a 1 would destroy the message. Since errors are a fact of life, any sensible system should have safeguards built in. The simplest way to built in safeguards is to repeat digits. If each bit is sent three times, then to communicate a message reading "101," the channel would actually carry "111000111." Then, if an error crept in, say "111000101," it would be possible to spot and correct that. The problem with this is that the message triples in length, so the one million zeroes becomes three million. Nor is it particularly reliable: just two errors could still destroy the message.

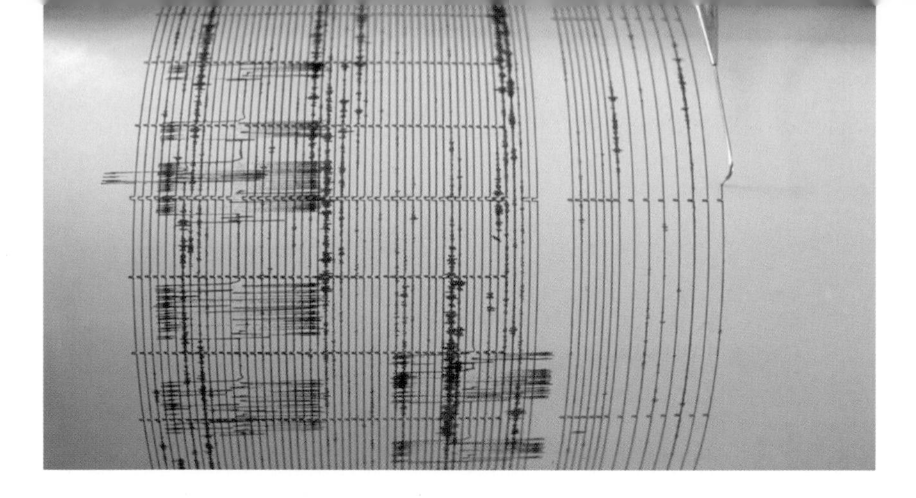

LEFT Information theory is increasingly used in the science of seismology. The concept of entropy can encode the predictability of seismic activity at a particular spot, helping scientists quantify the likelihood of an earthquake.

What was needed was a way to encode information efficiently for quick transmission, yet robustly in the face of any errors creeping in. In his groundbreaking paper of 1948 entitled "A Mathematical Theory of Communication," Claude Shannon addressed both of these questions. His answers were as unexpected as they were influential.

Entropy

How efficiently a message can be communicated depends on how predictable the string of symbols is. This unpredictability is measured by a number called the entropy of the information source. The higher the entropy, the more unpredictable the message. In English, for example, there is some amount of predictability: the letter "z" is considerably rarer than the letter "e," while the pair "cz" is much rarer than "th." A completely predictable source, such as one that only emits 1s, has an entropy of 0. Shannon estimated the entropy of the English language to be around 1.5 bits per symbol.

The two critical factors that determine how efficiently information can be transmitted are the entropy of the source (represented by H), and the bandwidth of the channel (meaning the number of bits per second it can carry, denoted as C). Shannon demonstrated how to encode information in such a way as to approach the maximum possible transfer rate: $\frac{C}{H}$. More remarkably still, he deployed error-correcting codes to show how a message can be communicated reliably even through a very noisy channel, perhaps where 99% of the message is corrupted, with only modest increases in the rate of transmission.

Shannon's ideas not only underpin today's internet and telecommunication systems, but they are also touching on an ever widening variety of other sciences, from energy transfer to predicting earthquakes, where the predictability of an information source is an important factor.

72 Arrow's impossibility theorem

BREAKTHROUGH ARROW'S IMPOSSIBILITY THEOREM SHOWED THAT NO CONCEIVABLE VOTING SYSTEM CAN ACCURATELY REFLECT THE PREFERENCES OF A POPULATION.

DISCOVERER KENNETH ARROW (1921–).

LEGACY ARROW'S THEORY WAS THE STARTING POINT OF SOCIAL CHOICE THEORY. WITH RESULTS SUCH AS HALL'S MARRIAGE THEOREM, THIS IS AN IMPORTANT TOOL IN POLITICS AND ECONOMICS.

The birth of democracy, meaning "government by the people," is usually ascribed to the city state of Athens around the sixth century BC. In the intervening millennia, numerous electoral systems have been tried. But which is the best and fairest form of democracy? In 1950, Kenneth Arrow proved the unnerving theorem: no perfect system can exist.

By modern standards, the ancient Athenian form of democracy was far from perfect, with neither women nor slaves entitled to vote. In more recent years, as democracy has spread around the world, various electoral systems have been tested out. The mathematics of elections, and other instances of social choice, are surprisingly subtle. In 1950, economist Kenneth Arrow produced his famous impossibility theorem, proving that a perfect democracy is essentially impossible—every system must be a compromise between competing factors.

Athenian government was largely by direct democracy, meaning that every motion was put to a public referendum. Nowadays, representative democracy is the norm. Instead of voting on individual laws, members of the public elect representatives to do so on their behalf. The central problem here is how to choose the candidates who best reflect the views of the populace at large.

OPPOSITE A 2008 electoral map of the USA, with blue representing counties voting Democrat and red those voting Republican. One anomaly in the US electoral system is that the votes of residents of less populated counties are de facto more valuable than those of crowded counties.

Electoral systems

The easiest election is a winner-takes-all system, known as simple majority voting, or in the UK (rather misleadingly) as "first past the post."

Arrow's conclusion was unsettling and unexpected: he found that every possible voting system must be unfair according to some measure.

Suppose there are three candidates: Alex, Bailey and Claude (A, B and C). In the election, A receives 40% of the vote, while B gets 35% and C 25%. In simple majority voting, A, being the candidate who gets the most votes, is thereby elected.

This may seem just and natural, but the trouble is that the chosen candidate is actually opposed by a majority of voters—60% in the above example. The more candidates there are, the more the problem is exacerbated. With 99 candidates, if one gets 2% of the vote and each of the others gets 1%, the chosen representative may be elected against the wishes of 98% of voters. It might also be that the elected candidate is the most unpopular one.

To overcome this scenario, voters in simple majority systems may decide to vote tactically. Someone may think, "My first choice is C, but they have no realistic chance of being elected, and I would much prefer B to win than A. So I will vote for B instead." In doing this, the voter is moving beyond their first preference and voting according to their second preference. For this to work, the voter has to have some knowledge of how other people are likely to vote.

Social choice theory

More sophisticated electoral systems already take into account voters' preferences beyond their first choice. Instead of voting for a single candidate, each voter ranks the candidates in order of preference. With this data, the system should produce an overall ranking of candidates according to the amalgamated rankings of all the voters. This is what is known as a social choice function.

There are innumerable ways that a social choice function might work. Simple majority voting simply tallies up people's first preferences and orders the candidates accordingly. Another way is for one person to decide the entire outcome: a dictatorship rather than a democracy.

Kenneth Arrow contemplated the properties that a system might need to be considered fair. His criteria were not complicated. Of course it should not be a dictatorship. Furthermore, it should have the property of Unanimity: if every individual voter prefers candidate *X* to candidate *Y*, then *X* should be ranked higher than *Y* overall. Slightly more complex is Independence of Irrelevant Alternatives: whether or not candidate *X* is ranked higher than candidate *Y* overall should depend only on voters' relative preferences of X and Y, not on their opinions of any third candidate Z.

Arrow's criteria seemed little more than common sense. But rather than using them to design the ultimate voting system, thereby settling centuries of political debate, Arrow's conclusion was infinitely more unsettling and unexpected: he found that there is no way any voting system can reliably satisfy these criteria. Every possible voting system must, therefore, be unfair according to some measure.

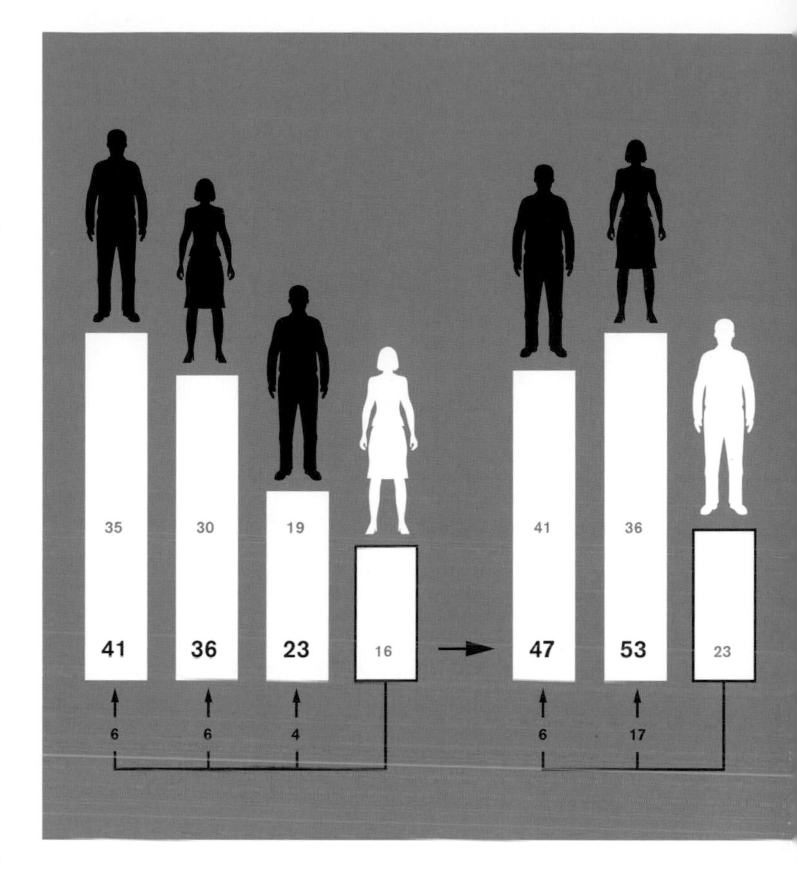

ABOVE Instant Run Off Voting or the Alternative Vote. In the first round, candidate D has received the least votes, and is eliminated. Her votes are then redistributed according to voters' second preferences. This continues until one candidate has over 50% of the vote, as happens for B in this example.

Hall's marriage theorem

In other contexts, mathematics brings more welcome news, showing that conflicting criteria can be met to the satisfaction of all. A famous instance was discovered by Philip Hall in 1935 and became known as Hall's marriage theorem. Suppose that a family of ten wants to share a box of ten chocolates. But people's tastes vary: perhaps mother is averse to nuts, while the children can't stand sweets containing alcohol, and the grandparents' teeth cannot manage hard toffee centers. Is there any way to please everyone?

Obviously, it may not be possible. If the violet cream is hated by everyone, that leaves only nine chocolates between ten people. Similarly, if the five children can only come up with a total list of four chocolates which any of them like, then there can be no way to satisfy even the children, let alone the whole family. A minimum requirement would be that each group of n people must be able to come up with a list of at least n chocolates which at least one of them likes. Surprisingly, Hall was able to prove that this criterion is all that is required to guarantee that there is a solution to the chocolate debate that is satisfactory to everyone.

73 Game theory

BREAKTHROUGH GAME THEORY IS THE MATHEMATICAL SCIENCE OF STRATEGY. A MILESTONE WAS THE INSIGHT THAT ALL GAMES OF A CERTAIN TYPE MUST CONTAIN EQUILIBRIA.

DISCOVERER JOHN VON NEUMANN (1903–57), OSKAR MORGENSTERN (1902–77), JOHN NASH (1928–).

LEGACY GAME THEORY TODAY HAS APPLICATIONS ACROSS SOCIAL AND POLITICAL SCIENCE, ECONOMICS AND ARTIFICIAL INTELLIGENCE RESEARCH.

Many situations in life require strategic thinking: from a hedge-fund manager deciding which businesses to invest in to a military general planning tactics for a war. Over the centuries, people have developed many games of strategy too, such as chess, checkers and Go. But strategy itself, in its purest form, only became a topic of study with the advent of game theory in the early 20th century.

One of the core concepts in game theory is that of a Nash equilibrium. The idea can be seen in that simplest of games, Scissors, Paper, Stone. If Ann and Bob are playing this game, Ann might plan her strategy by analyzing Bob's playing patterns. If she notices that he plays scissors more often than paper or stone, she can capitalize by increasing how frequently she plays stone. Indeed, she could try the strategy of playing stone every time; if Bob failed to amend his strategy in response, this would be very effective. But most likely, he would increase how much he plays paper.

However, it may be that Ann's analysis of her opponent doesn't reveal any chinks in his armor. So far as she can see, he is picking from the three gestures at random each time. Without any pattern to exploit, what should Ann do? In this case, her best hope is to ensure that her own play is similarly impregnable: she should play at random each time too. This is an example of a Nash equilibrium. In such a situation, even when each player is fully aware of how the other is playing, there is no way they can improve their own strategy.

OPPOSITE Games of strategy, including chess, checkers, backgammon and Go have a long history around the world. It was in these contexts that game theory first developed. Today it extends far beyond games, with relevance across the social sciences.

Games and conflict

Modern game theory began in 1944, with the publication of *Theory of Games and Economic Behavior* by Oskar Morgenstern and John von Neumann. One of the insights of this book was that equilibria arise unexpectedly, in a wide variety of situations. In 1951, John Nash extended this analysis and provided game theory with the standard formalization which continues to be used today. He lifted Morgenstern and von Neumann's theorem to new heights, proving that every possible non-cooperative game must contain an equilibrium.

In the context of the Cold War, game theory grew from a minor branch of mathematics into a notorious player on the international stage.

Around the time that Nash was doing his celebrated work, agencies outside mathematics were beginning to take notice of the potential offered by a science of strategy. Working for the US military, the RAND Corporation (standing for Research And Development) turned to game theory to investigate tactics of nuclear deterrence. In the context of the Cold War, game theory grew from a minor branch of mathematics into a notorious player on the international stage.

Among others, RAND employed the game theorists Merrill Flood and Melvin Dresher, who in 1950 provided one of the subject's most famous case studies: the prisoner's dilemma. This fictional story has a paradoxical feel to it, which served to correct a misunderstanding about equilibria, namely that they need be in any way optimal. Flood and Dresher's prisoners demonstrated that that need not be the case.

The prisoner's dilemma

Flood and Dresher contemplated two people (Ann and Bob again) who are arrested for a burglary which they did indeed jointly commit. They are held in separate police cells, where each ponders how to get him or herself off the charge, indifferent to the fate of their accomplice. Perhaps, Ann muses, she should cooperate with the police. If she helps them convict Bob, might they reward her with a reduced sentence? On the other hand, if she and Bob both keep quiet, perhaps neither will be convicted. What to do?

Meanwhile, the police have a problem. The current evidence is only enough to prosecute the pair for trespass. For burglary convictions to succeed, they need at least one suspect to cooperate. So the police make the following offer to each: if you confess and your accomplice does not, you will go free, while your accomplice will receive 10 years in prison. If you both confess, you will each be sentenced to 8 years. If neither of you confess, you will both be convicted of trespass, and each of you will receive 6 months in prison.

The two burglars are trying to minimize their individual jail time. From this perspective, the optimal strategy for the pair is for neither of them to confess. But this is not an equilibrium. If Ann expects Bob to stay silent, she has everything to gain by talking. So the Nash equilibrium—and likeliest outcome—is for them both to confess. But this leads to the largest total amount of jail time.

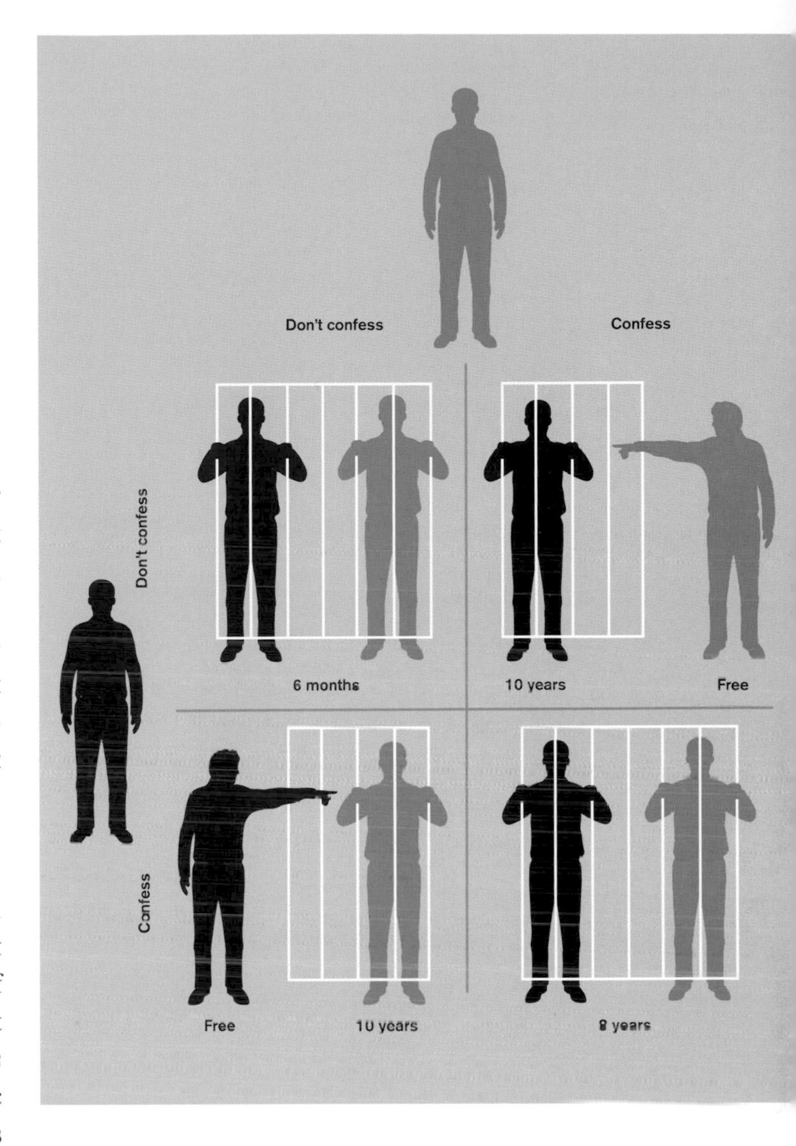

ABOVE A prisoners' dilemma. The optimal result for the pair of prisoners is the strategy of non-confession, landing them each with 6 months in prison. But the equilibrium point is where they both confess, getting them 8 years each.

Artificial intelligence

Strategy—the ability to plan ahead, thinking through different possible outcomes—is part of what makes us human. In recent years, artificial intelligence has become a major goal of scientific research, and game theory has been pulled into the mixture of theoretical and technological disciplines working toward it.

Claude Shannon, writing in 1950, noted that "chess is generally considered to require 'thinking' for skilful play" and suggested that if a computer could be programmed to play chess, it would mark progress toward other tasks requiring intelligence. For this reason, the defeat of grandmaster Garry Kasparov by IBM's Deep Blue in 1996 was a milestone in AI research. A more recent breakthrough was the unbeatable strategy for checkers derived in 2007 by a team of computer scientists led by Jonathan Schaeffer.

74 Exotic spheres

BREAKTHROUGH MILNOR FOUND WAYS TO REVEAL THE SECRETS OF THE SPHERES IN HIGHER DIMENSIONS, REVEALING SHAPES THAT ARE SPHERICAL FROM ONE PERSPECTIVE, BUT NOT FROM ANOTHER.

DISCOVERER JOHN MILNOR (1931–).

LEGACY MILNOR'S WORK LAUNCHED THE SUBJECT OF DIFFERENTIAL TOPOLOGY, WHICH HAS SINCE MADE DEEP DISCOVERIES ABOUT SHAPES IN HIGHER DIMENSIONS. HOWEVER, 4-DIMENSIONAL SPACE REMAINS MYSTERIOUS.

When are two shapes really the same? This question lies at the heart of many mathematical disciplines. Of course, any answer must depend on what is meant by the phrase "the same." In 1956, John Milnor discovered a shape that was the same as a sphere according to one definition, but not according to another. This exotic sphere spawned a whole new approach to geometry.

Every branch of geometry comes with its own particular notion of sameness. At the simplest level, in most contexts mathematicians do not distinguish between shapes that are identical but in different locations. But some approaches are much bolder in which shapes they class as being the same.

One of the widest notions of sameness arises in the subject of topology. Here squares, triangles and circles are all deemed to be the same, while cubes, octahedra and cylinders are all just spheres. The basic idea is that two shapes are topologically the same if one can be pulled, twisted or stretched into the form of the other (though without ripping or gluing).

OPPOSITE Spheres are among the most beautiful and symmetrical of all shapes. Understanding spheres, and the way that they can morph in to other shapes, has been an enduring mathematical theme over the last century. John Milnor's discovery of exotic spheres illustrated the huge depth and complexity of this question.

In between these two extremes, interesting situations can arise when two shapes fall between the gaps, being the same according to one definition, but not according to another. This is true of one of the most surprising shapes to be discovered in the history of mathematics: the exotic sphere found by John Milnor in 1956.

Morphing and smooth morphing

The perspective provided by topology has proved extraordinarily powerful since its origins in Leonhard Euler's work on the Bridges of Königsberg problem (see page 149). Today, topology is one of the most active branches of mathematics, encompassing many subdisciplines. But in 1956, John Milnor realized that a very slight refinement to the topological notion of sameness would make a spectacular difference to the entire subject.

In topology, the morphing that transforms one shape into another is permitted to be quite violent. It is legitimate to yank shapes around sharp corners, for example, as happens when a circle is pulled into the shape of a square. In the new subject of differential topology, the requirements were tightened. It was insisted that all morphing must be perfectly smooth. From this perspective, a circle is no longer the same as a square, since inserting sharp corners into a shape violates the requirement of smoothness. This is to be expected, though, since the square itself is not smooth, while the circle is. It will never be possible to bridge that gap with a smooth transformation.

BELOW Methods used by mathematicians to probe abstract shapes in high dimensions also have applications in studying and scanning more familiar shapes, such as in medical imaging technology and facial recognition software.

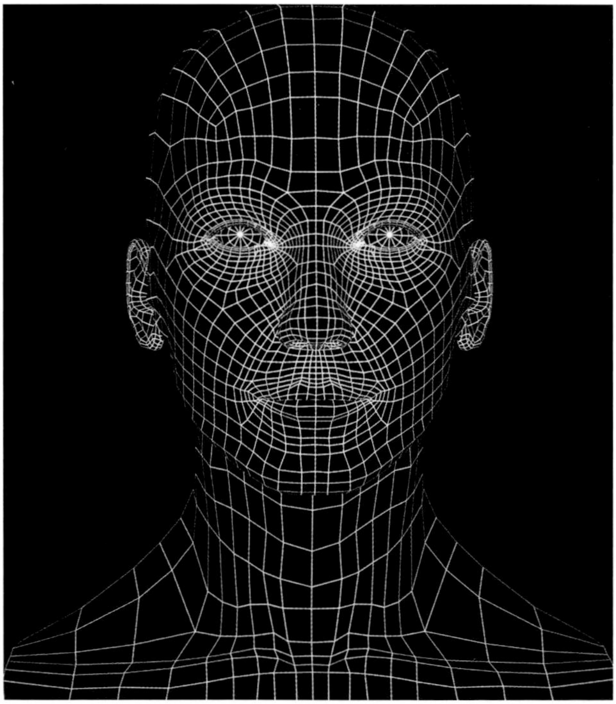

A deeper question asks whether it is possible to find two shapes that are both smooth, and that are the same from a topological perspective, but yet distinct from a differential perspective. It seems a strange scenario to contemplate, and indeed it is comforting to know that this sort of situation cannot arise in the geometrical settings which humans are most used to, namely shapes of one, two or three dimensions.

However, during the second half of the 20th century, mathematicians discovered that in the case of higher dimensions, these half-way shapes become a genuine possibility. The breakthrough was made by John Milnor in 1956. The smooth shape he found appeared at first sight to be a somewhat deformed example of a 7-dimensional sphere (the bigger brother of the familiar 2-dimensional surface). Indeed, from a topological perspective, that was exactly what Milnor had found, since a little pulling and twisting transformed his shape into ordinary spherical form. But this transformation was very violent, involving sharp folds and abrupt corners. The question became whether this process could be smoothed out. Could Milnor's shape be transformed into a sphere using only smooth processes? Shockingly, the answer was no. This, then, was a brand-new shape: topologically, it was a sphere, but from a differential perspective, it was not. It became known as an exotic sphere.

Exotic spheres and differential topology

Milnor's discovery fired the starting pistol for the new subject of differential topology. The immediate challenge was to understand whether this was a freakish one-off, or part of a larger pattern. Since 1956, numerous other exotic spheres have been found, in spaces of different dimensions. In 7-dimensional space, there are 27 exotic spheres in total, including the very first that Milnor had discovered. Each of them is topologically spherical, but differentially not so. In 11 dimensions, meanwhile, there are 991 exotic spheres. Yet the story is unpredictable: in 12 dimensions, for instance, there are no exotic spheres at all, just as there are none in one, two, three, five or six dimensions.

Since 1956, numerous other exotic spheres have been found, in spaces of different dimensions. In 7-dimensional space, there are 27 exotic spheres in total, including the very first to be discovered.

The toughest arena for differential topologists to battle with is 4-dimensional space. The biggest question in the subject asks whether or not there are any exotic spheres in four dimensions. As yet, no one has found one; yet nor has anyone proved the idea impossible.

On the other hand, it is now well understood that 4-dimensional space plays according to radically different rules from every other dimension. In the 1980s, using techniques from quantum physics, Simon Donaldson and Michael Freedman showed that 4-dimensional space itself comes in exotic varieties. Ordinary 4-dimensional space is known as R^4 (R representing the real numbers; see page 173). Donaldson and Freedman found spaces that are topologically identical to R^4, but differentially different from it. This was a truly astonishing result, since it was known that in every other dimension there is only ever one version of the ambient space. To complicate matters still further, in 1987 Clifford Taubes showed that there are actually infinitely many of these exotic R^4s. Making sense of the bewildering 4-dimensional world remains one of the biggest challenges to today's differential topologists.

75 Randomness

BREAKTHROUGH THE COMPRESSIBILITY OF A SET OF DATA, A CENTRAL IDEA FROM COMPUTER SCIENCE, WAS FOUND TO BE EQUIVALENT TO THE PHILOSOPHICALLY INTERESTING NOTION OF RANDOMNESS.

DISCOVERER RAY SOLOMONOFF (1926–2009), ANDREI KOLMOGOROV (1903–87), GREGORY CHAITIN (1947–).

LEGACY THE MODERN SUBJECT OF ALGORITHMIC INFORMATION THEORY ANALYZES THE INFORMATION CONTENT OF STRINGS OF NUMBERS AND PROBES THE LIMITS OF RANDOMNESS.

In the modern age, information is encoded through long strings of 0s and 1s, which flow along internet cables and ricochet off satellites. These binary stings encode everything from scientific data to the latest films and music. However, the theory behind such strings contains many surprises. How many 0s and 1s does one need to encode a piece of information? This question ties in with one of the most important and unexpected ideas in theoretical computer science: randomness.

When someone wants to email a heavy file, there are various pieces of software that can be used to reduce its size. But how can a file be made smaller without deleting anything? Data-compression programs work by exploiting patterns in the data, in the underlying string of 0s and 1s.

Patterns and compressibility

To take a simple example, if a string consists of a thousand 1s in a row (111111111111...), it is far more efficient to describe it and transmit the description than to waste time sending the whole original string. This procedure entails no loss of data; the receiver of the message will be able to reconstruct the original string exactly from the description.

This method works because the high level of structure within the data. The same approach will work for other strings such as 001001001001001001001... or 010010001000010000010000010... Each of these strings follows a predictable pattern which allows it to be dramatically compressed.

OPPOSITE Noise, such as TV static, is fundamentally unpredictable. The lack of any pattern in the data makes it hard to interpret or describe succinctly. The paradox of modern information theory is that this type of randomness is also hugely rich in information.

But what of a string which begins: 11001001000011111101101010100010 00100001011010001100001000110100 1...? It is far less clear whether this can be compressed. But in fact it can be whittled down to a single symbol, since these are the binary digits (or bits) of π. In 1960, Ray Solomonoff realized that the extent to which a string of bits could be compressed was an excellent measure of the total amount of information it contains. A string very rich in information cannot be compressed by much, while strings such as 1111111111... carry minimal information and can be compressed to almost nothing.

This same insight was expressed by two other researchers: Andrei Kolmogorov and Gregory Chaitin. We now talk about the Kolmogorov complexity of a binary string, meaning the shortest length to which it can be compressed.

Some numbers can be compressed into a few words of English but the English language contains only finitely many words, while the supply of numbers is infinite.

Berry's paradox

Suppose you wish to communicate a number to someone. The obvious approach is simply to send it digit by digit. But if the number is very large, this may not be efficient. For example, instead of sending "1606938044258990275541962 092341162602522202993782792835301376," it is far quicker to send "2 to the power of 200," meaning two multiplied by itself 200 times, which can be written very succinctly as 2^{200}.

It is not obvious, but there is a deep logical paradox lurking in these considerations. In 1906, the logician Bertrand Russell discussed this topic with a librarian at Oxford University by the name of G.G. Berry. The result of their conversation was the Berry paradox. Some numbers can be compressed into a few words of English: for example, "the result of multiplying the thousandth prime by the millionth Fibonacci number." But of course, the English language contains only finitely many words, while the supply of numbers is infinite. So at some point, we will reach numbers which cannot be so compressed. The formulation that Russell and Berry arrived at was "the smallest number that cannot be compressed into 12 words of English." On the one hand, this seems like an ingenious way to compress a very large number. Yet the paradox quickly reveals itself: the number described both can and cannot be described by 12 words of English.

The uncomputability of complexity

The Berry paradox has at its core the same phenomenon of self-reference as other classic paradoxes, such as Russell's own (see page 229). But

LEFT When compressing data such as digital photographs, *lossless* methods employ clever ways to compress the data without sacrificing accuracy. *Lossy* methods however entail a fundamental loss of information. Here the image is compressed at 300 dpi on the left, and 72 dpi on the right.

it has a consequence which is of profound importance in modern technology. It is by no means obvious, when presented with a very long number, whether or how much it can be compressed. So, the question is: is there some method which will start with a binary string and tell us the minimum length to which it can be compressed (which is to say exactly how much information it contains)?

In absolute terms, the answer is no; there are simply too many possible ways that compression could proceed. Indeed, if such a perfect compression program did exist, it would be able to defeat itself with a version of Berry's paradox. The upshot of this is something rather distressing: Kolmogorov complexity, the standard measure of information content, is a fundamentally uncomputable quantity.

Randomness and Chaitin's Ω

The simplest strings are those like 1111111..., which can be compressed to almost nothing. At the opposite end of the scale, what do information-rich, uncompressible strings look like? The answer is complete randomness. An absolutely incompressible string will also be absolutely unpredictable. At each stage, so far as the reader can predict, the chance of the next bit being a 0 will be exactly 50%. This type of string may emerge through a truly random process, such as tossing a fair coin.

It is hard to describe infinite random strings in any detail, since by definition they have no quick descriptions. Nevertheless, in the 1960s, Gregory Chaitin was able to identify an individual random string which became known as Ω (or omega). The number Ω can be interpreted as the probability that a randomly selected Turing machine (see page 281) will halt, rather than run forever.

76 The continuum hypothesis

BREAKTHROUGH PAUL COHEN DEVELOPED A WAY TO BUILD NEW MATHEMATICAL UNIVERSES TO ORDER. HE SHOWED THAT IT CANNOT BE RESOLVED ONE WAY OR ANOTHER WHETHER A CERTAIN LEVEL OF INFINITY EXISTS.

DISCOVERER GEORG CANTOR (1845–1918), PAUL COHEN (1934–2007).

LEGACY AMONG THE FIRST FORMALLY UNSOLVABLE PROBLEMS, COHEN'S WORK TRANSFORMED MATHEMATICIANS' ATTITUDES TO THE LOGICAL FOUNDATIONS OF THE SUBJECT.

The discovery by Georg Cantor in 1890 that there are many different levels of infinity was one of the most dramatic in the history of mathematics (see page 217). But like so many great scientific breakthroughs, it posed as many questions as it answered. One particular question was a torment to Cantor, and he died frustrated at his failure to produce a solution. This enigma became known as the continuum hypothesis, and it was not resolved until 1963. Its eventual solution would again rock the foundations of mathematics.

The continuum hypothesis was established as one of the great problems of the age at the start of the 20th century. David Hilbert was one of the most influential mathematicians of the age, and in 1900 he gave a speech outlining his priorities for the coming century. His list of 23 problems included some, such as the Riemann hypothesis, which remain just as mysterious today as they were when he spoke (see page 206). Others, such as Kepler's conjecture, eventually succumbed to the ingenuity of mathematicians armed with 20th-century technology (see page 373). The first problem on Hilbert's list was Cantor's continuum hypothesis.

OPPOSITE Georg Cantor showed that infinity is not a single entity, but an infinite hierarchy. The challenge he left behind was to understand how the different levels of infinity relate to each other.

An infinite mezzanine

Georg Cantor had split the infinite realm into different levels. Foremost among them is countable infinity. The name derives from the whole numbers—the primary tools for counting since time immemorial. Taken together, the whole numbers form a countably infinite set:

{1, 2, 3, 4, 5, …}. Meanwhile the set of real numbers, that is to say the collection of all possible decimal strings, also form an infinite collection. The level of infinity this occupies is the *continuum*. Cantor had proved, to the astonishment of his contemporaries, that the continuum is a genuinely bigger infinity than the countable level. But the question that had so frustrated him was this: is there an intermediate level between the two? Cantor invested a huge amount of time and energy attempting to rule out the possibility of any such halfway house; but he could not.

Cohen's theorem was that the continuum hypothesis is formally undecidable from all the usual rules of mathematics. In other words, no one had been able to provide either a proof or a disproof, because there could never be one!

As other thinkers began to investigate Cantor's work, his set theory assumed a foundational status in mathematics, as the bedrock on which the rest of the subject could be built. Yet the set theorists who achieved this feat were not able to resolve the question of the continuum hypothesis. No infinite set they could find fell in this gap. Every infinite set anyone could think of was either countable, or had the same size as the continuum, or was larger than either, sitting on a higher rung in the infinite ladder.

The first person to shine some light on the continuum hypothesis was Kurt Gödel. In 1940, Gödel showed that assuming the continuum hypothesis to be true did not introduce any contradictions into mathematics. This was a long way short of a proof, but it meant that it was at least consistent for the continuum hypothesis to be true. The search for a proof, so as finally to establish the truth of the matter, continued.

Cohen's forcing

In 1963, Paul Cohen finally solved the riddle of the continuum hypothesis, though not with the proof that logicians had long hoped for. A New Yorker by birth, Cohen was working as a professor at Stanford University in California. His mathematical background was not in logic, but in trigonometric analysis, where he had acquired a shining reputation. Turning his mind to set theory, Cohen was able to explain why an unsatisfactory state of affairs had endured for so long. Cohen's theorem was that the continuum hypothesis is formally undecidable from all the usual rules of mathematics. In other words, no one had been able to provide either a proof or a disproof, because there could never be one! The ordinary assumptions underpinning set theory were inadequate to resolve it one way or the other. It was an astonishing result, and Cohen demonstrated it in a very direct and dramatic way.

Earlier, Kurt Gödel was able to construct an entire mathematical universe in which the continuum hypothesis was true. Cohen's response was to construct another universe in which it was false. Both universes satisfied all the usual axioms and laws of set theory, so all conventional mathematical theorems would therefore appear identical in the two worlds. Yet in one universe, the continuum hypothesis was false: there was an intermediate level between the countable and the continuum. It was there because Cohen had deliberately constructed the universe around it. In the other universe, built by Gödel, there was no such intermediate level, and the continuum hypothesis was true.

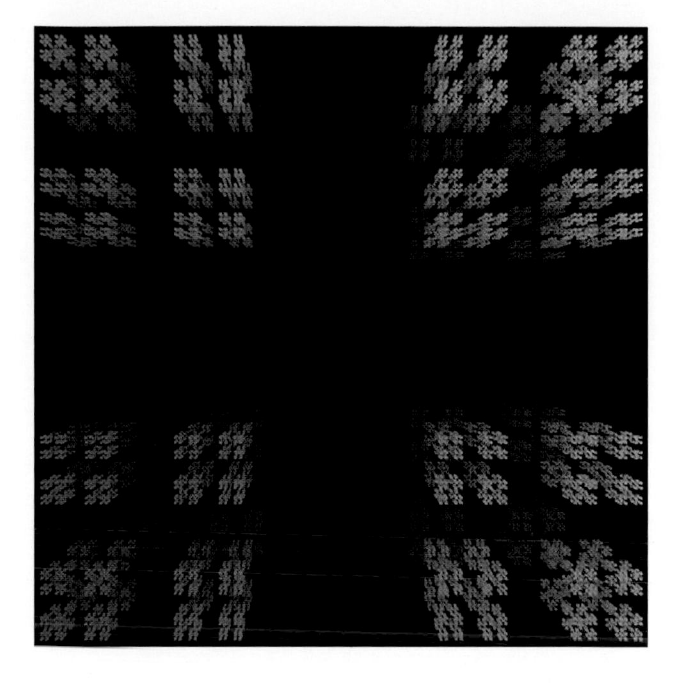

ABOVE Cantor dust is a fractal, formed by dividing a cube into 27 sub-cubes, then throwing away all but 8, and repeating this process ad infinitum for each remaining cube. The resulting shape contains continuum many points, but geometrically the shape has a volume of zero.

For his triumph, Cohen was awarded the Fields Medal, the top prize in mathematics. He remains the only logician to date to hold that distinction. President Johnson also awarded him the President's National Medal of Science for his "epoch-making results in mathematical logic." Cohen's proof of the independence of the continuum hypothesis threw mathematical logic back into the limelight. In particular, the method that he had used to construct his set-theoretic universes, known as forcing, became one of the most powerful tools available to logicians for probing the logical underpinnings of mathematics.

Since its invention, mathematicians have employed Cohen's forcing method to build a galaxy of different mathematical systems, each differing from the others in subtle ways. The continuum hypothesis was one of the earliest examples of a formally undecidable mathematical statement. Logicians now know of a great many examples, and Cohen's methods continue to be used to find ever more.

77 Singularity theory

BREAKTHROUGH SINGULAR POINTS ARE PLACES WHERE ORDINARY GEOMETRY BREAKS DOWN. HIRONAKA DEVISED A PROCEDURE FOR EXCISING SINGULARITIES FROM SHAPES.

DISCOVERER HEISUKE HIRONAKA (1931–), RENÉ THOM (1923–2002).

LEGACY TODAY'S MATHEMATICIANS AND PHYSICISTS CONTINUE TO ANALYZE THE MANY DIFFERENT TYPES OF SINGULARITY.

In mathematics, as in life, there are situations where the usual laws fail to apply. This phenomenon is also well known in physics. When a large star collapses to form a black hole, many of the familiar features of physics disappear, as matter, light, space and time are compressed down into a single point known as a singularity. The same was true at the very beginning of our Universe, at the time of the Big Bang from which everything we know emerged. But not all singularities are quite so explosive. This phenomenon has its roots in a much simpler setting, namely the geometry of curves.

Perhaps the simplest shapes are straight lines. Of course, these are perfectly smooth and regular, with no kinks, crossings or singularities in sight. At the next level of complexity come circles, along with the conic sections of Apollonius (see page 57). Though their geometry is subtler, these shapes are also perfectly smooth and regular, with no sharp corners or crossings to be seen.

OPPOSITE Optical caustics beneath shallow water. The surface of the sea is curved by waves and ripples, rather than lying flat. So when sunlight refracts, complex concentrations of light are formed at the seabed. Known as caustics, these are classic examples of geometric singularities in nature.

However, one more step up in complexity brings a surprise. This is the domain of the cubic curves. Algebraically, these are defined by equations containing an additional cubic term, that is an expression such as x^3 or x^2y.

Cusps and crossings

It was Isaac Newton who, in 1710, first analyzed the shapes described by these formulae. Just as there are three types of conic section, so

Newton's work led to the discovery that there are essentially 78 different cubic curves. Strikingly, many of these exhibited a fundamentally new phenomenon. Unlike a straight line, a circle or a conic curve, cubics sometimes cross over themselves, or come to a cusp (a sharp point), rather than flowing smoothly.

These crossings and cusps are both examples of what geometers term singularities: points on a shape where the usual description fails to apply. A curve is traditionally defined as resembling a straight line, at least when you zoom in closely enough. But at a crossroads or cusp this is not true. Near a crossroads, it appears not as one but as two lines. Near a cusp, the corner appears ever sharper the closer one looks, never smoothing out to a straight line.

Resolving singularities

Such singularities play havoc with all the usual techniques of geometry. To get around them, for many years geometers tried to resolve these singularities. They looked for ways of modifying the original shape to create another which is better behaved, having no singularities at all. Yet this must be done so that the new shape remains equivalent to the original, in a very precise sense.

Catastrophe theory became hugely famous and was subsequently seized upon to explain all manner of phenomena, from the destruction of fish stocks to the growth of medieval cities.

Newton himself found a way to resolve singularities in the specific case of 1-dimensional shapes, that is to say curves. One valuable method is to blow up a singular point, which means replacing it with several new points. According to this philosophy, a crossroads can be thought of as a double point, and can then be lifted to two new points. With this done, the original shape, with its crossroads, may appear as a shadow of a new curve which is entirely smooth at every point.

This blow-up procedure works well for curves, but in higher dimensions shapes are susceptible to more sophisticated types of singularities, such as folds and swallowtails. Not all of them are so straightforward to resolve.

Hironaka's theorem

In the early part of the 20th century, Oscar Zariski, one of the founders of modern geometry, worked on techniques for resolving the trickier types of singularities that can occur in 2- and 3-dimensional shapes. But it was Zariski's student Heisuke Hironaka who produced the seminal result in the subject in 1964, for which he was subsequently awarded the

Fields Medal (the mathematical equivalent of a Nobel Prize). Hironaka showed that all singularities, in shapes of all dimensions, can be resolved. So any shape defined by an algebraic equation must be equivalent to another that is perfectly regular, where all the singularities of the original have been removed.

ABOVE This shadow contains many singularities, where curves cut across each other forming crossroads. Yet the object itself (a slinky) has no singularities, but is perfectly smooth throughout. The standard approach to resolving singularities in some shape is to search for another smooth shape which has the original as its shadow.

Catastrophe theory

This was wonderful news for geometers. In many situations, Hironaka's theorem means that they no longer need to worry about singularities at all. Whenever a singularity appears, the whole shape can simply be traded in for an equivalent that is almost identical, but where all singularities have been resolved. In other contexts, though, the story is a little different. In many physical situations, the singularity is the most interesting part of the entire situation.

In the late 1960s, René Thom began to develop a particular approach to the subject known as catastrophe theory. Thom's catastrophes were singularities of a simple type, which he successfully classified into seven fundamental forms. This theory became hugely famous and was subsequently seized upon to explain all manner of phenomena, from the destruction of fish stocks to the growth of medieval cities. Even Salvador Dalí was inspired to depict a swallowtail catastrophe in his final painting.

Though some of these applications were perhaps somewhat fanciful, singularity theory has indeed proved to have numerous scientific applications, from the gravitation of a black hole to optical caustics—the patterns of dappled light that flicker across the seabed, whose unique geometry was first noticed in 1654 by Christiaan Huygens. Singularities continue to be studied intensively around the world today, as mathematicians attempt to understand the variety of ways in which geometry can go wrong.

78 Quasicrystals

BREAKTHROUGH QUASICRYSTALS ARE PATTERNS WHICH EXHIBIT VERY UNUSUAL SYMMETRIES. THE MOST FAMOUS EXAMPLES ARE PENROSE TILINGS.

DISCOVERER ROBERT BERGER (1938–), ROGER PENROSE (1931–), ROBERT AMMANN (1946–94).

LEGACY RECENTLY, THE GEOMETRY OF QUASICRYSTALS HAS BEEN PROPELLED INTO THE SCIENTIFIC MAINSTREAM, WITH DAN SHECHTMAN'S DRAMATIC DISCOVERY OF QUASICRYSTALLINE SOLIDS IN CHEMISTRY.

Different ways of using shapes as tiles have long delighted visual artists, and since the work of Kepler in the 17th century (see page 113) they have intrigued mathematicians too. Yet, during the 20th century, some extraordinary new tilings were discovered that seemed to defy the accepted laws of symmetry. With the discovery of physical structures called quasicrystals, these ideas would go on to cause a revolution in chemistry.

Patterns, such as those found on the tiled floor of a Roman villa or Moorish palace, can exhibit various types of symmetry. This means that twisting, sliding or reflecting the entire floor leaves it looking the same. Johannes Kepler was the first to analyze the relationship between the shape of the tiles and the symmetry of the resulting patterns. All of Kepler's tilings have translational symmetry. This means that if you take a step forward or backward or to the side, the floor will look identical to where you started.

Translational symmetry

All the most common tilings have this property of translational symmetry, so when Fedorov classified all the possible patterns with this property (see page 221), it seemed as if the study of 2-dimensional tilings and patterns had reached its end. But in the 1960s, interest arose in new ways of tiling a floor with much less symmetry.

It is not difficult to tile a floor in a way which has no translational symmetry. A common phenomenon is to start with a centerpiece, perhaps

OPPOSITE The ideas underlying Penrose tilings have previously been investigated by decorative artists of the Islamic world. By the 15th century, sets of shapes known as "girih tiles" were used to create aperiodic, Penrose-style patterns such as this one in Darb-e Imam in Isfahan, Iran.

the head of a flower, and then have tiles radiating out from there, like layers of petals. The resulting tiling will have rotational symmetry: if you turn the flower, the floor will look the same. But if the directions of the petals are all different, and there is only one center to the design, there will be no translational symmetry. Such patterns are called aperiodic.

In 1964, Robert Berger discovered a brand-new way to tile a plane, one which was rather unusual. The shapes he used had an extraordinary property, namely that the resulting pattern had no translational symmetry at all. The existence of aperiodic patterns was not new; what was truly unexpected was that Berger's tiles could *only* be used to create aperiodic patterns. The shapes resisted all attempts to squeeze them into a translationally symmetric pattern. However they were put together, if you started on a patch of floor and started walking forward or backward or to the side, no matter how far you went, you would never find yourself on a patch identical to where you started.

Linus Pauling acidly remarked, "There is no such thing as quasicrystals, only quasi-scientists."

Penrose tilings

Berger's astonishing discovery required tiles of over 20,000 different shapes. The obvious question was whether it was possible to recreate the same phenomenon with a smaller set of tiles. This was the moment when the physicist Roger Penrose entered the story. During the 1970s, Penrose discovered several beautiful tilings, and others in the same vein were found by Robert Ammann. Each of them shared the property that Berger had isolated, but for much smaller sets of tiles.

The most famous Penrose tilings require only two different tiles—for example, in the case of the rhombs tiling, two different rhombuses. Critically, though, the edges are notched to limit the possible ways they can slot together. The result is a beautiful and complex pattern, with no translational symmetry, but a striking fractal appearance. One of the major open questions in this area is whether it is possible to create the same effect with tiles of just one single shape.

Shechtman's quasicrystals

In 1982, a dramatic discovery was made which suddenly brought this subject into scientific prominence. Dan Shechtman is a materials scientist who in 1982 made a discovery that overturned the foundations of his subject.

LEFT The design of Ravensbourne College, London, is a homage to Penrose tilings. This section is adorned with an aperiodic tiling or quasicrystal constructed from three basic shapes: two irregular pentagons and an equilateral triangle.

It is a fundamental mathematical fact that any 2- or 3-dimensional repeating pattern is limited in the types of symmetry it can exhibit. This fact is known as the crystallographic restriction theorem, and it was critical in Fedorov's analysis of wallpaper groups (see page 221). It imposes limits on the type of rotational symmetry that such a pattern may have. A square is a shape with rotational symmetry. Turning it by 90° leaves it looking identical. Repeating this operation four times brings the shape back to the starting point, so a square has rotational symmetry of order 4. The crystallographic restriction theorem says that a repeating pattern may only have rotational symmetry of orders 1, 2, 3, 4 or 6. This fact has been used countless times by chemists, as it dramatically limits the possible structures of a crystalline solid.

So, when Dan Shechtman announced that he had discovered an alloy of aluminum with rotational symmetry of order 5 (like that of a pentagon), the chemistry community were not merely skeptical—many were openly contemptuous. The double Nobel Prize winning chemist Linus Pauling was particularly dismissive, acidly remarking, "There is no such thing as quasicrystals, only quasi-scientists." Yet Shechtman was eventually vindicated, winning the Nobel Prize for chemistry in 2011.

The mathematical explanation for his finding was that the structures he had seen did not have the translational symmetry of an ordinary crystal. This allowed them to bypass the restriction theorem and exhibit such unorthodox rotational symmetry. These were precisely the 3-dimensional analogs of Penrose tilings. In subsequent research, chemists found hundreds of examples of quasicrystals.

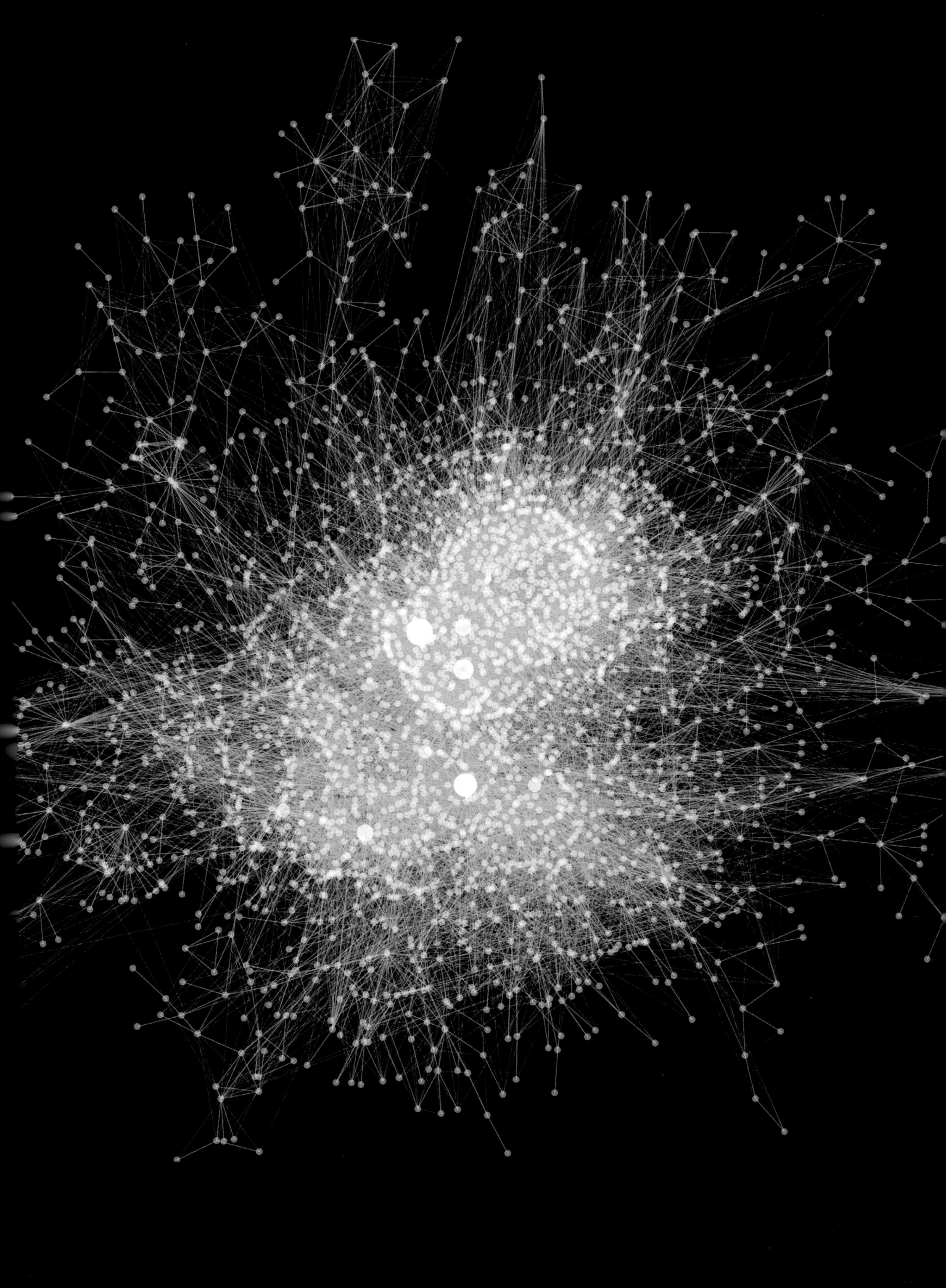

79 Friendship theorem

BREAKTHROUGH THE FRIENDSHIP THEOREM INVOLVES A DECEPTIVELY NATURAL CRITERION ABOUT NETWORKS OF FRIENDS. IT THEN SHOWS THAT ONLY NETWORKS OF A VERY SPECIAL FORM CAN SATISFY IT.

DISCOVERER PAUL ERDŐS (1913–96), ALFRÉD RÉNYI (1921–70), VERA SÓS (1930–).

LEGACY THE FRIENDSHIP THEOREM EXHIBITS THE ABILITY OF GRAPHS TO REPRESENT INFORMATION; IT ALSO STRIKINGLY ILLUSTRATES THE DIFFERENCE BETWEEN FINITE AND INFINITE MATHEMATICS.

Paul Erdős was one of the greatest mathematicians of the modern era. One of his most famous findings—proved, as so often happens, in collaboration with his friends—was a surprising result about the possible structures of networks of friends.

At a cocktail party, the resident mathematician notices that the group of people present has an unusual property. Each pair of guests have exactly one friend in common, who is also present at the party. Now, there is one simple way that this scenario could arise. It might be that the host of the party knows everyone else present, while the other guests are pairs of friends who know each other and the host, but no one else.

Friendship graphs

The surprising fact is that this is only one way that the friendship criterion can be met. This friendship theorem, as it became known, is a famous result in the subject of graph theory. A graph in this context is simply a collection of dots, some of which are joined together with edges, while others are not. Graph theory has its roots in Leonhard Euler's analysis of the bridges of Königsberg (see page 149), but it was during the 20th century that its power to distil a variety of situations to their logical essence became clear.

OPPOSITE An internet blog map, by Matthew Hurst. Blogs are shown as white nodes, and lines represent the links between them. Popular blogs have bigger nodes. The map reveals the community divided into two regions. The lower half, mainly socio-political blogs, is denser and more expansive than the upper half, which contains technical and gadgetry blogs.

To translate the cocktail party into a graph, we begin by representing each guest by a single dot. Then two dots are joined with an edge if the two people are friends. (We need to make the assumption that any two people are either friends or they are not; there are no degrees of friendship or one-way friends.)

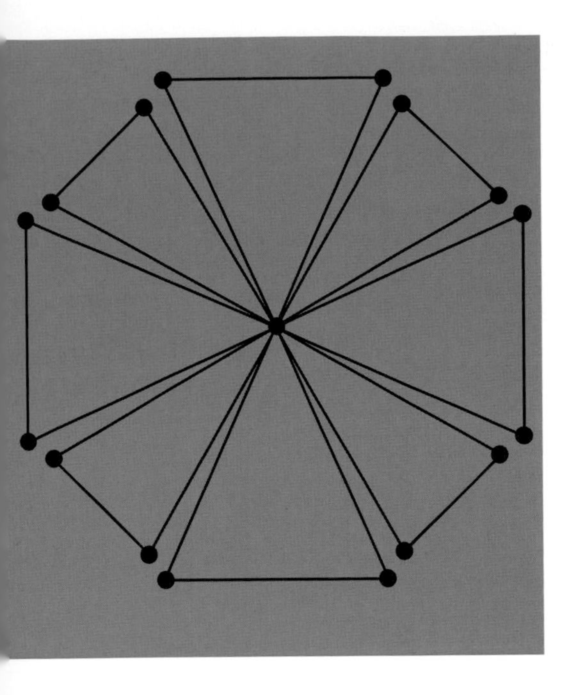

ABOVE A friendship graph connecting 17 people. The friendship theorem asserts, unexpectedly, that the only gatherings in which every pair of people have exactly one friend in common must take this form, with one person known to everyone, while the rest know each other in pairs.

In terms of the graph, the friendship criterion states that when you pick any two dots, there will be exactly one other dot which is connected to both of them. If you try experimenting with possible diagrams that satisfy this condition, you are soon forced to recognize that such a graph must have a very specific shape. There has to be one dot that is connected to all the others. Drawing this in the center, the remaining dots cycle around it in pairs, like the blades of a windmill. Yet actually proving that this is the required configuration was another matter.

This friendship theorem is a stark illustration of the difference between finite and infinite mathematical structures. As it stands, the theorem applies only to finite graphs (no matter how large). In the infinite realm, it fails spectacularly; it is possible to have an infinite friendship graph in which every two points share exactly one neighbor, but which is not centered on a single point. In fact, as Václav Chvátal and Anton Kotzig showed in 1974, there are a great many such graphs.

The result also illustrates the interconnectedness of different areas of mathematics. Since its first proof in 1966, the friendship theorem has been reproved several times, using a variety of algebraic, geometric and combinatorial techniques. But the original proof was by a trio of Hungarian mathematicians, including one of the 20th century's greatest minds, Paul Erdős.

Paul Erdős

Mathematicians, it has been suggested, fall roughly into two camps: problem-solvers and theory-builders. Erdős was the archetypal problem-solver. (Alexander Grothendieck, who developed algebraic geometry during the 20th century, is commonly cited as the ultimate theory-builder; see page 259.) Once he had a question in his mind, Erdős would wrestle with it, bringing to bear all the intellectual weaponry he could muster, until it eventually succumbed. Erdős was prolific in the extreme, publishing more papers in his lifetime than any other mathematician in history (the total exceeds 1416).

His exceptional mathematical ability went hand in hand with a highly unusual life. Erdős would travel the world, sleeping on the sofas of his colleagues, picking their brains and eating their food. His motto for this peripatetic lifestyle was "Another roof, another proof!"

Two of Erdős' most frequent collaborators were fellow Hungarians Alfréd Rényi and Vera Sós, with whom he proved the friendship theorem. (Rényi was not beyond amusing aphorisms himself, once remarking, "If I feel unhappy, I do mathematics to become happy. If I am happy, I do mathematics to keep happy.")

Erdős numbers

Paul Erdős was undeniably a hugely eccentric character, but as far as could be imagined from the stereotypical solitary genius. Indeed, he continues to hold the record for having the most number of co-authors of any mathematician: 509. This astonishing achievement is the basis of a piece of modern mathematical folklore, known as Erdős numbers. Appropriately enough, the joke takes the form of a piece of graph theory. Every mathematician is then represented by a dot, with Paul Erdős at the center. Two dots are joined with an edge, if the mathematicians in question have co-authored a paper together.

Erdős would travel the world, sleeping on the sofas of his colleagues, picking their brains and eating their food. His motto for this peripatetic lifestyle was "Another roof, another proof!"

Mathematicians are now ranked by their proximity to Erdős. The man himself has an Erdős number of 0, while his 509 co-authors each have Erdős number 1. People who have co-authored with one of these people, but not with Erdős himself have Erdős number 2 (there are thought to be 6593 such people), and so on. Someone's Erdős number is then nothing more than the length of the shortest path from them to the center. (Meanwhile mathematicians who have only written solo papers are said to have infinite Erdős numbers.)

80 Non-standard analysis

BREAKTHROUGH ROBINSON PIONEERED A NEW APPROACH TO MATHEMATICS, EMBRACING INFINITE AND INFINITESIMAL NUMBERS.

DISCOVERER ABRAHAM ROBINSON (1918–74).

LEGACY NON-STANDARD ANALYSIS CAUSED A STORM IN THE 1960S. TODAY IT IS A STANDARD PART OF THE MATHEMATICAL REPERTOIRE AND HAS BEEN EXPLOITED MANY TIMES TO PROVE THEOREMS IN MAINSTREAM MATHEMATICS.

Over the centuries, mathematicians investigated several different systems for geometry, number theory and algebra. In all cases, they have sought to probe the fundamental laws of the system being used. But in the early 20th century, mathematical logic began to shine some light on the strange relationships that are possible between a system and its underlying laws. In the 1960s, at the hands of Abraham Robinson, this developed into a brand-new approach to mathematics: non-standard analysis.

Most branches of mathematics involve a central object, known as a structure. This is where the action takes place. For algebra, the most important structure is the system of complex numbers (see page 105). For geometry, a crucial structure is the system of real numbers (see page 173). In arithmetic, it is the system of whole numbers. Other branches of mathematics, such as group theory and abstract algebra, consider other structures. In each case one of the pressing challenges is to understand the basic laws that describe the structure.

Understanding mathematical structures

The real numbers, for example, were finally understood through an analysis of the intermediate value theorem (see page 175). With this completed, an analysis of points and curves could finally move beyond the Euclidean paradigm and acquire a new mathematical sophistication. The intermediate value theorem was not simply a useful fact about real numbers; it was their main defining characteristic.

OPPOSITE An artistic representation of self-replicating nanorobots, designed to function at the molecular level. With the dawn of non-standard analysis, mathematicians were finally able to talk about *infinitesimally* small objects in a precise and meaningful way.

In the same way, in the study of the complex numbers, the defining moment was Gauss's proof of the fundamental theorem of algebra (see page 165). This, too, was more than merely a wonderful theorem. Gauss put his finger on exactly what it was that made the complex numbers so special: he discovered the fundamental law of the complex numbers.

Although seemingly more elementary, the system of whole numbers in which arithmetic is performed is far more difficult to axiomatize. This was witnessed by the famous incompleteness theorems of Kurt Gödel (see page 277). Nevertheless, some laws for arithmetic are known, although, of course, all are incomplete.

Model theory

In all these arenas, the general pattern was the same: to understand a mathematical object such as the real numbers, through its underlying laws or axioms. But this relationship, between a structure and its axioms, is itself something which can be studied. In the early 20th century, logicians such as Leopold Löwenheim and Thoralf Skolem began to do exactly that. Rather than focusing on one particular domain of mathematics, they considered this question in abstract, probing the relationship between general sets of laws and the structures which obey them.

The discovery of non-standard models was a fascinating development to logicians, but initially it was not of much importance outside logic. After all, it was the standard models which most mathematicians wanted to understand.

Löwenheim and Skolem's analysis uncovered a deep and unexpected fact: in general terms, a set of laws does not pin down a structure uniquely; a set of laws will admit a large variety of structures which obey it. These are called models of the axioms. Of course, among these models will be the object you most expect to see: the one you first thought of (the ordinary real numbers, for example). This is the standard model. But alongside this will be a large number of non-standard models, which resemble the real numbers in obeying the axioms, but differ in other, unexpected ways.

Infinitesimals return

The discovery of non-standard models was a fascinating development to logicians, but initially it was not of much importance outside logic. After all, it was the standard models which most mathematicians wanted to understand. Yet there was a suggestion from the history books that non-standard models might have a useful role to play.

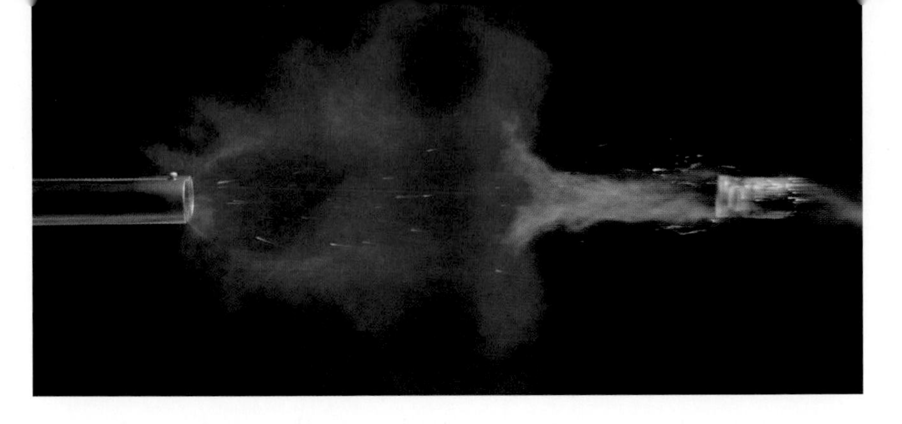

LEFT To calculate the velocity of a speeding object, early thinkers divided an infinitesimal distance by the infinitesimal time it took to travel. It was only with the emergence of non-standard analysis that this procedure received a theoretical justification.

During the earliest work on calculus, by Isaac Newton and Gottfried Leibniz (see page 133), and even in Archimedes' work (see page 53) thousands of years earlier, arguments were mostly phrased in terms of infinitesimals. These were notional infinitely small numbers. To calculate the speed of a traveling object at a single moment in time, the technique was to divide some infinitesimal distance traveled by the infinitesimally short length of time it took. Usually the result would be an ordinary real number, which quantified the object's speed at a specified instant.

With a more complete understanding of the real numbers, such arguments were consigned to the scrap heap. There are no such things as "infinitely small numbers": every number is either zero or has a definite positive size. To be more precise, it is certainly true that there are no infinitesimals among the real numbers. But non-standard models, such as the hyperreal numbers, do indeed contain infinitesimal numbers. Similarly, non-standard models of the whole numbers contain infinite numbers, while the standard whole numbers are all finite quantities by definition.

Non-standard analysis

In the 1960s, Abraham Robinson pioneered a revolutionary approach to mathematics. His idea was to use these exotic non-standard structures to shine some light on the models mathematicians really care about; the standard real, complex and whole numbers. His approach became known as non-standard analysis. At its center was a piece of logical trickery known as the transfer principle. If you can prove that a non-standard model exhibits some interesting behavior, in many cases the same thing will automatically follow for the ordinary model. This was essentially the idea exploited, albeit unwittingly, by Newton and Leibniz. Robinson turned it into a precise statement and a valuable weapon for the mathematicians who followed him. Today, Robinson's non-standard analysis is a common tool used by mathematicians to understand properties of ordinary mathematical structures.

81 Hilbert's tenth problem

BREAKTHROUGH DAVID HILBERT ASKED FOR AN AUTOMATIC WAY TO SOLVE EQUATIONS INVOLVING WHOLE NUMBERS. THE MRDP THEOREM ESTABLISHED THAT THERE COULD NEVER BE ONE.

DISCOVERER DAVID HILBERT (1862–1943), JULIA ROBINSON (1919–85), HILARY PUTNAM (1926–), MARTIN DAVIS (1928–), YURI MATIYASEVICH (1947–).

LEGACY FINDING A COMMON LANGUAGE BETWEEN THE SUBJECTS OF NUMBER THEORY AND LOGIC WAS A MILESTONE IN UNDERSTANDING THE LIMITS OF COMPUTATION.

Many of the biggest questions in number theory, such as Fermat's Last Theorem (see page 369) and Catalan's conjecture (see page 377), share the same basic format. Each begin with an equation. The challenge is to decide whether or not there are any whole numbers which satisfy it. In 1900, David Hilbert asked mathematicians to devise an automatic procedure for answering all such questions.

Fermat's Last Theorem asserts that a certain equation ($a^n + b^n = c^n$) is not obeyed by any whole numbers. Catalan's conjecture makes a similar claim for a different equation ($a^n - b^m - 1$). These are both Diophantine problems, taking their name from the Greek mathematician Diophantus. These have been a major strand of mathematical research since he contemplated them in his *Arithmetica* around AD 250 (see page 69). To study Diophantine equations is to unravel the possible relationships between whole numbers. History is very clear on one point: Diophantine problems, even those which seem simple, can be tortuously difficult to resolve. And to make matters worse, they tend to be difficult in different ways. Each requires its own set of sophisticated tools, which are almost useless when applied to other Diophantine problems.

David Hilbert found this piecemeal approach inadequate. He did not want to toil away solving Diophantine equations one at a time; he wanted a method to solve them all. In 1900, Hilbert famously presented a list of 23 problems for the mathematical community to focus on over the 20th

OPPOSITE An artwork inspired by Wang tiles. Given a collection of shapes, it is natural to ask whether they can be used to tile the plane. Work by Hao Wang and Robert Berger established that this, like Hilbert's 10th problem, is an incomputable question.

century. In the tenth, he called for a revolution in number theory: "Given a Diophantine equation ... devise a process according to which it can be determined by a finite number of operations whether the equation is solvable in rational integers [whole numbers]."

Algorithms and numbers

Hilbert was writing at the beginning of the 20th century, 40 years before the first digital computer. But his challenge would have major repercussions for the science of computation. In time, following the groundbreaking work of Alan Turing and Alonzo Church in the 1940s (see page 281), it became clear that the "process" Hilbert had envisaged should be an algorithm, which is to say a computer program.

Turing's work suggested a need for caution. He showed that computer programs are not all powerful—they have limitations. It might be that Hilbert was dreaming of the impossible. So the question became whether or not there could ever be a single computer program capable of solving all Diophantine problems. From being a call to arms for number theorists, Hilbert's tenth problem had evolved into a deep question for logicians, probing the limits of algorithms.

Turing's work suggested a need for caution. He showed that computer programs are not all powerful–they have limitations. It might be that Hilbert was dreaming of the impossible.

Computability and enumerability

The eventual solution to Hilbert's problem would hinge on a subtle distinction between different types of algorithm and cut to the heart of what it means to be "computable." Some collections of numbers can easily be described by a computer. One such is the set of square numbers: 1, 4, 9, 16, 25, 36, and so on. If I want to know whether or not 625 is included in the set, that question can easily be answered by an algorithm. A competent programmer could easily write a program to answer queries like this. Yes, 625 is in. No, 714 is out, and so on. So, this collection is deemed to be computable.

As well as being computable, this set is also Diophantine, meaning that it is described by an equation, namely $x = y^2$. If x and y are whole numbers satisfying this relationship, then x must be a square number. Many of the classes of numbers that mathematicians study are both Diophantine and computable. This was the first hint of a deep connection between number theory and logic. Might it be true that every Diophantine set is computable? If so, Hilbert's tenth problem would automatically follow:

it would be straightforward to write a computer program of the kind Hilbert wanted.

However, computer scientists discovered a subtle variant of computability. An *enumerable* set is a collection of numbers that a computer program can list: 5, 19, 804, 13, 22, and so on. At first glance, this seems a very similar definition. Yet the difference between a computable set and one which is merely enumerable cuts surprisingly deep. The distinction is this: in an enumerable set, you can never be sure when a number is out. If I want to know whether or not 625 is included, all I can do is wait until it appears in the list, at which stage I know that the answer is "yes." But if it fails to appear, there is no moment when I can be confident that the answer is definitely "no." No matter how long I wait, there is always a chance that 625 will be the next number on the list. This may seem like the sort of obstacle which an ingenious programmer should be able to bypass, but it is not so. Today's mathematicians are aware of many sets of numbers which are enumerable, but not fully computable.

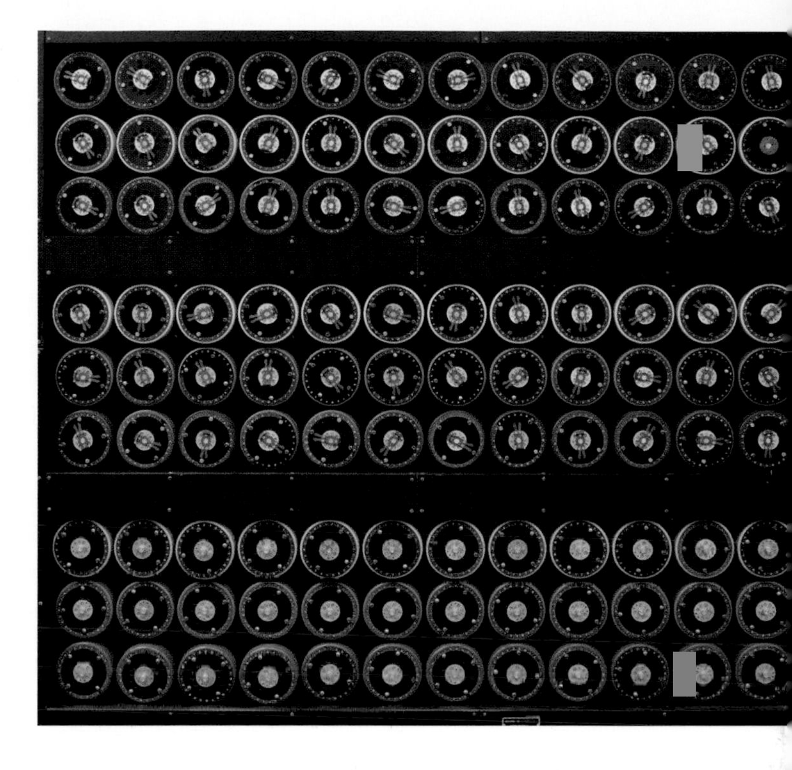

ABOVE Bombe decryption machine, designed by mathematician Alan Turing. This device was used during World War 2 to decrypt German communications enciphered with the Enigma machine. The bombe allowed messages to be deciphered, and was an early example of a laborious procedure being speeded up by automation.

The MRDP theorem

It is not hard to prove that every Diophantine set must be enumerable. But in the 1940s, a group of logicians began to believe that the opposite might also be true—that every enumerable set is actually Diophantine. This would be an immensely profound and unexpected result. If true, it would kill Hilbert's tenth problem, since it would imply the existence of Diophantine sets which are not computable, but merely enumerable. For over 20 years, Julia Robinson, Martin Davis and Hilary Putnam progressed toward a proof of this momentous theorem. The final hurdle was overcome in 1970 by Yuri Matiyasevich. The resulting MRDP theorem (after the surnames of the four researchers) meant that Hilbert's ambitious plans to automate a central plank of number theory can never be fulfilled.

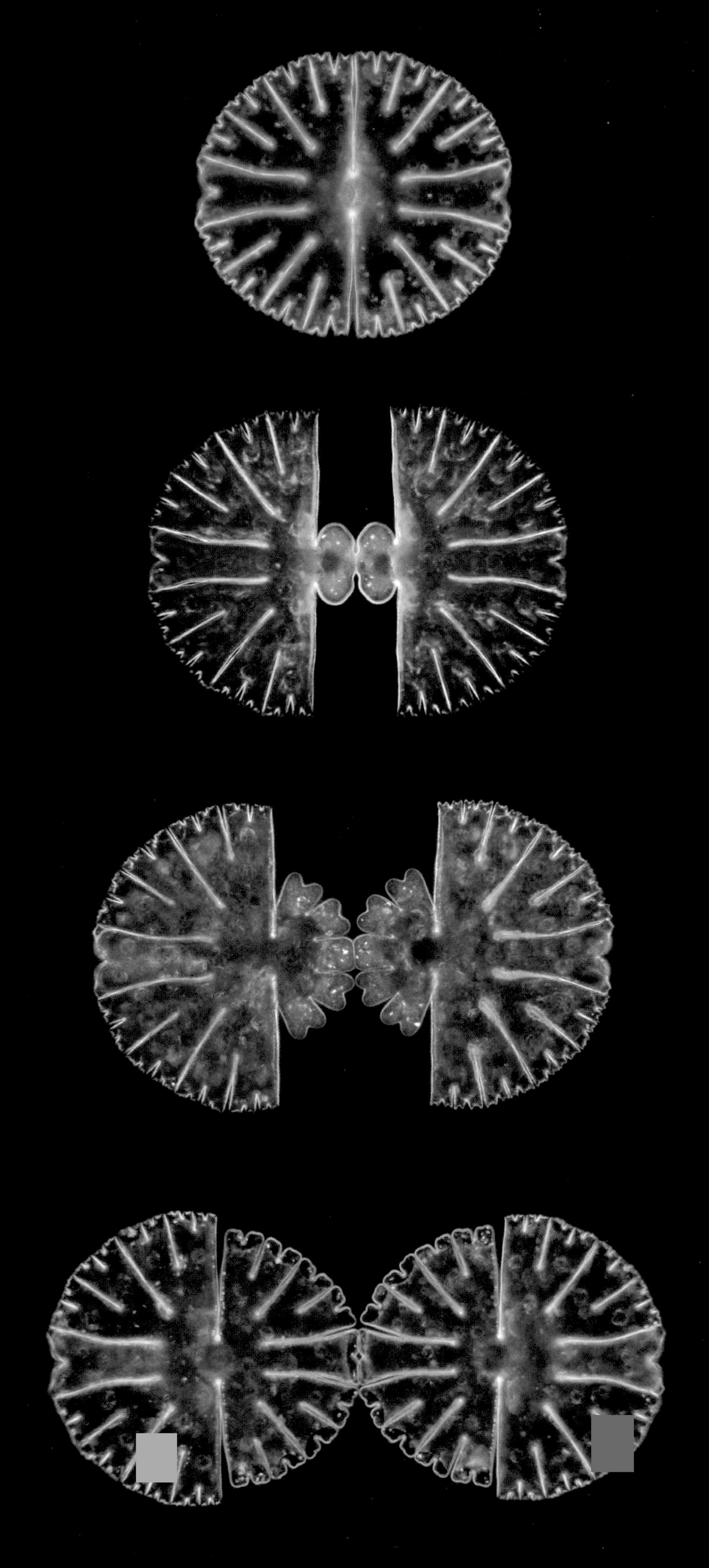

82 The game of Life

BREAKTHROUGH SEEKING TO UNDERSTAND HOW DEEP COMPLEXITY CAN EMERGE FROM THE SIMPLEST STARTING POINTS, JOHN CONWAY DEVISED A REMARKABLE SELF-PLAYING GAME KNOWN AS LIFE.

DISCOVERER JOHN CONWAY (1937–).

LEGACY LIFE WAS THE FIRST CELLULAR AUTOMATON. THE MANY VARIATIONS ON THE THEME WHICH HAVE SUBSEQUENTLY EMERGED ARE A MAJOR TOPIC OF STUDY.

In 1970, the mathematician John Conway set himself a challenge to probe the roots of complexity. He wanted to find the simplest possible set of rules that would nevertheless exhibit unpredictable, complex behavior. The result of his search was his most famous invention: the game of Life.

The rules to Conway's game are exceptionally simple. The action takes place on a grid, every cell of which is either "alive" or "dead." At the start of the game, the human player performs their only role: deciding which cells are initially set to "alive" and which to "dead." Then the game begins. Every second, each cell may change its state, depending on the eight cells surrounding it. The rules are:

A living cell will remain alive if it has 2 or 3 living neighbors.
Otherwise it will die.

A dead cell will come to life if it has 3 living neighbors.
Otherwise it will remain dead.

After the grid has been set up, the cells spring in and out of life according to these rules for as long as the game is allowed to run. Life's interest is in the relationship between the starting configuration and the long-term behavior of the grid. Sometimes the grid quickly settles down to a stable state. If the initial setting consists of a few isolated living cells, then in the second they will all die, and thereafter every cell in

OPPOSITE The stages of cell division of the Desmid (*Micrasterias thomasiana*), a green algae found in marshland. Conway's game of Life is one of the simplest yet most successful mathematical models of a biological process, and provided the starting point for much further research.

the grid remains stably dead. A 2×2 square of four living cells is also stable. Once that configuration is reached, it will remain for evermore. Other configurations may result in a repetitive loop: three living cells in a row will flicker back and forth between a horizontal and a vertical formation, for instance. But there are other simple configurations whose long-term development is hugely more complex and harder to describe.

Quantifying complexity

Conway realized that his game was complex; but how complex? In 1970, he asked whether any initial configuration could lead to indefinite growth. If only a limited number of cells are living initially, must it be true that there is some upper limit to the number alive at any subsequent stage? Or might some configuration result in an unlimited number of living cells? Conway offered a reward of $50 for an answer to this question. Later that same year, the prize was claimed by Bill Gosper, with his discovery of the "Gosper Glider Gun." This configuration shoots out a small pattern called "glider" every 30 seconds. These gliders never die, but glide away across the grid. As the total number of gliders grows, so does the number of living cells. This was the first suggestion that Life was not just an intellectual amusement, but contained genuine mathematical depth.

The school of thought called "digital physics" posits that ultimately the Universe itself is a cellular automaton, and that space, matter, gravity and all of familiar physics are emergent properties of some underlying computation.

This was spectacularly confirmed in 1982, when Conway achieved his ultimate aim: he showed that the game is Turing complete. This means that there is no upper limit on its computational abilities. In theory, Life is equal in computational ability to the most powerful modern computer (see page 365). As Life's cult following continued to expand, researchers would test this computational ability by finding ways to run intricate calculations, such as searching for prime numbers, or even constructing programmable computers within the grid.

Cellular automata

Life was conceived as a game, and is now recognized as the first example of a cellular automaton. But by altering the geometry of the grid to change the number of cells which neighbor each other, by increasing the number of states beyond just "living" and "dead," and by adjusting the rules by which cells move into and out of these states, an infinite number of variations on the theme are possible. By no means all of them share the very special property of being Turing complete. Many are simple systems, incapable of anything except unproductive stability or endless repetition of the same sequence. Others are chaotic systems, in which

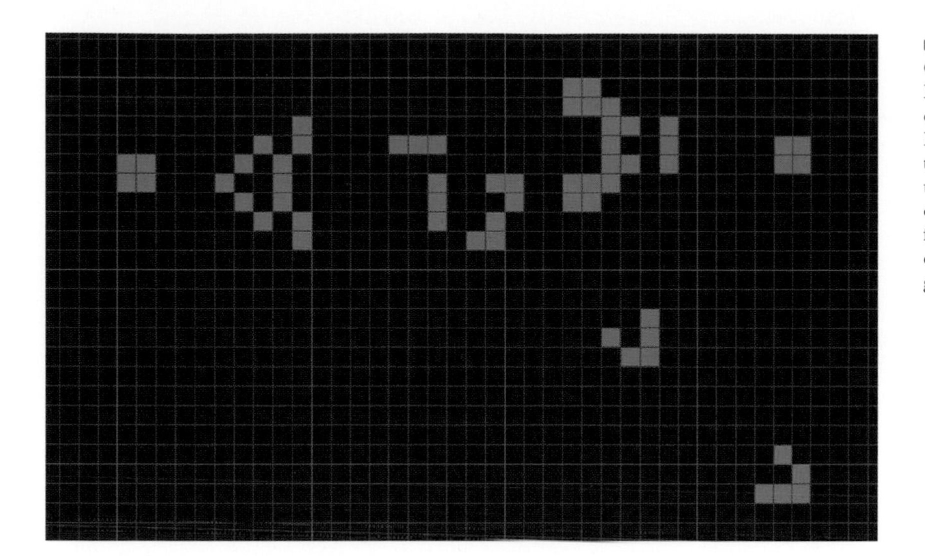

LEFT A Gosper Glider Gun in the game of Life. This pattern was discovered in 1970 by Bill Gosper, and was the first to produce an unbounded number of living cells in Life, firing out a never-ending stream of gliders (bottom right).

most initial configurations explode into wild unpredictability. These are too capricious for serious computation to be possible.

Yet between these extremes, other examples of Turing completeness have been found. The simplest is Rule 110, an automaton with only two states, where each cell has only two neighbors. It is a 1-dimensional automaton, a single line of cells which fizz between two states, according to laws even simpler than those of Life. Nevertheless, in 1985 Stephen Wolfram conjectured that Rule 110 should be capable of universal computation, and in 2002 Matthew Cook, while working for Wolfram Research, proved that this is indeed the case.

Computing the world

The computational power of Life and Rule 110 is as interesting philosophically as it is in technical, logical terms. Given that such systems are theoretically possible, an obvious question is whether any actually exist. The school of thought called "digital physics" posits that ultimately the Universe itself is a cellular automaton, or something similar, and that space, matter, gravity and all of familiar physics are emergent properties of some underlying computation. If so, this would transpose ideas such as algorithms and information from computer science to the very heart of physics. This idea continues to be explored, but the evidence is inconclusive. Meanwhile, researchers into artificial intelligence and artificial life have also latched on to the potential of cellular automata as efficient ways of modeling highly complex systems, from neurons firing within a human brain to organisms competing in an ecosystem.

83 Complexity theory

BREAKTHROUGH THE ABILITY OF COMPUTERS TO SOLVE PROBLEMS DEPENDS ON THEORETICAL SPEED LIMITS, KNOWN AS COMPLEXITY CLASSES. FOUNDATIONAL ANALYSIS WAS PERFORMED IN THE EARLY 1970S BY COOK AND LEVIN.

DISCOVERER STEPHEN COOK (1939–), LEONID LEVIN (1948–).

LEGACY COOK AND LEVIN POSED THE QUESTION OF WHETHER OR NOT $P = NP$. THIS NOW HAS A $1,000,000 PRICE TAG, COURTESY OF THE CLAY MATHEMATICS INSTITUTE, AND IS THE BIGGEST QUESTION IN THEORETICAL COMPUTER SCIENCE.

Over the centuries, mathematicians have proved themselves adept at finding solutions to all manner of tricky problems. But in some situations, a solution that is satisfactory to a mathematician may not be adequate for practical purposes. The codes that guard our bank accounts online, for instance, are easily crackable from a purely arithmetical perspective. But in computer science, the important distinction is between problems which are tractable on a manageable time scale and those which are not.

There is a well-known joke about a mathematician who wakes up to see his wastepaper basket on fire. Remembering that there is a bucket and a supply of running water in the bathroom next door, he thinks to himself, "A solution clearly exists"—and promptly goes back to sleep.

Of course, this is an unfair stereotype, but there is an element of truth behind it. If a mathematician wishes to solve a problem, he will usually be satisfied if he can find a method which is guaranteed to work. But in some situations this is not good enough. One is when your house is on fire, but a second is in computer science, where the fact that a problem has a theoretical solution is inadequate. It is also required that the solution can be implemented in the real world.

OPPOSITE An image from the Hubble Telescope's Ultra-Deep Field, revealing galaxies 13 billion light-years away, the deepest picture of the Universe so far. Previously, such time scales were the concern of astronomers alone, but the emergence of computational complexity theory has made them relevant in computer science.

Tractability and time scales

Suppose you wish to crack a ten-digit code. There is no problem in principle: just try 0000000000, then 0000000001, 0000000002, and

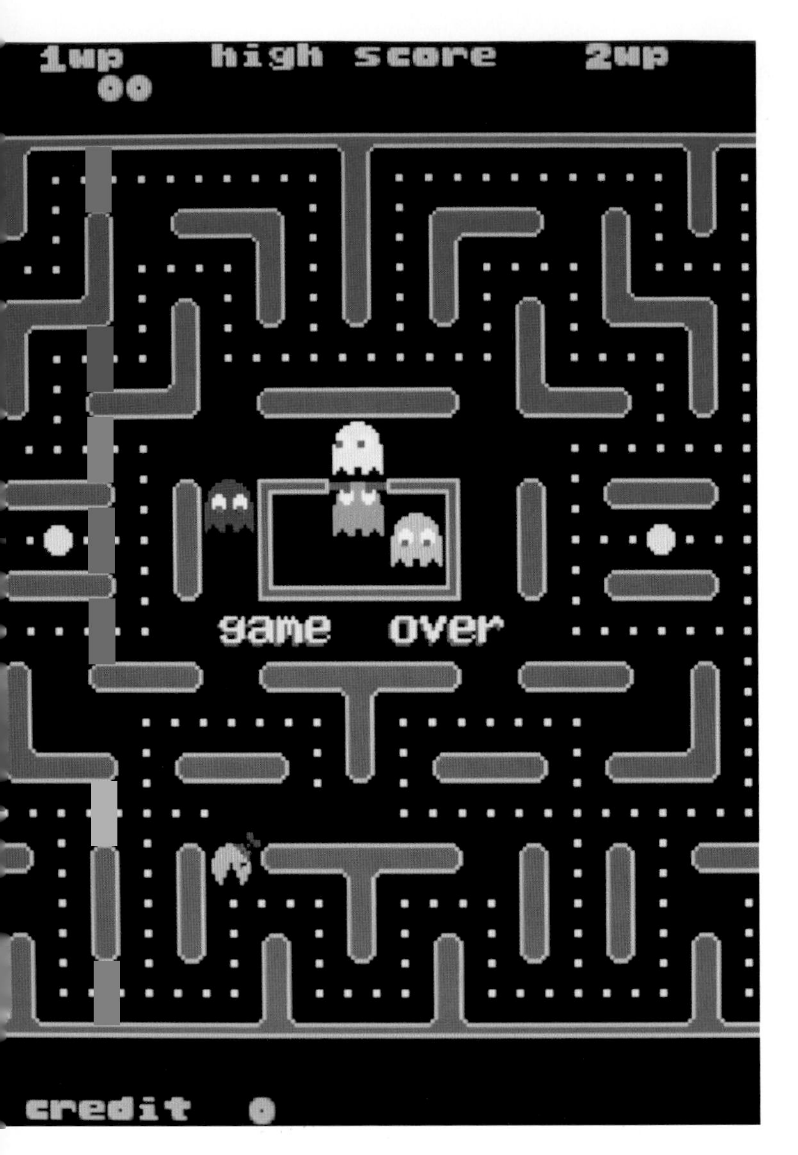

ABOVE In 2012, Giovanni Viglietta analyzed the computational complexity of several classic computer games. In particular, he established that NAMCO's Pacman is *NP*-hard, putting it in the same theoretical bracket as the traveling salesman problem.

so on. By the time you reach 9999999999, the code will have been cracked. Although this method is guaranteed to work perfectly, its drawback is obvious. If a computer can check 1000 possibilities a second, it will still take several months to complete the task. For a longer code of perhaps 100 digits, the procedure requires 10^{100} steps, which is to say 1 followed by 100 zeroes. At this stage, it ceases to be tractable at all, and will take longer than the lifetime of the Universe to complete (even with computers dramatically faster than those we have today).

This observation leads to a branch of study known as computational complexity theory, which emerged from Alan Turing's work on algorithms and Turing machines (see page 281). Essentially, it classifies mathematical tasks according to how long it will take a Turing machine (or, equivalently, a modern computer) to carry them out. Particularly significant work was carried out by Stephen Cook and, independently, Leonid Levin in the early 1970s, which posed the toughest question in the subject: whether or not *P* is equal to *NP*.

Computational complexity

The code-cracking procedure described above is very slow. For a ten-digit code, it requires 10^{10} steps to solve. For a 100-digit code, it will require 10^{100} steps, and for an n-digit code, 10^n steps. In the jargon of complexity theory, this problem requires exponential time. (On a quantum computer, it may be quicker: see page 365.)

Other problems can be solved more quickly. For instance, when your calculator adds together two numbers, for an input of two digits it might only need to carry out four steps, while an input of three digits might need nine. In general, an input of n digits might need n^2 steps. This is an example of a task which runs in polynomial time. The collection of all such tasks is known as *P*, and is the first of a great many complexity classes that theorists have discovered. As a general rule of thumb, *P* represents all those tasks which are tractable in the real world.

The subtlety of the whole subject comes from the fact that it can be deeply difficult to tell whether or not a given task is in *P*. Finding a quick algorithm certainly guarantees that the task is in *P*. But how can you establish that a problem is not in *P*? How can you rule out the possibility of someone finding a quick algorithm in future? This is really the central question in theoretical computer science.

N and *NP*

Cook and Levin each realized that the most pressing question concerned the relationship of *P* to another class known as *NP*. This essentially consists of tasks whose solutions can be verified quickly. Code-cracking is one example: it is quick to check a proposed solution. Another is the maximal clique problem.

In modern social media, people connect to each other online. One might ask the question: what is the largest clique one can find in such a network? A clique is a group of people where everyone is connected to everyone else. This seems a very difficult problem, since it is hard to know when one has found the best possible answer. However, if you ask whether or not a clique of size 100 exists, and then you are presented with a candidate group of 100 people, it is comparatively quick work to check that they do indeed form a clique. So this problem is in *NP*.

For a longer code of perhaps 100 digits, the cracking procedure requires 10^{100} steps. At this stage, the problem ceases to be tractable at all, and will take longer than the lifetime of the Universe to complete.

It is certainly true that every problem that is in *P* is also in *NP*: if a task can be completed quickly, it can be verified quickly. Going the other way, the expectation is that many problems are in *NP* but not in *P*, the clique problem being one candidate, and the traveling salesman (see page 341) being another. But as yet no one has proved this. Astonishingly, there is not a single example of a problem known to be in *NP* but not in *P*. Thus Cook and Levin's question of whether or not *P* = *NP* remains open today, and continues to be one of the largest in mathematics and computer science.

40
40

84 The traveling salesman problem

BREAKTHROUGH THE PROBLEM IS TO FIND THE SHORTEST ROUTE THROUGH A NETWORK OF CITIES. KARP PROVED THAT THIS IS *NP*-COMPLETE, MEANING THAT IT IS COMPUTATIONALLY DIFFICULT TO SOLVE, BUT QUICK TO VERIFY.

DISCOVERER RICHARD KARP (1935–).

LEGACY INSTANCES OF THE TRAVELING SALESMAN PROBLEM OCCUR IN OPTIMIZATION PROBLEMS THROUGHOUT SCIENCE AND ENGINEERING. SEVERAL APPROXIMATE ALGORITHMS HAVE BEEN FOUND, BUT IT REMAINS A MAJOR CHALLENGE.

A salesman is traveling around the country. He has 20 cities to visit in total, and between each pair of cities is a road he may take. But some roads are longer than others, and the salesman wants to arrange his trip to minimize the total distance he needs to cover. How should he proceed? This question has been considered by many mathematicians since the late 19th century, also arising in several branches of science, including the manufacture of microchips and DNA sequencing.

The obvious answer to the salesman's dilemma is to work out the various routes he could take, compare their lengths, and then pick the shortest. The trouble is that, with 20 cities, the total number of possible routes is over 100,000,000,000,000,000. So comparing them all is simply not feasible, even armed with a modern computer.

This traveling salesman problem, as it is known, is both mathematically interesting and of profound practical importance. Optimization problems appear in many spheres of work, such as in the design of public transport or the supply of gas and electricity to a town. Many optimization tasks are equivalent to a traveling salesman problem. Similar problems arise in the sciences, from genome-sequencing to microchip design, where a large body of material has to be organized or navigated in an efficient manner. In all these different scenarios, the fundamental mathematics is the same, with the roles of "cities" and "roads" played by various other entities. It may not be distance that needs to be minimized but some other quantity, the commonest being time.

OPPOSITE The traveling salesman problem poses a fundamental obstacle in many contexts including route-finding software. Finding the shortest route between two places is a tractable problem. For more complex journeys which visit several destinations it is usually necessary to settle for a route which is good enough, rather than optimal.

Graph theory and optimization

Today, the traveling salesman problem is formalized using the mathematics of graph theory. A graph is a collection of dots (or vertices), some of which are joined by edges. In the traveling salesman problem, the cities are the dots, and the edges are roads. Graph theory has its origins in the solution by Leonhard Euler of the Bridges of Königsberg problem (see page 149).

Richard Karp proved that the traveling salesman problem is fundamentally hard. It was only in the early 1970s that the technical framework was put in place to define what it truly means to be a "hard problem."

The question is whether there is a quicker way to proceed than a brute-force comparison of all possible routes. Perhaps an ingenious computer programmer could devise a method to home in on the shortest route without having to laboriously compare every available option? For many years, researchers contemplated this possibility. One promising approach is to begin at a city chosen at random, and then travel to its nearest neighbor. At each subsequent stage, the salesman should visit the closest city that he has not yet visited. It is a neat method, with the advantage that it is very quick to work out. The salesman only has to compare a small number of choices at each stage. But, as Karl Menger observed in the 1930s, this procedure usually does not produce the best possible result.

While some specific instances of the traveling salesman problem can be solved quickly, throughout the 20th century no efficient general method was found. This was despite the problem occurring in an increasing variety of contexts.

Karp's theorem

What researchers really wanted was a quick method for solving traveling salesman problems. But if none could be found, the second-best bet would be for someone to identify the fundamental obstacle. In 1972, the computer scientist Richard Karp did exactly that, by proving that the problem is fundamentally hard. Although this had been conjectured in the 1950s by Merrill Flood, it was only in the early 1970s that the technical framework was put in place to define what it truly means to be a "hard problem." In the jargon of computational complexity theory, Karp's theorem was that the traveling salesman problem is *NP*-complete.

This breakthrough placed the traveling salesman at center stage in modern computer science. Being an *NP*-problem meant that, despite being hard, there is nevertheless an angle from which the problem is tractable. If you are asked whether or not the salesman can improve on route of 100 kilometers, this might be a seriously difficult question to answer. Yet, if someone presents a putative shorter route, it is quick work to verify whether or not it actually works.

This tension between being hard to solve but still quick to check is at the heart of *NP*-completeness. It remains possible that someone may yet find a quick solution to the problem, though this is unlikely given the current state of our knowledge. Indeed, if someone could find a fast algorithm for traveling salesman problems, it would automatically translate to efficient methods for many other of the most intractable tasks in computer science. The only way this is possible is if *P* = *NP* (see page 337).

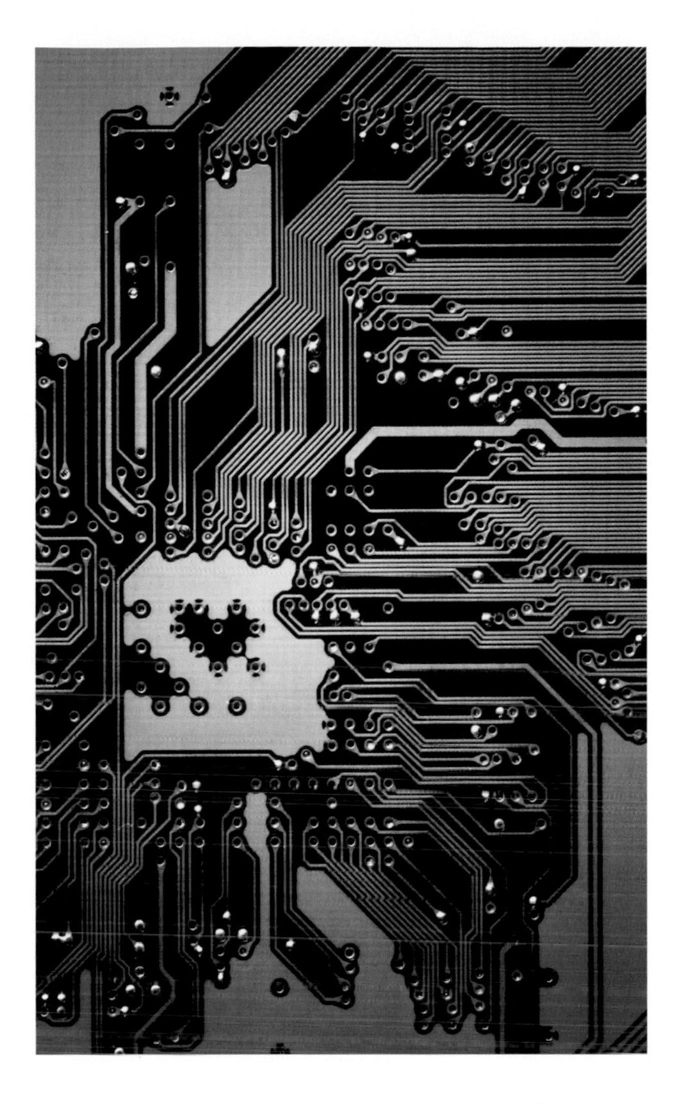

ABOVE Modern microchips contain many thousands of components to be connected. The circuits which connect them should be as short as possible to allow the chip to run faster. This leads directly to a traveling salesman problem for the chip's designers to grapple with.

Simulated annealing

Traveling salesman problems continue to appear in ever more contexts throughout science and engineering. Of course, these applications cannot wait for computer scientists to find a resolution to the *P* versus *NP* problem. Consequently various approaches have been adopted which rely on relaxing the goal from finding the best possible route to merely finding a reasonably good one. A significant development came in the 1980s, with the development of simulated annealing algorithms. The name comes from metalworking, whereby craftsmen improve the crystalline structure of a metal object by repeatedly heating and cooling it. In the mathematical analogy, the salesman would begin with a random route, and then automate a procedure to improve it. This works by encouraging the configuration to settle into a state of lower energy, as happens when larger crystals form within a metal.

85 Chaos theory

BREAKTHROUGH CHAOS THEORY IS THE ANALYSIS OF HIGHLY UNPREDICTABLE AND UNSTABLE SYSTEMS. THE FOUNDATIONS OF THE SUBJECT WERE ASSEMBLED IN THE 1970S.

DISCOVERER JOHN VON NEUMANN (1903–57), MITCHELL FEIGENBAUM (1944–), TIEN-YIEN LI (1945–), JAMES YORKE (1941–).

LEGACY UNDERSTANDING OF CHAOTIC SYSTEMS HAS GROWN GREATLY OVER RECENT YEARS, AND CHAOS THEORY NOW PLAYS A ROLE IN MANY BRANCHES OF SCIENCE, INCLUDING PHYSICS (SEE PAGE 000) AND POPULATION BIOLOGY.

Although it may not appear so to a student faced with a page of incomprehensible algebra, mathematics prizes simplicity. The cool curve of an ellipse and the crisp corners of a tetrahedron are far easier to describe than many of the patterns we see in the physical world. Events like the yearly fluctuations in a population of fish are hard to predict in advance, even when the relevant background is known. But the mathematics of chaos theory has shone considerable light on this topic. It shows that many systems remain totally unpredictable, even when their underlying rule is very simple. A particular breakthrough came in 1975, when Tien-Yien Li and James Yorke found a wonderful way to identify chaos, using a single, simple criterion.

How can one produce a random number? In the late 1940s, John von Neumann proposed a very strange answer to that question. He suggested that applying a simple algebraic rule a few times should do the job. The rule is to begin with some number, call it x, and then multiply x by $(1 - x)$, and multiply the result by 4. That is to say: $x \to 4 \times x \times (1 - x)$.

There does not seem to be anything especially "random" about this bit of algebra. Once the initial number is chosen, say $x = 0.1$, the result of applying the rule is then completely predetermined. But a little experimentation reveals von Neumann's insight. The sequence produced by this rule runs: 0.1, 0.36, 0.9216, 0.2890, 0.8219, 0.5854, 0.9708, and so

OPPOSITE Edward Lorenz coined the term "butterfly effect" to describe the extreme sensitivity to initial conditions that is characteristic of chaos. Whether or not a butterfly flaps its wings at a specific moment may cause major future differences in the weather, thanks to the chaotic nature of air-flow.

on (each number given to 4 decimal places). There does not seem to be much of a pattern here, and in fact that is no illusion. You can extend the sequence for as long as you like and no pattern will emerge. Someone who did not know the rule being used would find it virtually impossible to distinguish between this sequence and one produced by a genuinely random physical process such as radioactive decay.

The logistic map

The same rule was picked up in the 1970s by the biologist Robert May in his paper "Simple Mathematical Models with Very Complicated Dynamics." May thought of the rule as describing the population of an organism. If one year the population was x thousand fish, the next year it might be $4 \times x \times (1 - x)$ thousand. His question was: what would happen to the population over time?

The term "butterfly effect" reflects the fact that the equations governing our weather are also believed to be chaotic. So a butterfly flapping its wings could alter the weather on the other side of the world a year later.

Today, von Neumann's rule is known as the logistic map, and it is one of the simplest examples of mathematical chaos, a phenomenon which has been recognized in many different scientific situations, beginning with the three-body problem (see page 237). Yet, unlike the three-body problem, the logistic map is algebraically extremely simple, making it the perfect case study of chaos. What is more, in the late 1970s and early 1980s, it was shown that, in several important ways, the logistic map is typical of a much broader class of chaotic systems. This is especially true for the discoveries made by Mitchell Feigenbaum, Tien-Yien Li and James Yorke.

In von Neumann's pseudorandom number generator, everything rests on the number 4, known as the parameter. Changing that value completely alters the behavior of the system. If one replaces 4 with a new parameter of 2, the logistic map ceases to be chaotic. Instead, for any starting value, the sequence will quickly home in on a fixed value of 0.5. This is known as an attracting point of the system.

Increase the parameter from 2 to 3.4, and something new occurs. After a while, the sequence will endlessly flicker back and forth between two values around 0.84 and 0.45. This is known as an attracting 2-cycle. Raise the parameter a little higher to 3.5, and this is replaced with an attracting 4-cycle, and then at 3.55, an attracting 8-cycle, and so on. As the parameter increases, the length of the attracting cycle keeps doubling: 16, 32, 64, and so on. This behavior is what chaos theorists call a sequence of bifurcations.

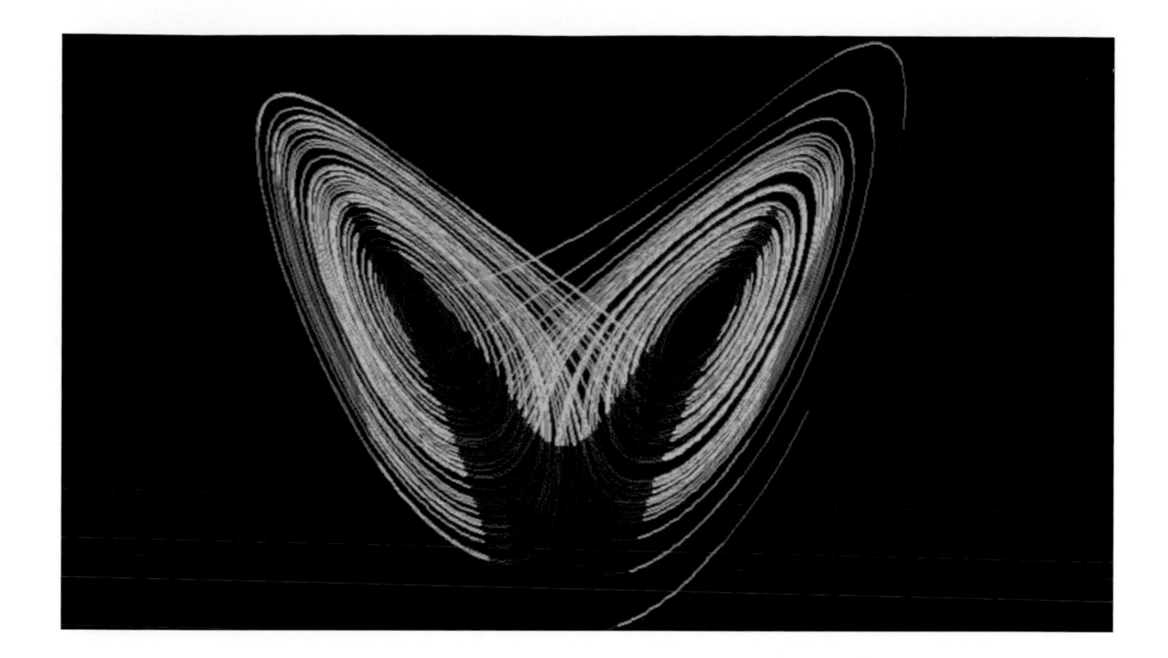

ABOVE A strange attractor. Over time, chaotic systems are often drawn toward a geometrical object, known as the system's attractor. When this object has an intricate fractal composition, as this one does, it is known as a strange attractor.

Yet these bifurcations do not continue indefinitely: only so long as the parameter remains below a critical value called the Feigenbaum point, around 3.5699. This is named after the chaos theorist Mitchell Feigenbaum, who proved that this sequence of bifurcations is typical of chaotic systems. (So much so that chaos theory is also known as bifurcation theory.)

Period three implies chaos

Once the parameter exceeds the Feigenbaum point, the system is truly chaotic. The sequence will never settle down into any predictable pattern, but will hop around seemingly at random forever. Sequences which start at very slightly different values, perhaps at 0.1 and 0.1000000001, will end up entirely unrecognizable from each other—a phenomenon commonly known as the butterfly effect. The name reflects the fact that the equations governing our weather are also chaotic. So a butterfly flapping its wings could alter the weather on the other side of the world a year later.

Yet even deep inside the chaotic portion of the logistic map, there are islands of stability. At a parameter of $1 + \sqrt{8}$ (around 3.83) the system has an attracting 3-cycle. This is a striking departure from the earlier cycles, which were all powers of two (2, 4, 8, 16, 32, and so on). Indeed, in 1975, James Yorke and Tien-Yien Li proved one of the great theorems in the subject: that the only systems that admit attracting 3-cycles are chaotic ones.

GULF OF MEXICO
CARIBBEAN SEA
GREATER ANTILLES
CAYMAN
JAMAICA
BAHAMAS
DOMINICAN REPUBLIC
PUERTO RICO
HAITI
ARUBA
COLOMBIA
ECUADOR
BELIZE
COSTA RICA
SAN FRANCISCO
SAN JOSE
LOS ANGELES
SAN DIEGO
TIJUANA
ACAPULCO
NEW ORLEANS
TAMPA
CHARLESTON
CAPE HATTERAS
MILLENNIUM

86 Four color theorem

BREAKTHROUGH THAT ANY MAP CAN BE PAINTED USING ONLY FOUR COLORS WAS PROVED IN 1976 BY APPEL AND HAKEN, IN ONE OF THE LONGEST PROOFS IN HISTORY.

DISCOVERER KENNETH APPEL (1932–) AND WOLFGANG HAKEN (1928–).

LEGACY APPEL AND HAKEN'S EXTENSIVE USE OF COMPUTERS TRANSFORMED ATTITUDES TO PROOF AND DEMONSTRATED THEIR POWER AS TOOLS FOR MATHEMATICAL RESEARCH.

In 1852, Francis Guthrie was examining a map of the counties of Britain. How many colors, he wondered, would he need to color it in? As he sat musing over this puzzle, Guthrie could not have known that his map-coloring question would grow into one of the most notorious problems in mathematics. Its eventual solution by Kenneth Appel and Wolfgang Haken in 1976 would alter the very nature of the subject.

Guthrie asked how many paint colors he would require if each county was to be colored differently from its neighbors. Being a student of mathematics, he was not really interested in what this might tell him about the geography of Britain, but rather in the underlying principle. Counting any abstract design as a "map," then how many paints would be needed if no two bordering "countries" may be the same color?

Map coloring

The square countries on a chessboard, for instance, can be painted using only two colors. (Countries touching only at a single point are allowed to be the same color.) It was easy to come up with more involved maps requiring three colors, and still more complex maps that could not be painted with fewer than four. But, try as he might, Guthrie was unable to find any map which needed five colors. However convoluted the pattern seemed at first, he could always find some way to color it using just four paints. So Guthrie came to the obvious conclusion: that four colors are sufficient for any map.

OPPOSITE Cartographers have appealed to deep mathematical theorems many times over the centuries to find new ways of making maps. But in the 19th century, a map-coloring riddle grew into one of the biggest questions in the subject.

RIGHT The Mercator Atlas, a map of Europe from the 1590s. It uses four colors (including the blue for the sea), though it was not until the 19th century that anyone investigated in detail how many were needed.

This was now a solid geometrical conjecture, demanding a rigorous mathematical proof. But Guthrie could not provide one. He mentioned the problem to his teacher, the eminent logician Augustus de Morgan. De Morgan was intrigued, and corresponded on the subject with several of the top mathematicians of the day, including the algebraist Arthur Cayley. In 1878, Cayley circulated the question at the London Mathematical Society and in the journal *Nature*, enquiring whether anyone knew of the answer to this deceptively simple riddle. With this, the map-coloring question was thrown wide open, becoming one of the major mathematical problems of the age. One to pick up the gauntlet was Cayley's former student Alfred Kempe, who triumphantly announced one year later that he had found the solution.

The five color theorem

Kempe's proof was published in the *American Journal of Mathematics* and his achievement won him great fame, even beyond mathematical circles. But 10 years later in 1890, Durham University lecturer Percy Heawood was reading Kempe's proof when he spotted something amiss. A critical step seemed to be missing from the argument. So, to Kempe's embarrassment, in 1890 Heawood published an article demonstrating that "there is a defect in the now apparently recognized proof."

Heawood was not able to close the gap fully, but he did make the first firm progress on the problem. By adapting Kempe's argument, Heawood was able to prove a weaker version of the conjecture: the five color theorem. Heawood spent much of his life studying the problem, attempting to refine his proof to demonstrate that four colors should be enough. Others, meanwhile, tried to concoct a counterexample: a pattern so complex that it required the full five colors.

To the frustration of everyone involved, neither approach was fruitful. This stalemate endured until 1976, when Kenneth Appel and Wolfgang Haken at the University of Illinois announced an extraordinary development. After all the twists and turns in the problem's history, they claimed they had finally proved the four color theorem. But when their colleagues came to read their proof, they were in for a shock.

A computer-assisted proof

Appel and Haken argued that if there is a counterexample (that is, a map that requires five colors) then there must be a smallest such, meaning one with a minimal number of countries. They then went on to probe what this minimal counterexample must look like, and demonstrated that it must contain one of 1936 specific configurations which they identified. They then went through the 1936 in turn, to prove it logically impossible. But each of these 1936 individual arguments required analyzing many thousands of broader patterns.

The proof was far and away the longest the world had ever seen. It was so immense, in fact, that it could not be digested by a single human being at all.

The proof was far and away the longest the world had ever seen. It was so immense, in fact, that it could not be digested by a single human being at all. To complete the argument had required over 1000 hours of computer-processing time. Appel and Haken's work was something completely new: a computer-assisted proof. This innovation would be even more important to mathematics than the theorem itself, but it was highly controversial in some circles. After all, the purpose of a mathematical proof is to advance human understanding. But if no human can understand it, what good is it?

But a mathematical proof is not an argument like any other. It is a sequence of steps, each logically deducible from those before, culminating in the theorem. Once a computer has been taught the rules of the game, it can distinguish between valid and invalid logical deduction. Since Appel and Haken's breakthrough, their proof has been simplified somewhat. But the case-by-case analysis is still intensive enough to require computer assistance. In the meantime, the use of the computer has revolutionized the subject of mathematics, as can be seen in the solution of other mathematical theorems such as Hales' work on the Kepler conjecture (see page 373).

87 Public key cryptography

BREAKTHROUGH BY EXPLOITING PRIME NUMBERS, PUBLIC KEY CRYPTOGRAPHY BREAKS THE SYMMETRY BETWEEN ENCODING AND DECODING, ALLOWING ANYONE TO ENCODE INFORMATION SECURELY.

DISCOVERER RON RIVEST (1947–), ADI SHAMIR (1952–) AND LEONARD ADLEMAN (1945–); JAMES ELLIS (1924–97), CLIFFORD COCKS (1950–) AND MALCOLM WILLIAMSON (1950–).

LEGACY AS THE PRIMARY PLANK IN MODERN INTERNET SECURITY, PUBLIC KEY CRYPTOGRAPHY REMAINS OF HUGE THEORETICAL AND PRACTICAL IMPORTANCE.

For as long as humans have existed, they have wished to keep secrets. Through wars and unrest, people have always wanted to communicate with their friends and allies, safe from the prying eyes of their enemies. For this purpose, soldiers, spies and secret societies down the centuries have employed codes and ciphers. But in the second half of the 20th century, this science of cryptography took on a new importance with the advent of the internet. With it came a brand-new way to encode data: public key cryptography.

The most familiar type of code is an alphabetic substitution, which entails jumbling up the alphabet. It might be that "a" is written as "G," "b" as "C," and so on according to the following key:

a	b	c	d	e	f	g	h	i	j	k	l	m	n	o	p	q	r	s	t	u	v	w	x	y	z
G	C	Q	F	Y	R	O	J	K	U	A	B	I	V	P	H	Z	S	L	W	D	M	E	T	N	X

To encrypt a message using this cipher, you begin with message in plain text, say "mathematics." Then you substitute each letter according to the key above, giving "IGWJYIGWKQL." For the recipient of the message to decipher it, she needs to know (or to find out) the rule that was used, and then merely swap the letters back. It is a simple idea, and it is notable for its symmetry. The process of deciphering the message is identical to that of enciphering it, but reversed. Critically, the two procedures require access to the same basic information, namely the key above.

OPPOSITE People interested in secrecy and security have often made use of mathematics, as have those trying to crack their codes. During the 20th century the stakes were raised even higher. In the internet age, we all rely on hi-tech security measures every day.

ABOVE Enigma was the code used by the German military during World War 2 and before. Both the encoder and decoder needed access to an Enigma machine and to know the day's settings. Using captured machines Polish and British cryptanalysts were able to crack the code, ushering the end of the war.

Over the years, numerous variations on this theme have been developed, many of them highly sophisticated and difficult for an interceptor to crack, a famous example being Nazi Germany's Enigma code. Nevertheless, the essential symmetry of the situation has always remained. Enciphering and deciphering a message were mirror-image procedures, relying on the same information. But in the second half of the 20th century, the computer revolution demanded a new approach. This arrived in the form of public key cryptography, where the symmetry between encoding and decoding was broken.

Public keys

In public key cryptography, the key comes in two parts: a public key and a private key. A bank trading online should make its public key freely available to everyone. Then, someone wishing to contact the bank securely can then use the public key to encipher their message. But deciphering cannot be done with just the public key. It needs the private key, which only the owner has access to.

In fact, the idea of public key cryptography is not new. It is the same idea used to send letters by mail. A bank's postal address is public knowledge, so anyone can deliver letters there (or employ a postman to do so). The address and letter box play the role of the public key. But to read the letters, an employee of the bank needs to be able to unlock the door, which means having the private key.

This old idea was given a modern mathematical twist in 1978 by the MIT researchers Ron Rivest, Adi Shamir and Leonard Adleman. The RSA algorithm (from the first letters of their surnames) used the arithmetic of very large numbers to produce a system for public key cryptography.

It remains widely used in internet security today. (It later transpired that James Ellis, Clifford Cocks and Malcolm Williamson had formulated a similar scheme in 1973 while working for the British intelligence services. It was used militarily, but remained classified as Top Secret until 1997.)

Numbers and codes

The heart of RSA encryption is number theory. One of the most ancient facts of mathematics is the fundamental theorem of arithmetic. This says that every whole number can be broken down into primes. So $36 = 2 \times 2 \times 3 \times 3$, for example. What is more, this happens in a unique way: however hard you try, you will never be able to break 36 down into any other collection of primes besides two 2s and two 3s. In relation to small numbers like 36, the "fundamental theorem" hardly seems to merit its grandiose title. But when applied to bigger numbers, it becomes invaluable. Starting with a large number, such as 62,615,533, it is by no means obvious how to break this down into primes. Even with all the powers of modern computers, mathematicians cannot do much better than an exhaustive search of all eligible primes, testing 2, 3, 5, 7, 11, and so on in turn. This is very time consuming, and with much bigger numbers (those of 100 digits long, say) it is completely unfeasible.

The RSA algorithm used the arithmetic of very large numbers to produce a system for public key cryptography. It remains widely used in internet security today.

However, if you are told the answer (7907×7919) it is the work of a moment to check that it is indeed correct. At this stage, the fundamental theorem of arithmetic guarantees that this is the only valid answer. This asymmetry between the time it takes to find the answer and the time needed to check it is the cornerstone of RSA encryption. The large number (62,615,533) acts as the public key, though which messages are encoded. But to decode the message requires the private key, which consists of the prime decomposition 7907×7919.

The idea is beautifully simple, and the details were worked out by Rivest, Shamir and Adleman in their seminal paper of 1978. Of course, the security of the entire system relies on potential interceptors not being able to work out the private key from the public key. The hope is that multiplying two primes together is what is known as a one-way function, meaning something very quick to do, but slow to undo. The evidence so far suggests that this is the case, but mathematicians in the intervening years have failed to produce any guarantees. Indeed, if large-scale quantum computing becomes viable, then RSA codes will be crackable (see page 369).

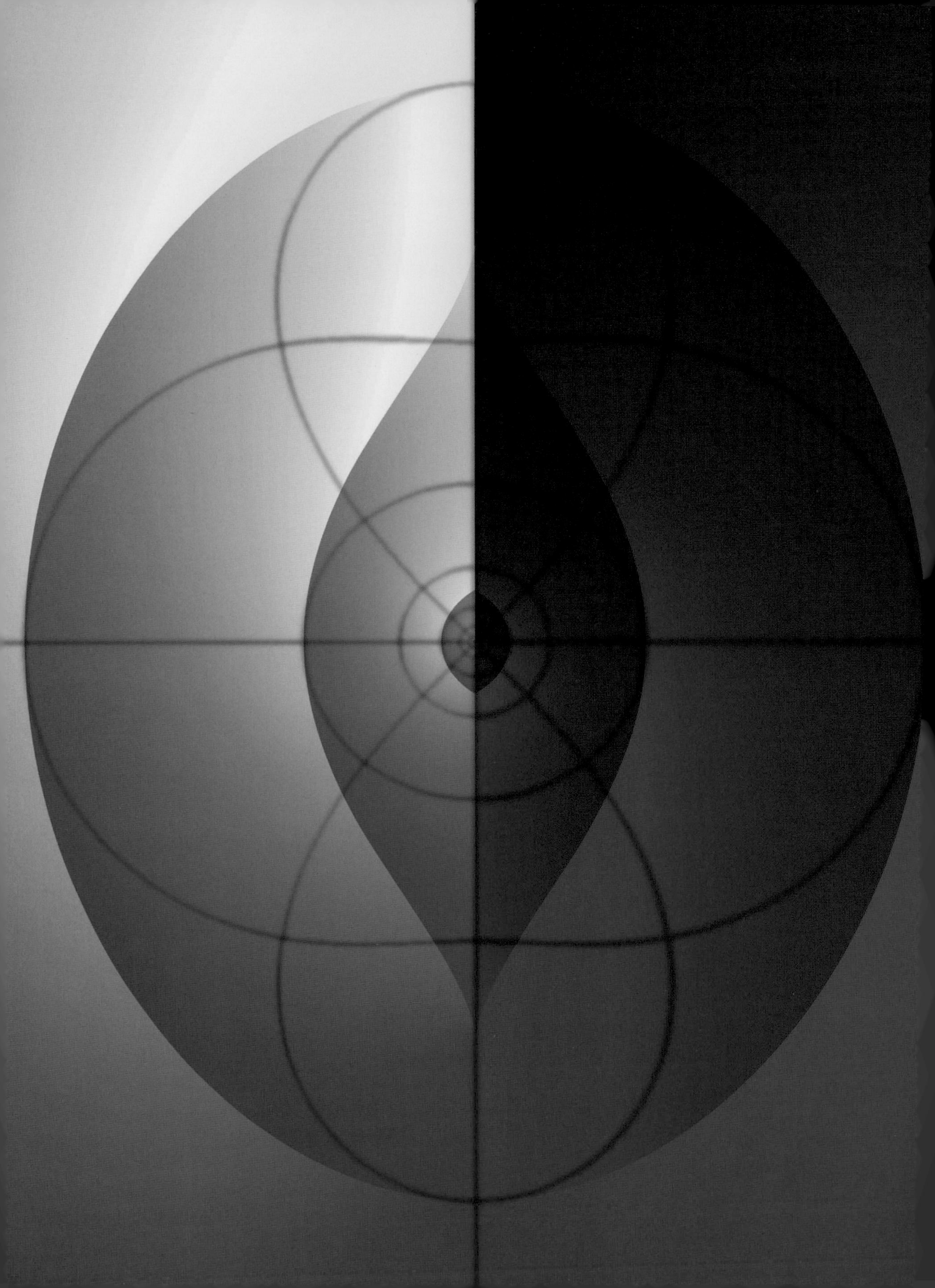

88 Elliptic curves

BREAKTHROUGH OVER THE 20TH CENTURY, A CONFLUENCE OF NUMBER THEORY AND GEOMETRY DREW ATTENTION TO UNUSUAL OBJECTS KNOWN AS ELLIPTIC CURVES.

DISCOVERER LOUIS MORDELL (1888–1972), GERD FALTINGS (1954–), BRYAN BIRCH (1931–), PETER SWINNERTON-DYER (1927–).

LEGACY ELLIPTIC CURVES REMAIN A MAJOR FOCUS OF MODERN MATHEMATICS. BUT ALL EFFORTS TO PROVE BIRCH AND SWINNERTON-DYER'S MONUMENTAL CONJECTURE HAVE SO FAR FAILED.

Since the work of Diophantus, number theorists have wanted to understand the possible relationships between whole numbers. These relationships are expressed by equations, and these same equations can also be interpreted by geometers as shapes. During the 20th century, the spotlight fell on a particularly interesting type of equation describing a very special shape called an elliptic curve.

Throughout mathematical history, there have been two major approaches to mathematical objects. Number theory has always been interested in studying the system of whole numbers, while geometers have wanted to understand shapes. Yet these two subjects are more closely connected than might be expected, as Diophantus first noticed (see page 69).

Geometry and number theory

Geometrically, that grand old man of geometry Pythagoras' theorem says that the three sides of a right-angled triangle are connected by the equation $x^2 + y^2 = z^2$. There is no particular constraint on the type of numbers needed here. But as a number theorist, Diophantus was interested in knowing which whole numbers obey this equation. These are known as Pythagorean triples, such as 3, 4, 5 ($3^2 + 4^2 = 5^2$).

This illustrates an idea which would become fundamental during the 20th century. The equation $x^2 + y^2 = z^2$ describes a geometrical object—in fact, in this example, it is a conical surface. The number

OPPOSITE Artwork based on a lemniscate. The lemniscate (resembling the symbol for infinity, ∞) was first described in 1694 by Jakob Bernoulli. Attempts to calculate the length of stretches of this shape led to some of the earliest analysis of elliptic curves.

theorists' question is whether there are any points on the surface whose coordinates are all given by whole numbers. In this case, as Diophantus realized, the answer is yes, the point (3, 4, 5) being an example. But in the case of the surface given by $x^3 + y^3 = z^3$ the answer is no. As Pierre de Fermat famously observed (see page 369), no whole numbers can obey this equation, which means that no points on the corresponding shape have whole number coordinates.

Curves and Faltings' theorem

This idea led 20th-century number theorists such as Louis Mordell to ask a deceptively simple question: which geometrical objects contain points with whole number coordinates, and which do not? Extending this slightly, a major topic developed around rational points on shapes, meaning places whose coordinates are all rational numbers, or fractions of whole numbers.

Faltings' theorem was undoubtedly one of the great breakthroughs in 20th-century number theory. Yet it still did not answer all the number theorists' questions about curves.

The first and most important shapes are 1-dimensional curves. It was clear from the outset that there were not going be any easy answers to this question: some curves have no rational points at all, while others have infinitely many. In between are curves with a limited, finite number of rational points. It was a major technical challenge, therefore, to understand which was which. This would shine a light on the possible and impossible relationships between whole numbers.

Luckily, there is a natural way to differentiate between curves. It involves assigning them a number known as their genus. The most familiar curves, such as circles, straight lines and conic sections (see page 57) have genus 0. For curves like this, it is not too difficult to determine the distribution of rational points: the answer will always be either that the curve has none at all or that it has infinitely many.

In 1922, Mordell made a bold conjecture about the behavior of more complicated curves—those with genus 2 or higher. He said that they should only ever contain finitely many rational points. This idea would have major consequences for our understanding of the whole numbers. Yet Mordell was not able to prove his conjecture, and through the 20th century it stood as a monument to our poor understanding of the interaction of number theory and geometry. It was not until 1983 that Gerd Faltings broke the deadlock, when he was finally able to provide a proof to Mordell's claim.

Elliptic curves

Faltings' theorem was one of the great breakthroughs in 20th-century number theory. Yet it still did not answer all the number theorists' questions about curves. Between the curves of genus 0 (which are easy to understand), and the curves of genus 2 and higher (which were tamed by Faltings) lie the troublesome curves of genus 1. These are known as elliptic curves, and are highly unusual objects in several ways.

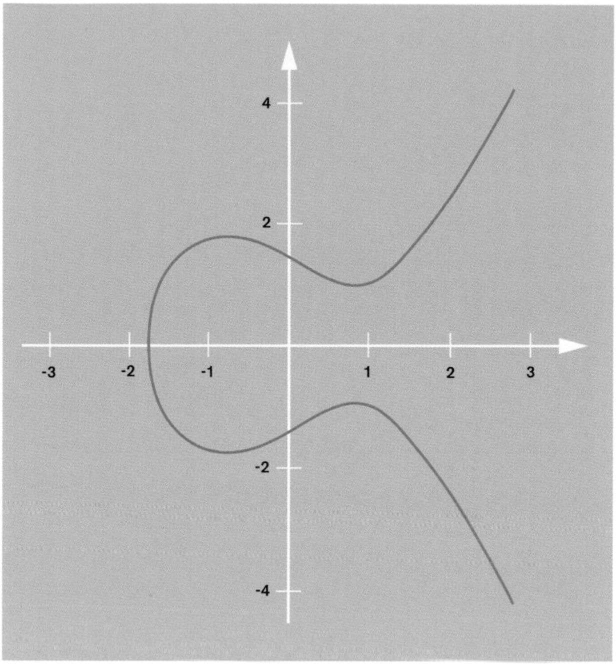

To start with, elliptic curves are not merely curves. They are also groups (see page 389), meaning that there is a way to add points together. If you pick any two points on an elliptic curve, there is a coherent system for combining them to give a third point, just as adding two whole numbers produces a third. The resulting object is very interesting to algebraists, and in recent years it has also been exploited to great effect by cryptographers. Because elliptic curve addition is computationally difficult to undo, it is today used as a basis for efficient public key cryptography (see page 353).

ABOVE An elliptic curve. Despite their name, elliptic curves do not closely resemble ellipses, being infinitely long. Elliptic curves are described by cubic equations, this one being given by $y^2 = x^3 - 2x + 2$.

Birch and Swinnerton-Dyer conjecture

For today's number theorists, the major question about elliptic curves remains the matter of rational points. But elliptic curves are extremely delicate objects: some contain infinitely many rational points, and others only finitely many. How can we tell which is which? In 1965, Peter Birch and Henry Swinnerton-Dyer addressed this question. They assigned a mathematical device to the curve, called an *L*-function. In fact, it is a close cousin to Riemann's zeta function, used to study the prime numbers. This device, they claim, encodes the information about how many rational points the elliptic curve contains.

As of the year 2000, the Clay Foundation is offering a $1,000,000 reward for a proof or counterexample of this striking claim. But, so far, Birch and Swinnerton-Dyer's function is no more yielding to the ingenuity of today's number theorists than Bernhard Riemann's.

89 Weaire–Phelan foam

BREAKTHROUGH AN EFFICIENT NEW METHOD FOR DIVIDING UP 3-DIMENSIONAL SPACE, INSPIRED BY MOLECULAR CHEMISTRY AND THE PHYSICS OF FOAMS; THE DISCOVERERS ALSO USED SOPHISTICATED COMPUTER MODELING TECHNIQUES.

DISCOVERER WILLIAM THOMSON (LORD KELVIN) (1824–1907), DENIS WEAIRE (1942–), ROBERT PHELAN (1971–).

LEGACY THE DISCOVERY OF SUCH A COMPLEX WAY TO DIVIDE UP SPACE EFFICIENTLY CHALLENGES THE NOTION THAT THE SIMPLEST ANSWER IS ALWAYS THE BEST ONE.

In the late 19th century, the eminent mathematician and physicist Lord Kelvin asked himself a deceptively simple question. Suppose you want to divide a large room into cells, each of exactly the same size. What shape should the cells be, if you want to minimize the amount of material needed to build their walls?

The most obvious approach is to divide the whole space into cubes. But there are numerous other possibilities: perhaps tall thin cells would be more efficient? Or cells with more exotic shapes, such as irregular octahedra (cells with eight triangular faces)? Nor do the faces of the cells need to be flat. Taking all of these into account, which of these many options requires the least amount of material to fabricate its walls?

Kelvin declared that the problem was "solved in foam," meaning that foam and froth, such as made by soap or whisked egg whites, would tend to minimize the amount of material needed. With this in mind, Kelvin and the many scientists who followed spent hours examining the geometry of foam in minuscule detail.

OPPOSITE In 1993, analysis of foam provided a breakthrough in Kelvin's conjecture, a classical geometrical question. Here, interesting optical effects are produced as light reflecting off the back and front walls of detergent bubbles interferes.

Pappus' hexagonal honeycomb

A similar question to Kelvin's had been considered over a thousand years earlier by the geometer Pappus, around AD 320. He had asked: if you want to divide up a page of paper into cells of the same size, what shape should they be to minimize the amount of ink needed to draw them? Pappus compared the three most obvious candidates: a square grid, a hexagonal grid, and a triangular grid. He found that hexagons provide the best answer. What Pappus could not manage, however, was to rule out the possibility of a grid of a more exotic shape, which would be more efficient than hexagons.

The tubes that bees use for storing honey have hexagonal cross-sections, rather than squares, triangles or any more complicated shape, and this choice minimizes the amount of wax they need.

Extraordinarily, this gap persisted until 1999, when Thomas Hales finally resolved the hexagonal honeycomb conjecture (as it had become known). If the time it took to answer Pappus' question was a surprise, Hales' final answer was not. Pappus had been right: the most efficient pattern is indeed a hexagonal grid. Indeed, this fact has been exploited by bees in designing their nests for millions of years. The tubes that bees use for storing honey have hexagonal cross-sections, rather than squares, triangles or any more complicated shape, and this choice minimizes the amount of wax they need.

Kelvin's conjecture

Kelvin's question echoed that of Pappus, but it concerned three dimensions, where the situation is even murkier. All the same, Kelvin thought that he had found the answer: a shape which later became known as the Kelvin cell. It is built from eight hexagonal and six square faces, where the hexagons are very slightly curved. Kelvin carefully constructed a wire model of this arrangement, now affectionately known as "Kelvin's bedspring." As ever, though, actually proving the optimality of his bedspring design was another matter. To his frustration, Kelvin was unable to do it.

Weaire–Phelan foam

In retrospect, Kelvin did not stand a chance. For, unlike Pappus's hexagons, Kelvin's conjecture was wrong. This was established in dramatic style in 1993, when the Irish physicists Denis Weaire and Robert Phelan found a better design, inspired by chemistry, and using advanced modeling techniques.

Weaire–Phelan foam uses 0.3% less material than Kelvin's and is a more intricate construction, with a strikingly organic appearance. Their original motivation was not to outdo Kelvin, but to study the physics of real-life wet foams (rather than dry foams, as the theoretical idealizations are known). They were inspired in their work by an unusual chemical compound known as a clathrate, in which molecules of one chemical are contained within a "cage" formed by another.

Weaire and Phelan examined the shapes of these chemical cages and wondered if the design could be modified to create foam. They used a computer program called Surface Evolver to finesse the curvature of their cells, and were as stunned as everyone else when the result turned out to refute Kelvin's conjecture.

In fact, their foam is built not from just one type of cell, as Kelvin's was, but from two of different shapes (though their volumes are equal). One cell is composed of 12 pentagonal faces, and the other from two hexagons and 12 pentagons. One Weaire–Phelan unit consists of two of the former, and six of the latter.

As well as occurring on the molecular scale in crystals, Weaire–Phelan foam has also subsequently been realized by humans on a monumental scale. In 2008, the Aquatic Center at the Beijing Olympics was constructed from a single giant piece of Weaire–Phelan foam. However, it was not until 2011 that a research team at Trinity College Dublin, which included Denis Weaire and the physicist Ruggero Gabrielli, succeeded in realizing the Weaire–Phelan structure as an actual foam, produced from ordinary detergent solution.

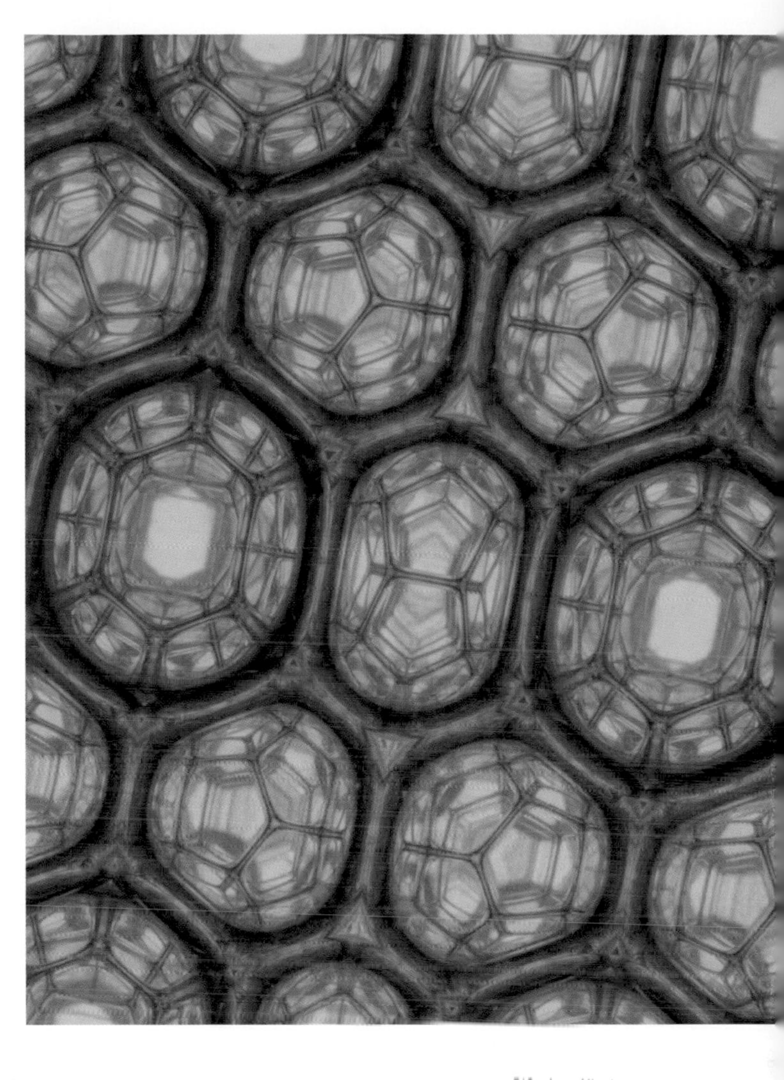

ABOVE Weaire–Phelan foam, as photographed in the laboratory of Trinity College Dublin in 2011. Producing the foam involved painstakingly creating plastic containers whose walls were shaped in precisely the right way.

The huge question, of course, is whether Weaire–Phelan foam is the best possible answer to Kelvin's problem, or whether there are still more efficient foams waiting to be found. Either way, a lesson of Weaire and Phelan's breakthrough is that mathematicians can no longer rely on their own ingenuity in such matters. Analysis of the many and varied structures thrown up by nature, along with sophisticated computer modeling techniques, are now prized weapons in the modern geometer's arsenal.

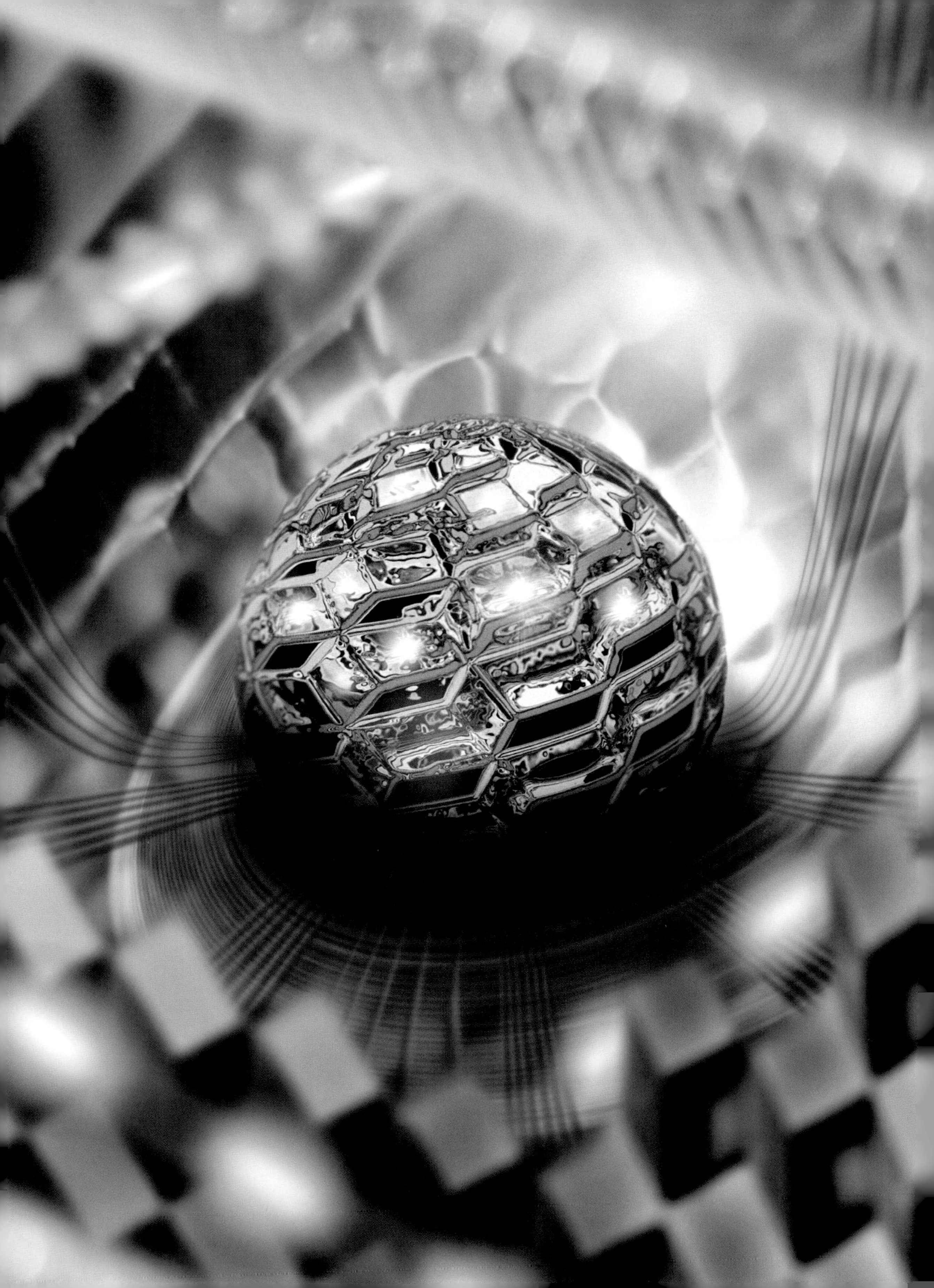

90 Quantum computing

BREAKTHROUGH ONE OF THE CHALLENGES FOR TODAY'S MATHEMATICIANS IS TO UNDERSTAND THE POSSIBILITIES AND LIMITATIONS OF QUANTUM COMPUTERS. IN 1994, PETER SHOR FOUND A METHOD FOR FACTORIZING NUMBERS QUICKLY.

DISCOVERER PETER SHOR (1959–)

LEGACY SHOR'S ALGORITHM WAS A MILESTONE IN THEORETICAL COMPUTER SCIENCE. THE PRACTICAL IMPACT WILL BE SEEN WHEN (OR IF) A WORKING QUANTUM DEVICE IS BUILT.

One of the central facts in mathematics asserts that every whole number is constructed from primes. So 28 for example can be broken down as $2 \times 2 \times 7$. This has been known since Euclid's time, at least, and has a majestic title to reflect its importance: the fundamental theorem of arithmetic. But in 1994, this ancient and venerable theorem was given a modern, quantum twist by Peter Shor.

In fact, the fundamental theorem of arithmetic goes a little bit further: every whole number can be broken down into primes, in exactly one way. So each whole number corresponds to the unique collection of primes from which it is built: it cannot also be true that 28 is equal to $2 \times 3 \times 5$, or 2×13, or any combination apart from $2 \times 2 \times 7$. When applied to small numbers such as 28, this all seems obvious. But when considering much larger numbers, things start to become a little more mysterious.

The factorization problem

When presented with a larger number, such as 62,615,533, mathematicians are assured by the fundamental theorem of arithmetic that it can be broken down into primes. Yet actually finding these basic constituents is another matter. In principle, there is a simple procedure which will work: just test each prime in turn. It is easy to see that 2 does not divide 62,615,533, and nor do 3 or 5. We could try 7, 11, 13, 17, and continue in the same vein. But at this rate, it could take an unreasonably long time to arrive at the answer. And for an even bigger number, say

OPPOSITE An artist's impression of a quantum computer's crystal core, under high magnification. The design of tiny crystals, built in laboratories atom by atom, is a current avenue of research in efforts to build a working quantum computer.

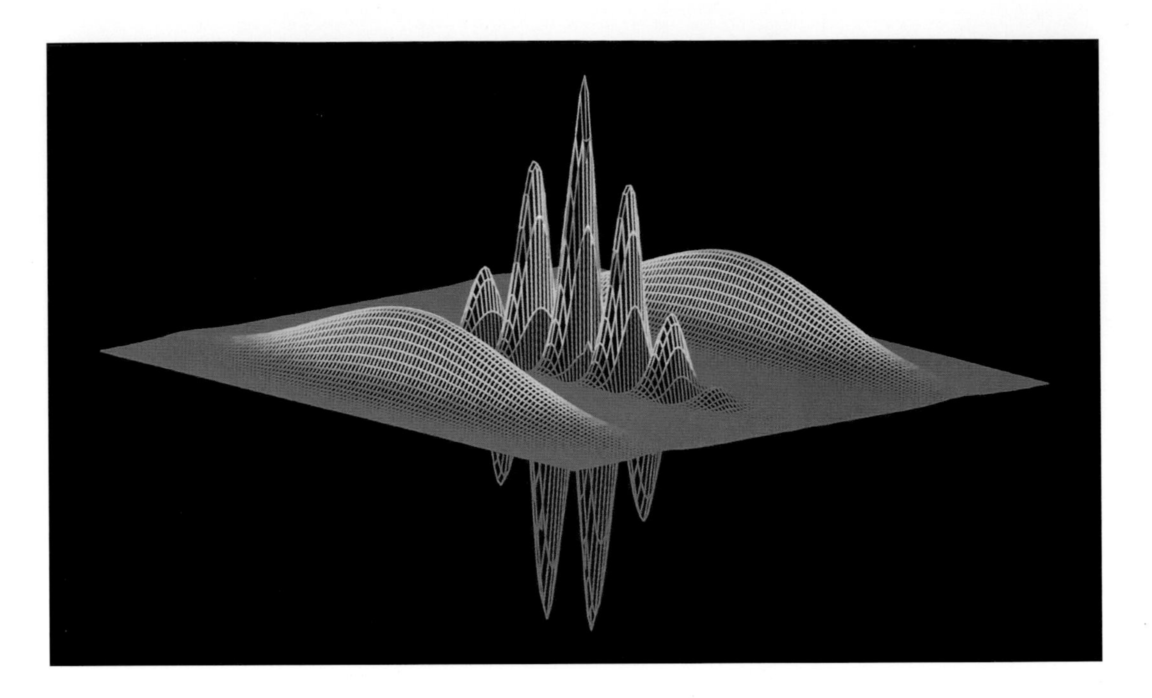

ABOVE Conventional computers store data in bits as either 0 or 1. The qbits of a quantum computer have some probability of being in either state. The tallest peak represents a qbit at 1, and the deepest trough one at zero. The lesser peaks and troughs represent qbits with intermediate probabilities.

one 100 digits long, the search could take longer than the lifetime of the Universe. For all practical purposes, then, this factorization problem is impossible by this method.

But perhaps there is a quicker way? For years, mathematicians have searched unsuccessfully for some trick to speed this process up. But this is no longer just a riddle for mathematicians to ponder. In the age of internet shopping and banking, the factorization problem has acquired a status of the highest practical importance. The codes which protect our personal details online, such as credit-card numbers and home addresses, are ingenious systems; but behind them is a simple idea. Each system employs some large number to encode its messages. This is known as the public key. However, to decode the message, you also have to know the number's prime constituents. These form the private key, which only the intended recipient can access (the bank or the email account holder). Crucially, to crack the code you have to know—or to find out—the prime constituents of the public key. The security of the entire system relies on people outside the security loop not being able to do so (see page 353).

According to standard mathematics and computer science, this is a safe procedure, although there is no guarantee. It is conceivable that someone will devise a clever computer program to factorize integers far more speedily than is currently possible.

Shor's algorithm

In 1994, a mathematician at Bell Laboratories did exactly this. Peter Shor produced a revolutionary new method for solving the integer factorization problem. What is more, it could do so incomparably quicker than any other known method. The catch was that Shor's algorithm cannot run on a conventional computer. It was an entirely new type of object: a quantum algorithm, designed for a quantum computer.

Feynman realized the potential power that would be unleashed if this scenario could be exploited for the purpose of computation.

In quantum physics, subatomic particles do not have exact positions or velocities in the way that larger, more familiar objects such as people do. Instead, each particle is smeared across a range of possible states, some likelier than others. In this condition, it resembles a wave rather than a particle, with the peaks of the wave corresponding to more probable positions of the particle, and troughs to less likely ones. But when someone or something on the macroscopic scale intervenes, such as by measuring the particle's position, the wave collapses down to a single position. It is an extraordinary, counterintuitive idea, but one backed up by almost a century of evidence.

In a sense, before the measurement is taken, the particle is doing infinitely many things at the same time. In 1982, the Nobel Prize winning quantum physicist Richard Feynman realized the potential power that would be unleashed if this scenario could be exploited for the purpose of computation. Instead of proceeding one step at a time, a quantum computer could be spread across a range of states. By skillfully manipulating that distribution, and the interactions between the different possible states, a quantum programmer might coax the machine into having a high probability of landing in a situation which it would take a traditional computer far longer to reach.

This was exactly the principle behind Shor's algorithm. The mathematics behind it was unequivocal: Shor's method could factorize large numbers very quickly. This sparked huge interest in the theory of quantum computation, now a major topic in computer science. It also raised the stakes, as the development of a workable quantum machine would now be a world-changing event. So far, however, only tiny quantum devices have been constructed successfully. They are able to run Shor's algorithm, but 15 is the biggest number yet factored. As things stand, a machine powerful enough to run Shor's algorithm for large numbers still remains a distant prospect.

91 Fermat's Last Theorem

BREAKTHROUGH IN 1995, WILES PRODUCED THE LONG-AWAITED SOLUTION TO A NOTORIOUS PUZZLE POSED BY PIERRE DE FERMAT. IN SO DOING, HE IDENTIFIED A DEEP CONNECTION BETWEEN TWO MATHEMATICAL WORLDS.

DISCOVERER PIERRE DE FERMAT (1601–65), ANDREW WILES (1953–).

LEGACY WILES' DISCOVERIES EN ROUTE TO HIS PROOF REVOLUTIONIZED NUMBER THEORY, PROVIDING A NEW RAFT OF TECHNIQUES FOR STUDYING THE WHOLE NUMBERS.

The mid-17th-century thinker Pierre de Fermat was a seminal figure in the history of number theory. Yet, for all his great accomplishments, his name will forever be associated with one particular conundrum he did not solve. His so called "last theorem" has become emblematic of the depths that can lie beneath even seemingly simple mathematical puzzles. Its eventual solution by Andrew Wiles in 1996 is perhaps the most famous tale of mathematical brilliance and tenacity of the last few decades.

The roots of Fermat's enigma lie in an even greater landmark of the mathematical world. For thousands of years, Pythagoras' theorem has occupied a central place in geometry. It says that when you square each side of a right-angled triangle (that is, multiply each of the three lengths by itself), the two smaller squares add together to equal the largest. In algebraic terms, $a^2 + b^2 = c^2$ (which is short for $a \times a + b \times b = c \times c$).

A natural question is how this ancient piece of geometry interacts with the oldest and most important mathematical entities: the whole numbers. The answer to this question is not obvious. Typically, if two of a, b and c are whole numbers, then the third will not be. A right-angled triangle whose two shorter sides are 1 and 2, for example, has a longest side of $\sqrt{5}$, an irrational (and therefore non-whole) number.

OPPOSITE Fermat's Last Theorem poses restrictions on how cubes can relate to each other. He wrote "It is impossible to separate a cube into two cubes." This means that no two cubes can be combined to give a third, if all three have sides given exactly by whole numbers.

Pythagorean triples

Yet early number theorists realized that there are special cases where all the lengths are whole numbers. The first has lengths of 3, 4 and 5.

These numbers fulfill Pythagoras' theorem, since $3^2 + 4^2 = 9 + 16 = 25 = 5^2$. Diophantus of Alexandria recognized how important facts such as this are. Equations such as Pythagoras' theorem describe the possible (and impossible) relationships between whole numbers. Diophantus was fascinated by Pythagoras' equation, and in his most famous work, *Arithmetica*, published around AD 250, he presented a method for generating Pythagorean triples like (3, 4, 5). Further examples are (5, 12, 13), (9, 12, 15) and (7, 24, 25).

All further investigations suggested that Fermat had been right: that these equations are insoluble. But Fermat's claimed proof was never discovered, and nor could anyone else provide one.

When Diophantus' *Arithmetica* was rediscovered in the early 17th century, it inspired a new generation of number theorists. In the centuries that followed, a central goal of number theory was to understand the whole numbers, through the Diophantine method of determining which equations can and cannot be satisfied within them. This approach especially captured the imagination of Pierre de Fermat, the French lawyer, government official and sensational amateur mathematician.

Fermat's powers

As he leafed through his copy of *Arithmetica*, Fermat pondered a variation of the problem. What if he replaced the squares in Pythagoras' theorem with cubes? Could he find three whole numbers, a, b, c, where $a^3 + b^3 = c^3$ (that is, $a \times a \times a + b \times b \times b = c \times c \times c$)? His experiments produced no positive results, and Fermat began to believe that it could never be done. What about fourth powers: $a^4 + b^4 = c^4$? Again, he could find no whole numbers obeying this equation. Nor could he solve $a^5 + b^5 = c^5$. Fermat concluded that for all values of n bigger than 2, the equation $a^n + b^n = c^n$ has no solutions among the whole numbers. In his copy of *Arithmetica*, he penned the infamous lines:

> It is impossible to separate a cube into two cubes, or a fourth power into two fourth powers, or in general, any power higher than the second into two like powers. I have discovered a truly marvelous proof of this, which this margin is too narrow to contain.

These words would tantalize and torment mathematicians for hundreds of years. All further investigations suggested that Fermat had been right: that these equations are insoluble. But Fermat's claimed proof was never discovered, and nor could anyone else provide one. Fermat's "last theorem" grew into perhaps the most famous riddle in mathematics, until British researcher Andrew Wiles locked his sights on it in the 1980s.

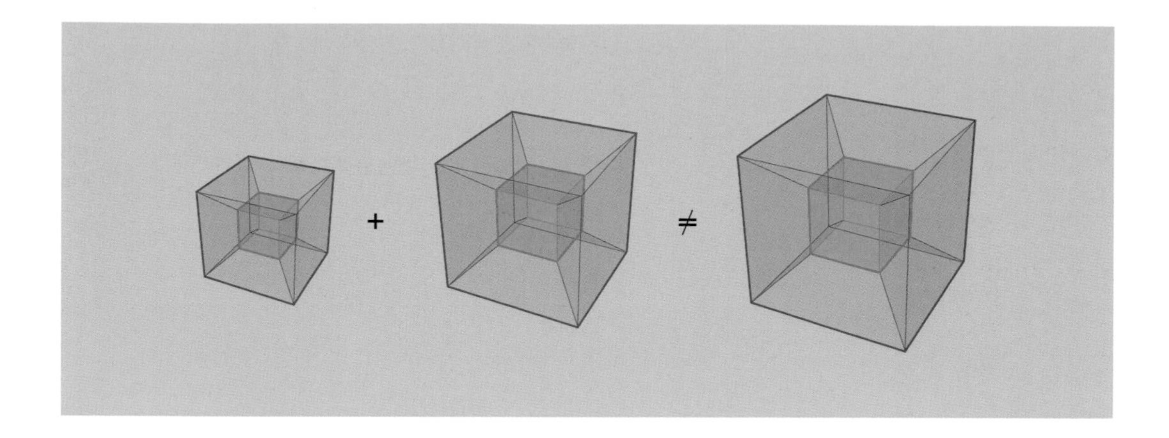

ABOVE Three hypercubes. Supposing that the dimensions of all three are whole numbers, it can never be that the hypervolume of the smaller two add up to that of the largest.

Wiles had first seen the statement as a child, and became determined to produce the proof that had eluded so many.

Putting the pieces together

A crucial development came in 1986, when Ken Ribet proved a result known as the epsilon conjecture. This breakthrough brought Fermat's Last Theorem within sight of known mathematical methods for the first time. Yet there were still immense obstacles to overcome. Wiles' approach was a proof by contradiction. He imagined that Fermat was wrong, that there was some triple of whole numbers a, b, c which satisfied $a^n + b^n = c^n$ (where n is at least 3). The first step was to use these three numbers geometrically to define an object called a Frey curve. Next, he needed to show that this curve could be transported to a completely different mathematical world. Back in the 1950s, two Japanese mathematicians, Yutaka Taniyama and Goro Shimura, had argued that there should be a deep connection between elliptic curves like Frey's and subtle objects called modular forms which live among the complex numbers. But no one had managed to prove it. Their modularity conjecture became Wiles's goal. If that could be proved, Ribet's epsilon theorem would do the rest.

Retreating into a near solitary existence for seven years, Wiles worked unceasingly on the proof, eventually announcing its completion in 1993. It was met with a flurry of excitement, but the experts who set about understanding the work quickly found a gap in the proof. Wiles had to return to the drawing board. A year later, with the assistance of his former student Richard Taylor, he had succeeded in navigating around the roadblock. They proved a restricted version of Shimura and Taniyama's modularity conjecture, enough to establish the impossibility of Frey's curve. With that, the proof of Fermat's Last Theorem was complete.

92 Kepler's conjecture

BREAKTHROUGH JOHANNES KEPLER POSED A QUESTION ABOUT THE BEST WAY TO STACK SPHERES. IN 1998, THOMAS HALES PROVIDED A "99% CERTAIN" PROOF OF KEPLER'S CONJECTURE.

DISCOVERER JOHANNES KEPLER (1571–1630), THOMAS HALES (1958–).

LEGACY ONGOING ATTEMPTS TO VERIFY HALES' RESULT REQUIRE IMPROVEMENTS IN COMPUTERIZED PROOF-CHECKING, A CHALLENGE FOR RESEARCHERS INTO ARTIFICIAL INTELLIGENCE.

In the late 16th century, the seafarer Sir Walter Raleigh was engaged by Queen Elizabeth I to explore and colonize the New World. During one of these voyages, he pondered the way that cannonballs were stored on his ship. He could not have known that his line of thought would turn into one of the outstanding geometrical questions of the coming centuries.

Traditionally, cannonballs were piled up in pyramids, just as oranges are still stacked at fruit stalls today. Raleigh wondered many balls each pyramid contained. If he had, say, 35 cannonballs, how high should the resulting pyramid be? He put these questions to the mathematician and astronomer Thomas Harriot, who was on board as navigator, accountant and all-purpose scientific consultant. With a few simple calculations, Harriot was able to answer Raleigh's queries. But they sparked in his mind some much subtler considerations.

On the bottom layer, the cannonballs were arranged in hexagons, with each ball touching six others. The next level was similar, with one ball not directly above another, but each resting as low as possible. Then the third layer sat directly over the first. Harriot wondered how efficient this method was. Might there be another system, which could pack more cannonballs into the same space? Harriot was an early convert to the theory of atoms, and for him this question had implications far beyond the transport of naval ammunition.

OPPOSITE Kepler's conjecture stated that the most efficient way to pack spheres together was in layers like this, laid on top of each other. In fact, this arrangement is a standard method used by greengrocers stacking fruit.

RIGHT Kepler's conjecture was first stated in his paper "On the Six-Cornered Snowflake." In pondering why snowflakes have six-fold symmetry, Kepler considered that circles and spheres can be packed together in hexagons, and proposed that this is the most efficient method possible.

Harriot discussed the matter with his friend, the German scientist Johannes Kepler. It turned out that the cannonball packing is very efficient, with a packing density of around 74% (or more precisely $\frac{\pi}{\sqrt{18}}$), meaning that the spheres occupy 74% of the available space, with only 26% being left empty. In contrast, arranging the spheres in a cubic lattice (with each touching four others on the same level, plus one above and one below) only fills around 52% of the space.

Neither Harriot nor Kepler was able to find any other arrangement of spheres that improved on the packing density of 74%, though slight variations could match that figure. In his seminal paper on crystals, entitled "On the Six-Cornered Snowflake," Kepler conjectured that this arrangement of spheres is indeed the most efficient possible. Of course, neither the sailors on Raleigh's ships nor any modern-day greengrocer would be remotely surprised by this revelation. Yet Kepler was not able to come up with a rigorous proof.

Gauss's lattices

Despite the seemingly obvious answer, the Kepler conjecture, as it became known, went on to defy generations of mathematicians. No one could find a more efficient way to pack spheres; nor could they prove that the cannonball packing arrangement is indeed optimal. A breakthrough came in 1831, when Carl Friedrich Gauss proved that this is the best possible regular configuration of spheres. That is to say, so long as the balls are arranged according to a repeating pattern, there is no way to improve on the density of 74%. All that remained for Kepler's conjecture to be solved was to rule out some strange, haphazard and yet highly efficient way of packing spheres.

Gauss also proved an analogous result for circles. Suppose you are challenged to fit as many identical coins on to a tabletop as possible. The obvious approach is to mimic the bottom layer of the cannonball packing and arrange the coins hexagonally, with each touching six others. With its packing density of around 91% (more accurately, $\frac{\pi}{\sqrt{12}}$) Gauss proved that this is the most efficient regular packing. It fell to later thinkers, Axel Thue in 1890 and László Fejes Tóth in 1940, to rule out any possible irregular configuration of circles.

Hales finally unveiled his extraordinary proof. The written part of the proof amounted to 250 pages, but this was dwarfed by the 3 gigabytes of computer code required to settle the 100,000 individual subproblems into which Hales had divided it.

Hales' theorem

In 1953, Fejes Tóth also outlined a strategy for proving the original Kepler conjecture. He realized that the problem could be reduced, in principle, to a single calculation. This entailed finding the minimum value achieved by a certain algebraic expression. The trouble was that the expression was extraordinarily long and complex, making the calculation completely impractical. With great foresight, however, Fejes Tóth commented that the situation might change, given "the rapid development of our computers."

Computers were indeed central to the next chapter of the story. It was Thomas Hales, a mathematician at the University of Michigan (now at Princeton University), who was able to turn Fejes Tóth's insight into a computer program to verify Kepler's conjecture. In 1998, assisted by his graduate student Samuel Ferguson, Hales finally unveiled his extraordinary proof. The written part of the proof amounted to 250 pages, but this was dwarfed by the 3 gigabytes of computer code required to settle the 100,000 individual subproblems into which Hales had divided it.

The monumental proof was handed to a panel of 12 experts, led by László Fejes Tóth's son, Gábor. After four years of work, however, the experts surrendered. In 2003, they announced that they were "99% certain" of the truth of the calculation, but unable to certify it completely. That remains its status at the time of writing. Like the proof of the four color theorem before it (see page 349), Kepler's conjecture has challenged the very nature of mathematics in the computer age, and it is in this spirit that the final chapter in the story may come to be written. Hales is currently working on what he terms the Flyspeck project, standing for Formal Proof of Kepler (or FPK). The goal is a new proof whose logical deductions can be formally verified by proof-checking software.

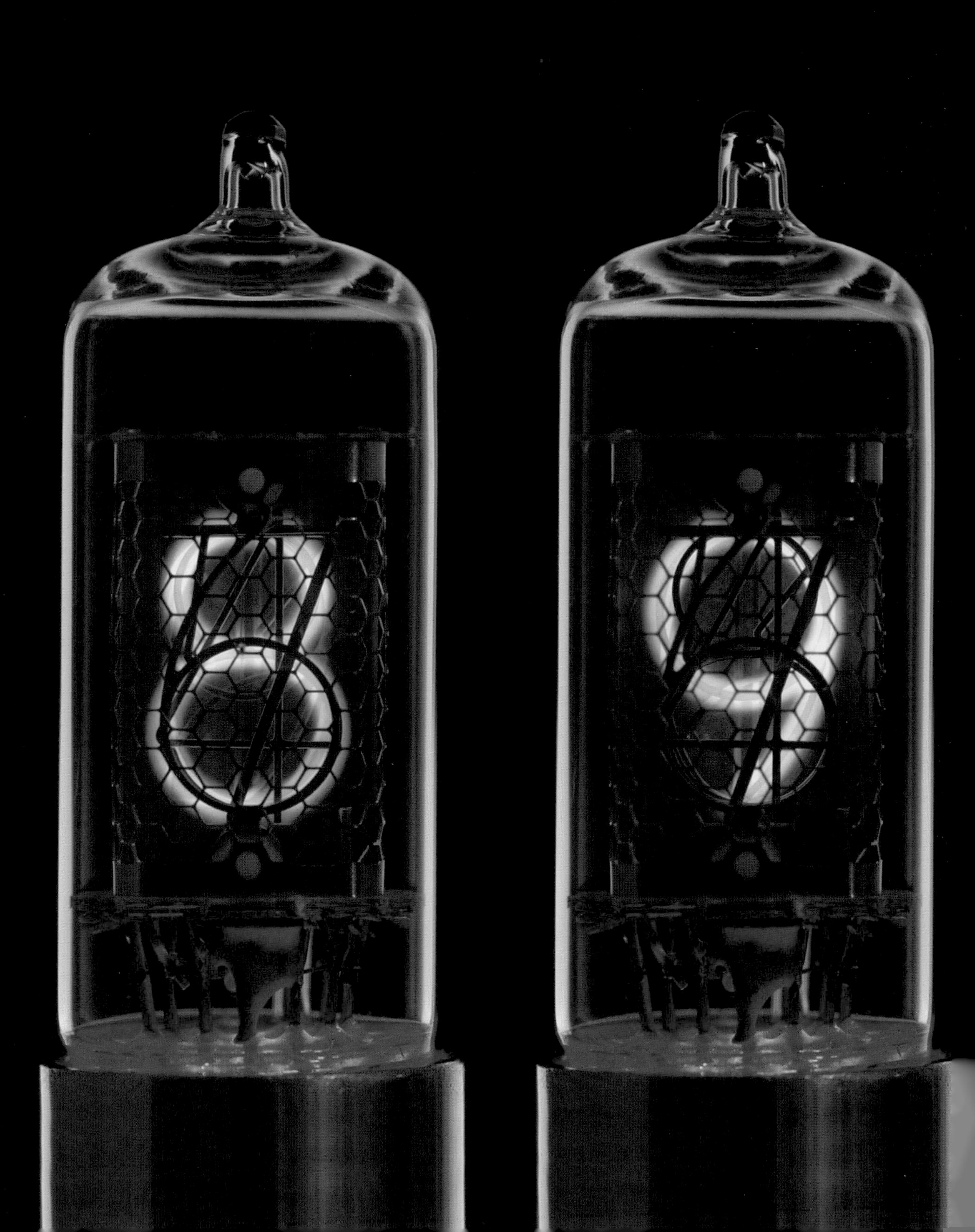

93 Catalan's conjecture

BREAKTHROUGH IN 1884, EUGÈNE CATALAN NOTICED THAT 8 AND 9 ARE THE ONLY NEIGHBORING POWERS AMONG THE WHOLE NUMBERS. MIHĂILESCU EVENTUALLY PROVED THIS IN 2002.

DISCOVERER EUGÈNE CATALAN (1814–94), PREDA MIHĂILESCU (1955–).

LEGACY MIHĂILESCU'S THEOREM WAS A MAJOR ACHIEVEMENT IN NUMBER THEORY. BUT MANY RELATED QUESTIONS REMAIN UNSOLVED.

To non-mathematicians, it can seem extraordinary that the whole numbers, that is 1, 2, 3, 4, 5 and so on, should hold any mysteries at all. After all, doesn't every eight-year-old child have a fairly good grasp of them? Yet even simple observations about whole numbers can hold the keys to deep and difficult problems. One of the most striking examples was a conjecture made by the Belgian mathematician and republican activist Eugène Catalan in 1884. It concerned the numbers 8 and 9.

Both 8 and 9 are powers of other numbers. A power is a number multiplied by itself a number of times. In these cases, 8 is a power of 2, specifically $2 \times 2 \times 2$ (or 2^3). Similarly 9 is 3×3, or 3^2. It hardly needs to be mentioned that 8 and 9 are also neighbors among the whole numbers. But, Catalan realized, this is exactly what makes them so special.

Neighboring powers

Search as he might, Catalan could find no other instances of neighboring powers. So he made the obvious conjecture: that there are no other neighboring powers—that 8 and 9 are the only place where this happens.

Catalan was a respected mathematics researcher and lecturer whose career was often complicated by his forthright republican views. On more than one occasion his political activism clashed with his employment, such as when he refused to swear allegiance to Emperor Napoleon and

OPPOSITE Strange neighbors. Catalan's conjecture tells us that 8 and 9 are the only pair of neighboring powers among all the whole numbers. Though simple to state, it took over a century for a proof to be provided.

was promptly fired from his teaching position. Despite these setbacks, Catalan made several important contributions to mathematics, in both number theory and geometry. He could not, however, provide a proof of his conjecture about the numbers 8 and 9.

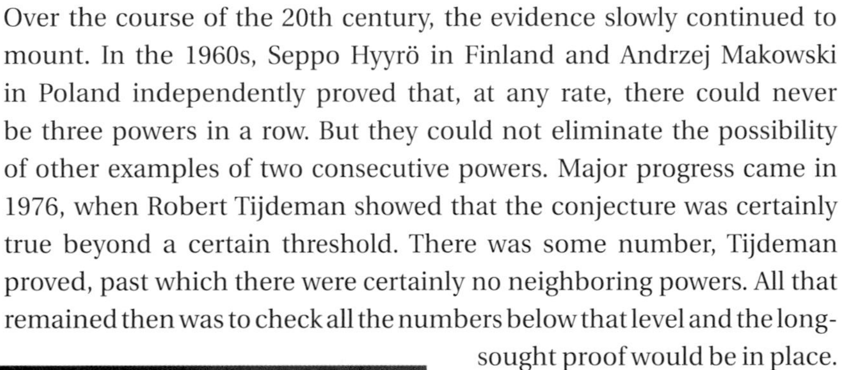

Five hundred years earlier, the rabbi Levi ben Gershon, also known as Gersonides, had made a similar observation. He had successfully proved that 8 and 9 represent the only case of a power of 2 sitting next to a power of 3. While this did not rule out other powers neighboring each other, it suggested that Catalan's hunch was right.

Over the course of the 20th century, the evidence slowly continued to mount. In the 1960s, Seppo Hyyrö in Finland and Andrzej Makowski in Poland independently proved that, at any rate, there could never be three powers in a row. But they could not eliminate the possibility of other examples of two consecutive powers. Major progress came in 1976, when Robert Tijdeman showed that the conjecture was certainly true beyond a certain threshold. There was some number, Tijdeman proved, past which there were certainly no neighboring powers. All that remained then was to check all the numbers below that level and the long-sought proof would be in place. But unfortunately Tijdeman couldn't put an exact figure on the threshold. Researchers in the 1990s managed to come up with a figure, but it was so absurdly large that the strategy was not feasible, even with the arrival of the computer age.

BELOW A portrait of Eugène Catalan by Emile Delperée in 1884. As well as a distinguished mathematician, Catalan was also an outspoken republican, who lost his job for refusing to swear an oath of allegiance to the Emperor Napoleon.

It was not until 2002 that Catalan's question was at last resolved. Using Galois theory (see page 179), the Romanian mathematician Preda Mihăilescu provided the long sought proof. Catalan had been correct: 8 and 9 are indeed the only consecutive powers.

Neighbors but one

Although Catalan's conjecture has now been settled, closely related questions remain open. For example, one could enquire about powers that, instead of being immediate neighbors, are 2 apart. Examples are 25 (which is equal to 5^2) and 27 (that is 3^3). The same question can be posed for powers 3 apart, as $125 = 5^3$ and $128 = 2^7$ are. In 1936, Subbayya Sivasankaranarayana Pillai conjectured that there should only be a limited supply of powers which are 2 apart, and 3 apart, and so on. In other words, for every number (call it k), the pairs of powers which are exactly k apart should form a strictly finite collection.

So far, the only instance of Pillai's conjecture which has been proved is where $k = 1$, which is to say Catalan's original conjecture. All the other cases remain defiantly open. Their truth would follow automatically from one of the biggest questions in number theory: the *abc* conjecture.

Major progress came in 1976, when Robert Tijdeman showed that there was some number beyond which there were certainly no neighboring powers. But unfortunately Tijdeman couldn't put an exact figure on the threshold.

As easy as abc

The *abc* conjecture is a tremendously subtle statement, contemplated by the number theorists Joseph Oesterlé and David Masser in 1985. They considered three whole numbers, a, b and c which are related by $a + b = c$, where a and b have no factors in common. An example is $a = 4$, $b = 5$ and $c = 9$. Now, take all the prime factors of a, b and c (ignoring any repetitions) and multiply them together to form a new number, d. In this case, $d = 2 \times 5 \times 3 = 30$. The crux of the matter is to compare c and d. Oesterlé and Masser noticed that most of the time d is bigger, as happens in this case. But occasionally c might be bigger, as is true for the triple $a = 1$, $b = 8$, $c = 9$. What the *abc* conjecture asserts is that these exceptional cases are rare and break the rule by only a small amount. The technical statement is that if e is any number bigger than 1 (even by a tiny amount), then there will only be finitely many triples of a, b, c where $c > d^e$.

At first sight, this seems fearfully arcane. But if it were proved, the repercussions around the mathematical world would be huge. Not only would all the outstanding cases of Pillai's conjecture automatically follow, but so would a proof of Waring's problem (see page 241) and a raft of other outstanding problems in number theory. In August 2012, Shinichi Mochizuki released a purported proof of this momentous conjecture. Whether or not it is correct remains unclear at time of writing.

94 Poincaré's conjecture

BREAKTHROUGH POINCARÉ NOTICED THAT, FROM A CERTAIN PERSPECTIVE, SPHERES ARE THE ONLY SHAPES WITHOUT ANY HOLES. THIS CONJECTURE WAS FINALLY PROVED BY PERELMAN IN 2002.

DISCOVERER HENRI POINCARÉ (1854–1912), GRIGORI PERELMAN (1966–).

LEGACY PERELMAN'S WORK SURPASSED EVEN POINCARÉ'S ORIGINAL OBSERVATION, LEAVING US WITH A WHOLLY NEW UNDERSTANDING OF 3-DIMENSIONAL SPACES.

At the turn of the 20th century, one of the greatest mathematicians in the world, Henri Poincaré, turned his mind to the contemplation of holes. Which shapes have holes? And which do not? No one could have predicted that these questions would develop into one of the deepest and most difficult subjects in mathematics.

Poincaré realized that the best perspective for analyzing holes comes from topology. Topology was already growing into a fashionable theory, and with Poincaré's backing, it would go on to influence almost every branch of mathematics over the 20th century. Like geometry, topology is the study of shapes. It differs from geometry, however, in its approach. Geometers fret about the fine details of a shape—the lengths of its lines, its angles, the degree of curvature in each region. Meanwhile, a topologist is only interested in the much broader features of a shape, aspects which can survive a violent regime of pulling, stretching and twisting. The result is that families of shapes which a geometer would consider to be different are viewed by topologists as essentially the same.

Poincaré realized that the most important data that survives topological stretching and twisting is the number and type of holes within the shape. The surface of a sphere, for example, contains no holes. To a topologist, a sphere is the same thing as a cube or a cylinder, and it is no coincidence that these shapes also have no holes. The classic example of a surface with a hole is a donut, or, as it is known to mathematicians, a torus. This is topologically different from a sphere. Different again are the double

OPPOSITE The breakthrough which opened the door to the long-awaited proof of Poincaré's conjecture was treating curvature as something mutable, even flowing, rather than static.

torus, with its two holes, and the triple torus, and so on. However, the mathematical lexicon contains far weirder surfaces than these, a famous example being the Klein bottle. Does the Klein bottle have a hole? At this stage, Poincaré realized that the old, informal notion of a hole was not good enough. To determine whether or not strange shapes like the Klein bottle contain holes, he needed a proper, precise definition.

Contracting loops

If you draw a loop on a sphere, it can move and shrink away to a single point. Try the same trick on a torus, though, and the loop will get stuck around the hole. That was Poincaré's test for whether or not a surface has a hole. If every possible loop contracts, then there are no holes. So what of the Klein bottle? Some loops do get stuck. So the Klein bottle, like the torus, has a hole.

Poincaré's question asked whether, topologically speaking, the 3-hypersphere is the only 3-dimensional shape with no holes. He believed that it was, but was unable to prove it.

In fact, it soon became clear from a topological perspective that the sphere is the only surface with no holes. Surfaces are 2-dimensional shapes. What, Poincaré asked, of shapes in higher dimensions? The difficulty is that these are impossible to visualize. Humans can just about cope with ordinary 3-dimensional space. But there are other 3-dimensional spaces that curve around on themselves, just as 2-dimensional surfaces like the sphere, torus and Klein bottle do.

Hyperspheres

In fact, each dimension has its own version of the sphere. Like its little brother, the 3-dimensional hypersphere is a shape with no holes. Poincaré's question asked whether, topologically speaking, the 3-hypersphere is the *only* 3-dimensional shape with no holes. He believed that it was, but was unable to prove it.

The 20th century was a boom time for topologists. But through it all, Poincaré's conjecture about the 3-hypersphere stood defiantly open. The latest powerful techniques were not enough to prove it, while even the most imaginative of mathematicians were unable to concoct a new hole-less shape to refute it. Surprisingly, in yet higher dimensions, the situation was more tractable. In 1961, Steven Smale proved that in dimensions 5, 6, and upward, the analogous rule holds true. In each case, there is only one shape without holes, namely the resident hypersphere. In 1982, Michael Freedman was able to deduce the same thing for the 4-dimensional hypersphere. So for two decades, the only

LEFT A Klein bottle (center) and variations on the theme, made by glassmaker Alan Bennett in 1995. In topology, the most important data about a shape is the number and type of holes in contains. Poincaré's conjecture asserted that spheres are the only shapes containing no holes.

remaining mystery was the one Poincaré had originally enquired about: the 3-dimensional hypersphere.

Ricci flow

There was a plan of attack, though. Richard Hamilton came up with an idea to exploit a very different area of mathematics, that of flows. Mathematicians had long studied the flow of heat and of liquids (see page 181). Hamilton invented the Ricci flow as an ingenious way of turning curvature into something fluid, instead of static. Hamilton's plan was to begin with a shape with no holes, and then allow its curvature to flow naturally around the shape. He believed that in time it would settle down into the form of a hypersphere, proving Poincaré's conjecture.

It was a brilliant and beautiful idea; but it didn't quite work. The trouble was that the flow sometimes hit a bottleneck, pinching off a piece of the shape. This obstacle was finally overcome by a reclusive Russian mathematician by the name of Grigori Perelman. Perelman was able to dissect every shape carefully, in such a way that the Ricci flow could run freely around each piece. By applying his technique and allowing the curvature to flow, it finally became clear that any hole-less shape would indeed eventually reveal itself as a hypersphere.

Perelman's triumph was one of the greatest mathematical feats of recent times. But in the months and years that followed, he retreated further into isolation, refusing both the $1,000,000 prize he was awarded by the Clay Institute, and mathematics' top prize, the Fields Medal.

95 Constellations of primes

BREAKTHROUGH CONSTELLATIONS ARE PATTERNS WITHIN THE PRIME NUMBERS. ASKING WHETHER AND HOW OFTEN CERTAIN PATTERNS APPEAR IS A DEEP AND DIFFICULT QUESTION, BUT A MAJOR BREAKTHROUGH CAME IN 2004 WITH THE PROOF OF THE GREEN-TAO THEOREM.

DISCOVERER TERENCE TAO (1975–), BEN GREEN (1977–).

LEGACY THE GREEN–TAO THEOREM, AND DIRICHLET'S THEOREM BEFORE IT, WERE MAJOR MATHEMATICAL EVENTS. NEVERTHELESS, MORE REMAINS MYSTERIOUS THAN IS KNOWN IN THIS AREA.

Even today, the prime numbers are a central concern in modern mathematics. A particularly pressing question is: what patterns are hiding within the infinite sequence of prime numbers? Partial answers to this question were provided in 1837 by Johann Dirichlet, and in 2004 in the wonderful theorem of Terence Tao and Ben Green. But the whole subject remains shrouded in mystery.

In 1837, Johann Dirichlet analyzed the relationship between the prime numbers and another simpler type of sequence, starting with a number such as 2, and then repeatedly adding on another number, perhaps 5. This produces a sequence known as an arithmetic progression: 2, 7, 12, 17, 22, 27, 32, 37, and so on.

Dirichlet proved a wonderful theorem describing how the primes cut across these various sequences: he showed that every such arithmetic progression must hit infinitely many prime numbers. (The only exceptions are sequences like 6, 10, 14, 18, 22, and so on. In this case every number in the sequence will be even, and so non-prime. In more technical terms, the obstacle here is that the two initial numbers share a common factor, specifically 2.)

OPPOSITE The Pleiades, part of the constellation of Taurus, and among the brightest stars in the night sky. As the star-gazers of the past searched for meaning in the patterns they saw in the night sky, so today's mathematicians hunt for constellations among the prime numbers.

Progressions of primes

Dirichlet's theorem was a leap forward in understanding, but there were many questions that it left unanswered. Such as: how many primes may themselves be in arithmetic progression? For example, the sequence

ABOVE The Arecibo message, broadcast in 1974 toward the star cluster M13. Humanity's message to the Universe comprised 1679 binary digits. This number is two primes multiplied together (23×73), which means there are only two ways to arrange the data as a rectangle, from which the message can be seen.

3, 5, 7 is a special run of primes, in that each is the same distance from the previous one (in this case, they are 2 apart). Similarly, 11, 17, 23, 29 are four primes, each separated from the last by the same distance (namely 6).

How long can sequences like this be? The search quickly becomes hard, as the individual numbers involved become very large, too. The longest currently known arithmetic progression of primes consists of 26, beginning with 43,142,746,595,714,191 and then increasing in steps of 544,680,710. It has long been conjectured that there should be arithmetic progressions of primes of every possible length. This idea dates back at least to 1770, to the work of Edward Waring and Joseph-Louis Lagrange. But the conjecture resisted all attempts at proof until 2004, when Ben Green and Terence Tao collaborated to prove their stunning theorem.

If you want a list of 100 primes, each exactly the same distance from the last, the Green-Tao theorem guarantees that there will be such a list somewhere. It does not, however, provide much useful information about where to start looking!

Constellations of primes

A closely related question is how many times a certain pattern of primes should appear. The most famous instance of this question is the twin prime conjecture. Twin primes are primes which are two apart, such as 3 and 5, and 11 and 13. Millions of such pairs have been found; at time of writing the largest known pair are 100,355 digits long! The twin prime conjecture asserts that, like the primes themselves, the list of twin primes actually continues forever—there are infinitely many such pairs. Unfortunately, however, all efforts to prove this have so far failed.

Twin primes are one example of what are sometimes called constellations of primes. A constellation is a pattern: this particular pattern consists of just two numbers, 2 apart. A slightly broader class of constellations is addressed by de Polignac's conjecture. In 1849, Alphonse de Polignac asserted that, for every even number, there are infinitely many pairs of primes that are a fixed distance apart. So there must be infinitely many primes which are 4 apart (starting with 3 and 7), then infinitely many like 5 and 11, which are 6 apart, and so on. Of course, de Polignac's conjecture includes the twin primes conjecture as a special case. No cases of de Polignac's conjecture are yet proved for certain.

The Hardy–Littlewood conjecture and hypothesis H

De Polignac's conjecture is itself subsumed into one of the outstanding questions in the whole of mathematics: the Hardy–Littlewood conjecture. Just as the prime number theorem (see page 207) provides accurate estimates of how many individual prime numbers there are beneath a certain limit, so G.H. Hardy and John Littlewood proposed an extension of that theorem to describe how many times a certain constellation of primes should appear. For example, their conjecture can be used to give an estimate for the number of twin primes below one million, say. (The estimate is 8248, while the exact answer is 8169.)

If you want a list of 100 primes, each exactly the same distance from the last, the Green–Tao theorem guarantees that there will be such a list somewhere. It does not, however, provide much useful information about where to start looking!

All the evidence is that these estimates work just as well as those produced by the prime number theorem. However, unlike in that case, it has not yet been proved that they are bound to do so. If it could be proved, then Dirichlet's theorem and the Green–Tao theorem would be subsumed by a single result, implying the truth of a host of conjectural facts about the prime numbers, including the twin prime conjecture, de Polignac's conjecture and several other claimed results.

In 1958, Andrzej Schinzel and Wacław Sierpiński pushed the bounds of the Hardy–Littlewood conjecture even further to formulate a sweeping statement which would essentially settle the entire question of constellations of primes. Their extended conjecture, known as hypothesis H, would also imply results such as the fact that there are infinitely many primes of the form $n^2 + 1$ (an example is 5, which is equal to $2^2 + 1$).

Patterns among the prime numbers have been a major preoccupation over the centuries. The fact that the twin prime conjecture remains unsolved is indicative of the deep difficulty of such questions. In the Hardy–Littlewood conjecture and hypothesis H, mathematicians have found vast generalizations of this problem. Whether such sweeping conjectures are within the scope of this generation of number theorists remains to be seen. If they are, it would finally bring some light into this dark corner of mathematics.

96 The classification of finite simple groups

BREAKTHROUGH PROVING THAT EVERY FINITE SIMPLE GROUP MUST BE ONE OF A KNOWN TYPE. THIS INVOLVED TAMING THE NOTORIOUS MONSTER GROUP.

DISCOVERER OVER A HUNDRED MATHEMATICIANS AROUND THE WORLD, INCLUDING DANIEL GORENSTEIN, JOHN CONWAY, MICHAEL ASCHBACHER AND SIMON NORTON.

LEGACY A MILESTONE IN ALGEBRA WHICH HAS BEEN OF INCALCULABLE VALUE TO RESEARCHERS EVER SINCE.

In 1972 in Chicago, Daniel Gorenstein hatched one of the most ambitious plans in the history of mathematics. Gorenstein was a researcher into the subject known as group theory. From its beginnings in the early 19th century, group theory had grown into a central strand of mathematics. Now, Gorenstein thought, the techniques were in place to mount an assault on the biggest questions in the subject. Over 30 years later, with the input of over 100 mathematicians, the project reached its triumphant conclusion: a complete classification of finite simple groups.

Groups have their origins in the work of Évariste Galois and Niels Abel in the early 19th century (see page 177). These thinkers were interested in algebra, but they came to recognize the relevance of a notion from a different branch of mathematics: geometry. That idea was symmetry. Just as rotating a square by 90° interchanges its edges but leaves it looking identical, so Galois realized that solutions to an equation could be swapped around without altering the equation's fundamental appearance.

Symmetries of shapes

That similar rules should govern such different situations indicated that something profound was afoot, and in 1854, Arthur Cayley formulated the abstract laws of a group. The same group can appear as the symmetries of a shape such as a square or a cube, and as the symmetries of an equation, and in innumerable other mathematical scenarios. But a

OPPOSITE Cubic houses in Rotterdam, Netherlands. All the symmetries that we see in the natural world, and those of art and architecture, are expressible in the language of group theory. That theory has been developed abstractly over the last century, with the classification of finite simple groups being a crowning glory.

group theorist will not commit themselves to any one setting, preferring to understand the group's structure internally.

Finite groups

Some groups are infinite. A circle, for example, has an unlimited number of symmetries. You can rotate it by 0.52°, or by 311.43°, or any value in between, and it will look identical. A square, meanwhile, has only finitely many symmetries. Rotations by 90°, 180°, 270°, or 360° are true symmetries of the shape, but rotation by 33.19° is not—it leaves it looking wonky.

Both finite and infinite groups exist in a wide variety of forms. The collection of whole numbers constitutes an infinite group, as do other number systems such as the real and complex numbers, and the many structures that have been discovered during investigations into abstract algebra (see page 165). Finite groups are, by their nature, more limited. Nevertheless, the variety of finite groups discovered during the 20th century still seemed so wide as to defy understanding. So finite group theorists decided to focus their efforts on the most important type.

Certain groups are analogous to prime numbers, in that they can't be broken down into any smaller groups. These indecomposable groups are known as simple.

It was Galois who identified a critical difference in certain groups analogous to prime numbers, in that they can't be broken down. These indecomposable groups are known as simple. This observation became all the more important when Camille Jordan and Otto Hölder realized in the 1860s that the analogy with prime numbers runs even deeper. Just as every whole number can be broken down into primes, so every finite group can be deconstructed into a collection of finite simple groups.

Gorenstein's program

Jordan and Hölder had shown that simple groups are the bedrock on which our understanding of all finite groups must depend. So it was on these special groups that Gorenstein set his sights in 1972. What he wanted was a classification of finite simple groups, meaning a complete listing of all of them. Classification theorems like this play pivotal roles in mathematics, starting with the ancient geometers' fascination with the Platonic solids (see page 37). In that case, Theaetetus had not only come up with a list of five beautiful shapes, but had then sealed the result with the wonderful theorem that these five types are all there are: no sixth Platonic solid will ever be found. Gorenstein hoped to mimic this result, and called on group theorists around the world to help carry out his 16-step plan for classification. Hundreds of experts rallied to the cause, but even so, it was a daunting prospect. To begin with, it was far from clear that the currently known list of examples were all there were. It seemed highly likely that complicated new finite simple groups were waiting to be discovered. What was more, unlike the five Platonic solids, the currently known collection of simple groups already formed an infinite list.

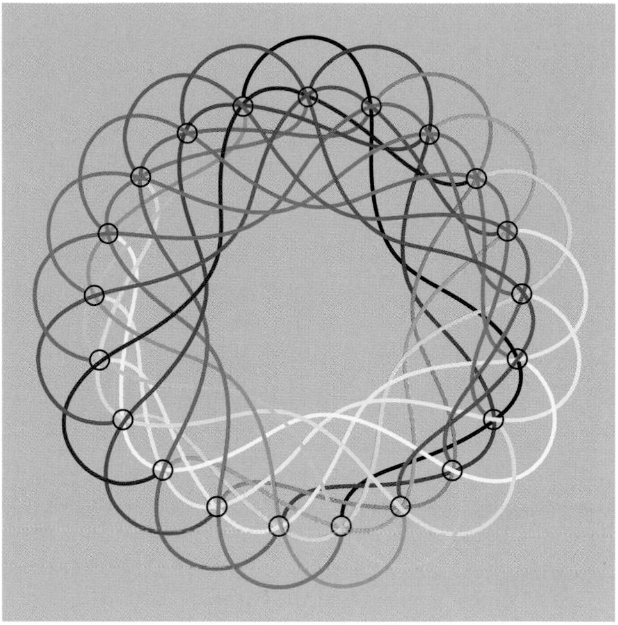

Families and sporadics

The classification demanded that the finite simple groups be organized into families. The experts identified 18 of these. Each family comprised symmetries of certain types of geometrical object. The simplest family consists of the groups of rotations of a regular triangle, pentagon, heptagon, and generally a polygon with prime number of sides. (The rotations of a square or other non-prime polygon form a non-simple finite group.) The other families related to more abstract geometries.

ABOVE A projective plane, in which each of the 21 points lies on 5 lines, and each of the 21 lines passes through 5 points. The symmetries of objects such as this give rise to families of finite simple groups listed in the classification.

Inconveniently, though, these 18 families did not quite cover every possibility. Some awkward one-off groups were also found which failed to fit any of the patterns. These sporadic groups posed a major challenge to Gorenstein's program. In their case, an exhaustive list was necessary. It turned out that there are 26 sporadic groups in total, the largest being *the monster*, which has size:

808,017,424,794,512,875,886,459,904,961,710,757,005,754,368,000,000,000.

Though initially declared complete in 1981, holes in the classification were subsequently discovered. It was not until 2004 that these gaps had been patched, at which stage the wonderful theorem could finally be stated: every finite simple group must belong to one of the 18 families, or be one of the 26 sporadic groups.

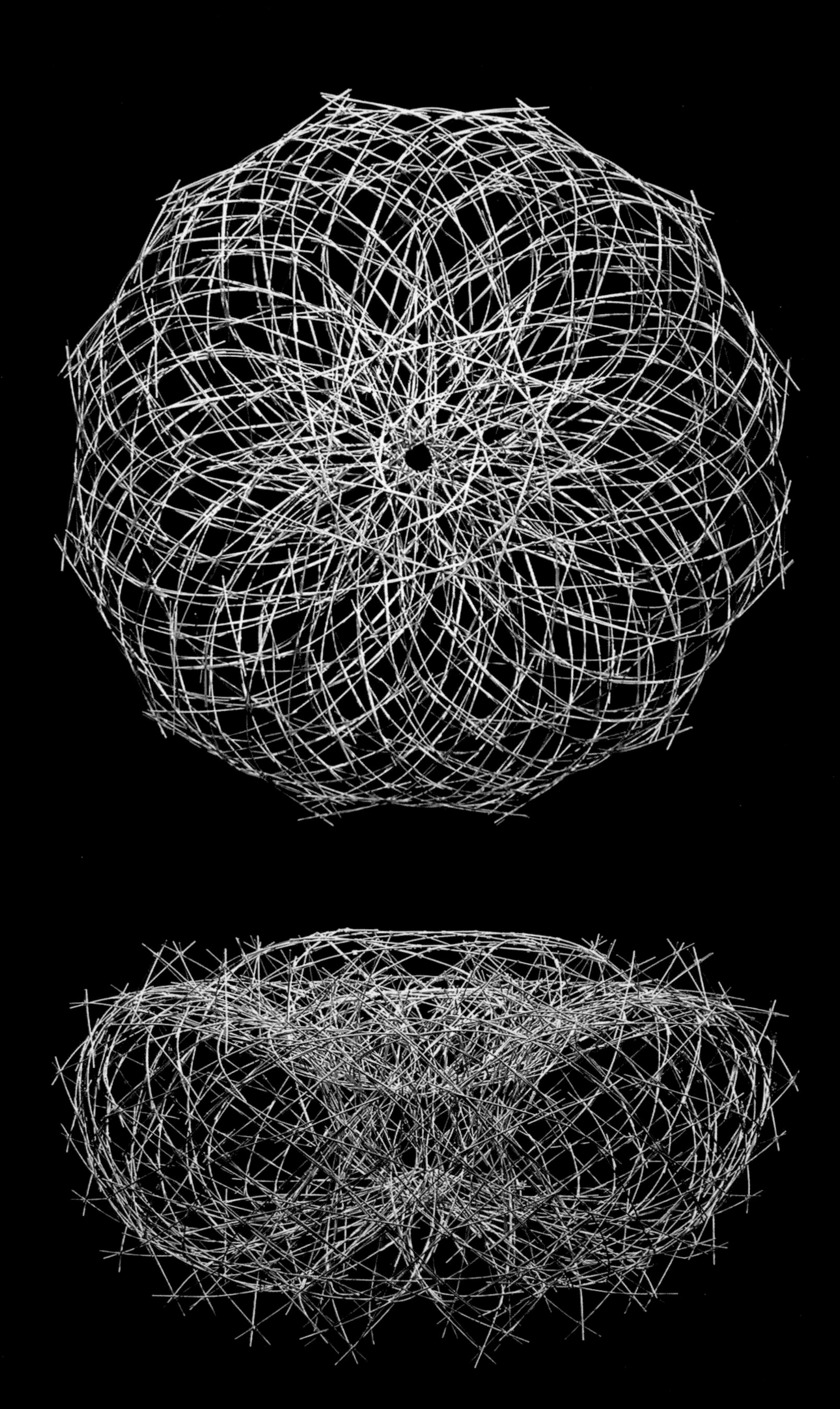

97 Langlands program

BREAKTHROUGH ROBERT LANGLANDS PUT TOGETHER A NETWORK OF CONJECTURES TO UNIFY SEVERAL TOPICS IN MATHEMATICS. A BREAKTHROUGH WAS MADE IN 2009, WITH NGÔ'S PROOF OF THE SO-CALLED FUNDAMENTAL LEMMA.

DISCOVERER ROBERT LANGLANDS (1936–), LAURENT LAFFORGUE (1966–), NGÔ BẢO CHÂU (1972–).

LEGACY ALTHOUGH SEVERAL OF LANGLANDS' CONJECTURES HAVE SUBSEQUENTLY BEEN PROVED, THERE REMAINS A DAUNTING AMOUNT STILL TO DO.

The research program outlined by Robert Langlands in 1967 is undoubtedly one of most ambitious ventures in the history of mathematics. Yet, in recent years, there has been significant progress, Ngô Bảo Châu's proof of the "fundamental lemma" in 2009 marking a particular breakthrough.

For several decades, number theorists have been investigating the properties of a particularly difficult type of geometrical object: the elliptic curve (see page 357). There has been interest in these objects since the time of Newton, but it was only during the 20th century that elliptic curves began to appear in some of the biggest stories in mathematics.

Modular forms

When Andrew Wiles proved Fermat's Last Theorem (see page 369), the key step in his proof involved elliptic curves. In particular, he deduced a very surprising fact: that elliptic curves are also modular forms. Modular forms are objects from a completely different area of mathematics, that of complex analysis (see page 154). They are highly symmetrical structures living among the complex numbers. That elliptic curves should actually be modular forms in disguise was first conjectured by Yukuta Taniyama and Goro Shimura in the 1950s. In 1986, Gerhard Frey dragged this arcane topic firmly into the spotlight, with a sensational observation: if Fermat's Last Theorem was false, this would give rise to a non-modular elliptic curve. So a proof of the Taniyama–Shimura conjecture would automatically imply Fermat's Last Theorem.

OPPOSITE "Torus" by Manuel A. Baez. As well as being a geometrical object a torus has an algebraic structure, seen by analyzing the different routes which can be taken around it. Algebraic tori in different dimensions play a central role in Langlands' program.

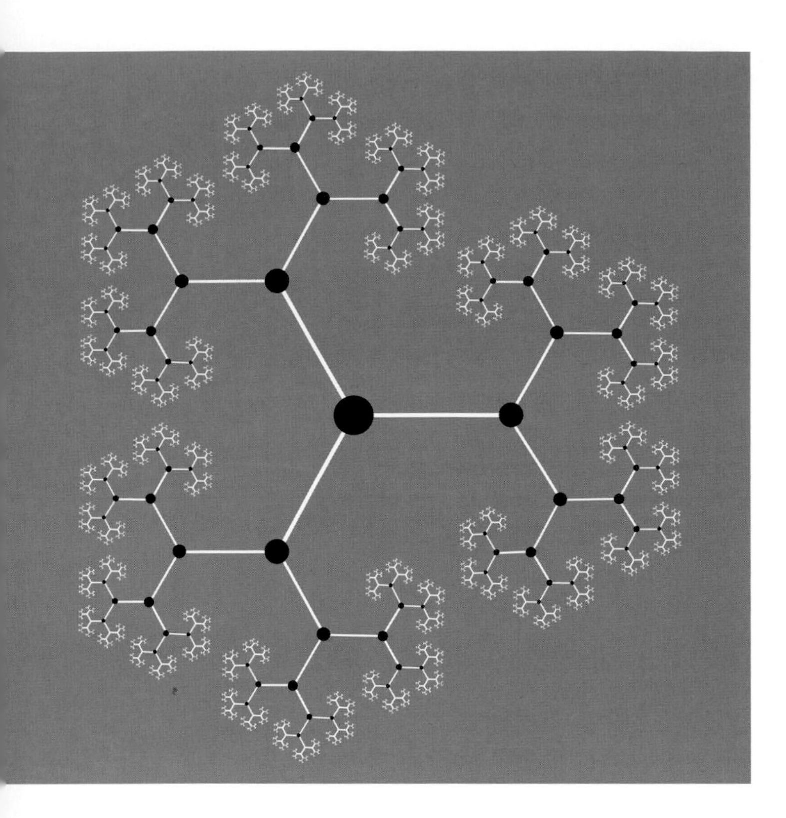

ABOVE A mathematical tree, in which every node has three neighbors and there are no loops. Objects such as this can be extracted from algebraic scenarios. Understanding the symmetries of the resulting trees plays a role in Langlands' fundamental lemma.

This was exactly Wiles' approach. However, he did not prove the full Taniyama–Shimura conjecture, but rather a limited version which applied only to "semi-stable" elliptic curves. This was enough for Fermat's Last Theorem to follow. A complete proof of the whole modularity theorem, as it became known, followed in 2001 at the hands of Christophe Breuil, Brian Conrad, Fred Diamond and Richard Taylor.

Robert Langlands

The modularity theorem was a startling result, unexpectedly forging a connection between two seemingly distant areas of mathematics. But the theorem itself was only the tip of the iceberg, going on to form a small part of one of the biggest challenges in modern mathematics, known as the Langlands program.

The Langlands program began inauspiciously. Writing to the famous number theorist André Weil in 1967, Langlands started on an unpromising note:

> In response to your invitation to come and talk I wrote the enclosed letter. After I wrote it I realized there was hardly a statement in it of which I was certain. If you are willing to read it as pure speculation I would appreciate that; if not—I am sure you have a waste basket handy.

Contained in the letter was a unifying vision of two very different areas of mathematics: algebra and complex analysis. On the side of complex analysis, the central objects are automorphic forms—a broad expansion of modular forms. In both scenarios, through very different means, one can concoct a mathematical gadget called an L-function, a close cousin of the Riemann zeta function (see page 205). Essentially, the Langlands program involves establishing that the same L-functions arise on each side of the divide. These L-functions thereby create a bridge between the algebraic and analytic worlds.

Langlands' conjectures apply in multiple different settings, and more progress has been made in certain areas than in others. The most intractable situations are those relating to number theory. Yet, in more geometrical areas, success has been possible. In the specific setting of so-called functions fields, Laurent Lafforgue was able to prove correct a large swathe of Langlands' conjectures in 2002, winning a Fields Medal for his work in the same year.

Ngo's proof of the fundamental lemma

None of the high-powered mathematicians whom Langlands approached were able to supply the required proof. This raised the stakes considerably, and Langlands' minor lemma was promoted to the rank of the fundamental lemma.

Monumental though they were, Langlands himself was able to make a start on the maze of conjectures. One of the earliest successes was the discovery of a device known as the Arthur–Selburg trace formula, which extracts number-theoretical information from a geometrical scenario. Langlands wanted to put this formula to work, but whenever he tried to do so, he encountered an obstacle. As he added together certain collections of numbers, the answers came out slightly differently from how he expected. This was irritating, but did not seem a major blow to his hopes. So he set one of his graduate students the job of proving that the answer he had obtained was really the same as the number he needed. Langlands initially thought this should be nothing more than a lemma—mathematicians' name for a minor mathematical result, several ranks below a true theorem.

The graduate student was not able to oblige, however, and when Langlands tackled the problem himself, he also found it considerably more difficult than he had expected. He discussed it with his peers, but none of the high-powered mathematicians whom Langlands approached were able to supply the required proof. It remained a frustrating obstacle to progress, to the extent that several researchers in the area simply began to assume the lemma's truth. This raised the stakes considerably, and Langlands' minor lemma was promoted to the rank of the fundamental lemma.

The breakthrough arrived in 2009, when Ngô Bảo Châu deployed a sophisticated idea from mathematical physics, that of a Hitchin fibration, to finally prove the fundamental lemma, and thereby establish the truth of the many results which depended on it. Ngô's theorem also earned him a Fields Medal, and it represents a major step toward making sense of Langlands' grand vision. Even so, a great deal of work is needed before the Langlands program can be fully realized.

98 Reverse mathematics

BREAKTHROUGH REVERSE MATHEMATICIANS INVESTIGATE HOW THE FACTS AND THEOREMS OF MAINSTREAM MATHEMATICS RELATE TO THE UNDERLYING LOGICAL SYSTEMS BEING USED.

DISCOVERER GERHARD GENTZEN (1909–45), HARVEY FRIEDMAN (1948–), STEPHEN SIMPSON (1948–).

LEGACY FRIEDMAN'S SEEMINGLY SIMPLE STATEMENTS ABOUT WHOLE NUMBERS WHICH REQUIRE THE STRONGEST LOGICAL AXIOMS THREATENS TO USHER UNPROVABILITY IN TO THE HEART OF MATHEMATICS.

Gödel's incompleteness theorems (see page 277) turned the world of mathematical logic upside-down. It meant that the long-held goal of a single set of laws underlying the arithmetic of the whole numbers had to be abandoned. In the years since, proof theorists have compared the strengths of different logical laws for arithmetic, while reverse mathematicians have asked which laws are required to prove specific theorems. One of the most spectacular discoveries involves statements that are not provable within any of the usual logical frameworks.

The standard laws for describing the whole numbers are those of Peano arithmetic, named after its inventor, the 19th-century logician Giuseppe Peano. This is a list of seven simple axioms, which largely seem to be statements of the obvious. For example, if x and y are unequal numbers, then so are $x + 1$ and $y + 1$. It had once been hoped that Peano's laws might provide a solid underpinning for the whole of mathematics. But in 1931, Gödel's incompleteness theorem killed that idea stone dead. Despite being incomplete, it seemed adequate for most practical purposes.

Even if it was technically incomplete, one might at least hope that Peano arithmetic should be consistent, meaning that it doesn't contain any hidden contradictions. If it is to be any use at all, then following the rules of Peano arithmetic should never produce a nonsensical outcome such as $1 + 2 = 4$.

OPPOSITE A reverse mathematician aims to reconstruct a logical framework by examining its mathematical consequences. Just as the patterns within a stained-glass window produce a complex interplay of colored light, so the relationship between the logical laws of numbers and the theorems that mathematicians prove about them is far from straightforward.

Proof theory

Unfortunately, Gödel produced an unwelcome surprise here too. His second incompleteness theorem stated that no logical framework for arithmetic could ever prove itself consistent. Yet Gödel left open a subtle possibility, which Gerhard Gentzen seized upon only a few years later. After the upsets caused by Gödel, Gentzen's theorem was more reassuring: he showed that Peano arithmetic is indeed consistent. He avoided Gödel's second theorem by working not within Peano arithmetic itself, but in an alternative logical system. Gentzen's new system was in some ways stronger than Peano's, but in other respects weaker. It could not prove itself consistent, but it was enough to prove Peano arithmetic consistent.

Gentzen's work sparked a flurry of interest in comparing the strengths of different axiomatic systems. This new subject became known as proof theory, and it continues to be a preoccupation of logicians today.

In the years since Gentzen's seminal result, proof theorists have analyzed a large variety of different logical systems. An interesting question is how these relate to the theorems of mainstream mathematics. Given a classical mathematical theorem, what are the underlying logical assumptions needed to prove it? Such questions have surprising answers. For instance, it turns out that the Jordan curve theorem (see page 209) corresponds to a logically stronger system than the intermediate value theorem (see page 173).

Concrete incompleteness

Gödel's theorem guaranteed the existence of unprovable statements—facts about numbers which the given laws are inadequate to prove. Fascinating though they were, Gödel's examples were highly artificial, and unlikely to worry researchers in mainstream mathematics.

Given a classical mathematical theorem, what are the underlying logical assumptions needed to prove it?

One of the earliest examples of concrete incompleteness was discovered in 1977 by Jeff Paris and Leo Harrington. They identified a slight variation of Ramsey's theorem (see page 273), which it seemed clear that all numbers should satisfy. This was a natural statement, exactly the sort of thing that might appear within ordinary mathematics. Nevertheless, Paris and Harrington established that the result was unprovable within Peano arithmetic.

Since this breakthrough, several further instances of concrete incompleteness have been found. Particularly shocking are those discovered by the logician Harvey Friedman, one of the founding fathers of reverse mathematics. It is not too difficult to extend Peano arithmetic in such a way as to render the Paris–Harrington theorem provable. But in recent years, Friedman has identified patterns among the whole numbers whose existence cannot be proved in *any* ordinary logical system.

ABOVE König's lemma says that if a mathematical tree has infinitely many nodes, but each node only has finitely many neighbors, then the tree must have an infinitely long branch. This fact cannot be proved in the simplest logical frameworks, and is an axiom of certain stronger systems.

Large cardinal axioms

Any unprovable statement can be turned into a provable one by adding the appropriate new axioms to the rule book. But which new rules are needed to tame Friedman's patterns? With none of the usual systems for mathematics seeming to be up to the task, the answer came as a major shock.

The logically strongest laws of mathematics have their origins in Georg Cantor's theories of infinity (see page 217). Following Cantor, modern set theorists accept the existence of many levels of infinity, known as cardinal numbers. In fact, Cantor's methods lead directly to the conclusion that there are infinitely many such levels. But one can ask whether there are still higher levels, so outrageously high up that they remain forever out of reach, and cannot be constructed, or even glimpsed, by any standard mathematical procedure.

These entities, if they exist, are called large cardinals. But, like the continuum hypothesis (see page 309), their existence cannot be deduced from any established principle of mathematics. The only way to get large cardinals is to assume their existence as a new law. It had been widely assumed that the existence or not of large cardinals is irrelevant to the everyday concerns of most mathematicians. Yet it turned out that large cardinal axioms are exactly the extra laws needed to render provable Friedman's statements about the whole numbers.

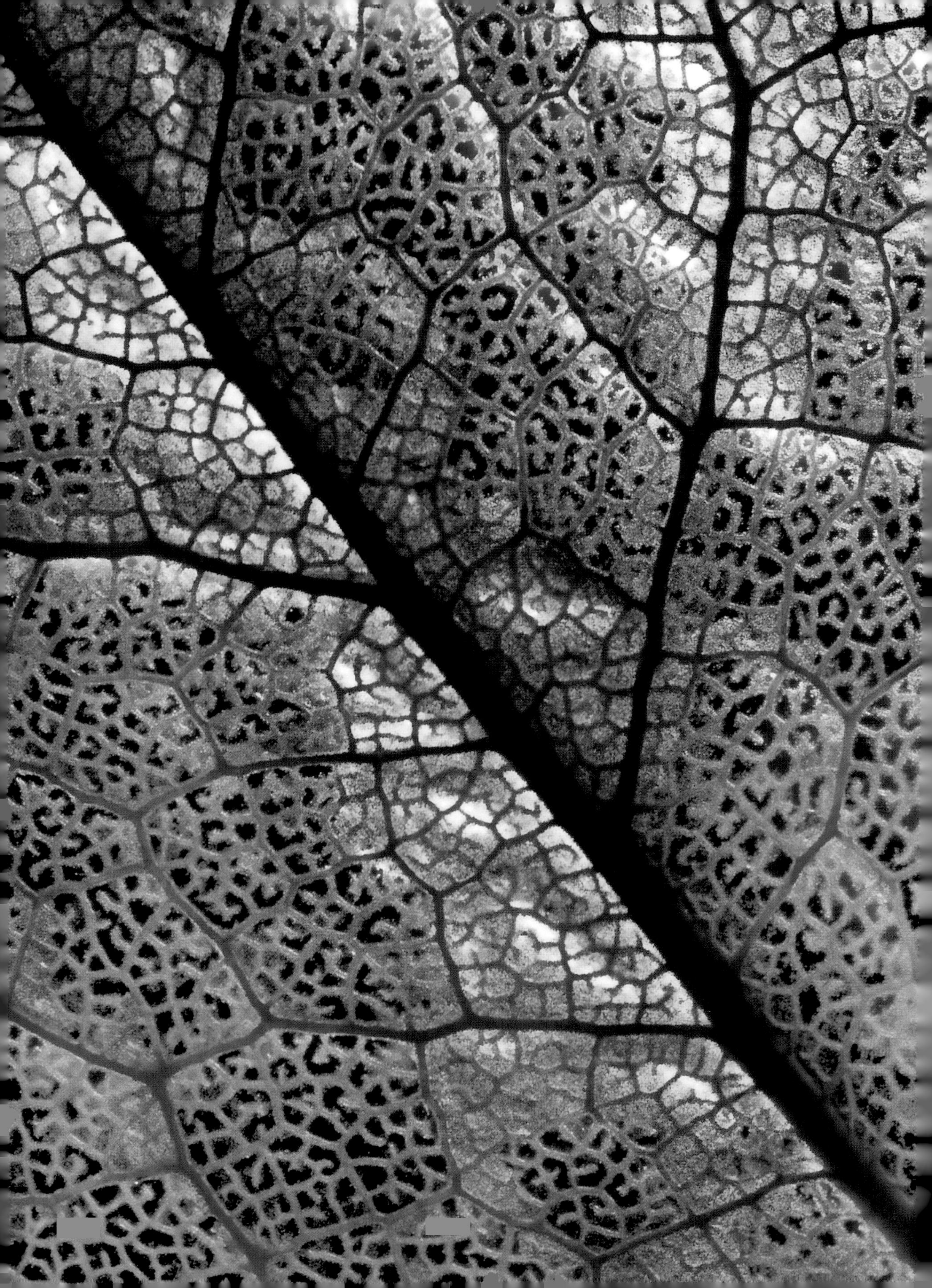

99 Partitions

BREAKTHROUGH A PARTITION IS A WAY OF SPLITTING A SET OF OBJECTS INTO SMALLER SUBSETS. IN 1918, HARDY AND RAMANUJAN PROVIDED A GOOD ESTIMATE FOR THE NUMBER OF POSSIBLE PARTITIONS. THIS PAVED THE WAY FOR THE WORK OF ONO'S TEAM IN 2011.

DISCOVERER G.H. HARDY (1887–1947), SRINIVASA RAMANUJAN (1887–1920), KEN ONO (1968–), JAN HENDRIK BRUINIER (1971–), AMANDA FOLSOM (1979–), ZACHARY A. KENT (1977–).

LEGACY PARTITIONS ARE SIMPLE OBJECTS WHICH HAVE NEVERTHELESS DEFIED UNDERSTANDING FOR MANY YEARS. THEIR CONQUEST MARKS A MILESTONE IN OUR UNDERSTANDING OF NUMBERS.

The deceptively simple notion of partition has inspired some exceptionally deep number theory since the early 20th century. It began with mathematics' greatest mathematical double act, that of Srinivasa Ramanujan and G.H. Hardy. Later, in 2011, a team led by Ken Ono was finally able to lay many of the questions surrounding partitions to rest.

How many ways are there to write the number 4 as a collection of smaller numbers added together? A little experimentation reveals five possibilities: 1 + 1 + 1 + 1, 1 + 1 + 2, 2 + 2, 3 + 1, and finally just 4. These are the five partitions of 4. (In the context of partitions, it is critical that 3 + 1 is counted as being the same as 1 + 3.)

Partitions are a simple and natural idea, but the mathematics behind them is far from straightforward. For small numbers, there is no difficulty: the number of partitions of 1 is 1, for 2 it is 2, and for 3 it is 3. But as the sequence continues, the pattern becomes ever less clear: 1, 2, 3, 5, 7, 11, 15, 22, 30, 42, 56, 77, 101, and so on. Mathematicians over the centuries have searched for the underlying rule to this sequence. If someone needs to know the number of partitions of 100, say, is there a formula that will quickly give the answer? Or is there no choice but to begin counting the partitions one by one?

The first person to devote serious thought to this question was Leonhard Euler, who discovered what is known as a generating function for

OPPOSITE The skeleton of a leaf, dividing the surface into regions of different sizes. How larger numbers can be divided into combinations of smaller ones is captured by the idea of a partition.

partitions. In principle, Euler's method could provide the answer: that there are 190,569,292 different partitions of 100. In fact, though, it was a slow method which meant crunching through all the intermediate numbers first. For large numbers, it was an impractical approach.

Hardy & Ramanujan

A more direct approach to the partition problem arrived with the partnership of Srinivasa Ramanujan, a phenomenal self-taught talent from rural India, and the pre-eminent British mathematician of the early 20th century, G.H. Hardy.

By any measure, Ramanujan was one of humanity's greatest geniuses. Growing up in Tamil Nadu in southern India, he quickly devoured all the mathematical books available to him at his school in Kumbakonam, and by the age of 13 was already proving theorems of his own. Working alone and in extreme poverty, Ramanujan rediscovered several famous mathematical results for himself, including several relating to diverging series (see page 97) and formulae for solving algebraic equations (see page 85). Ramanujan wrote his work in a highly idiosyncratic style, and, even at his most brilliant, never fully mastered the notion of a rigorous mathematical proof. Despite this, he produced a stream of highly original research, which he claimed was revealed to him in dreams by his family's goddess, Namagiri.

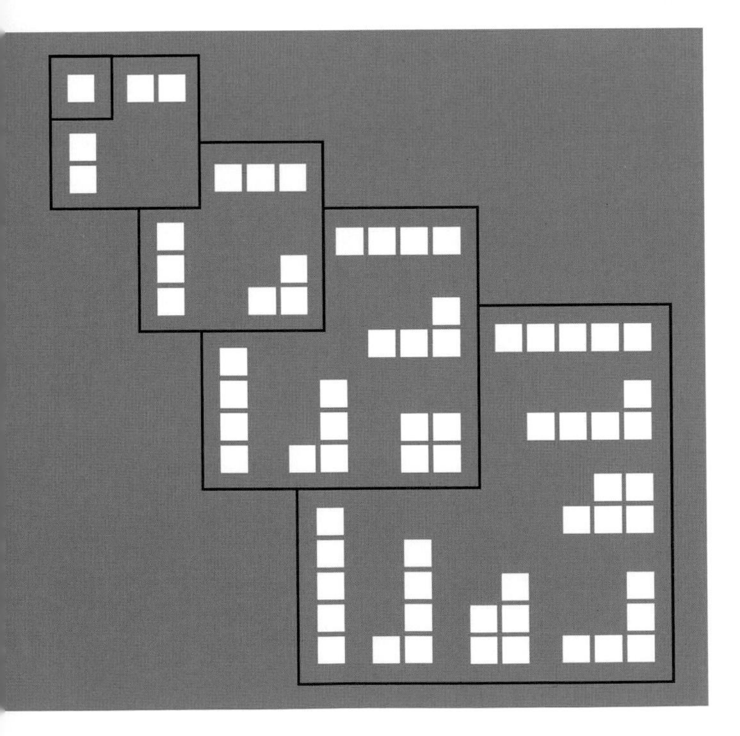

BELOW The partitions of the numbers 1–5 as represented by Ferrers diagrams.

Ramanujan's efforts to develop his career in India were hampered when he was refused entrance to the University of Madras, having failed the non-mathematical portions of the entrance exam. Forced to support himself as a clerk, in 1913 he wrote a fateful letter to the Cambridge University professor G.H. Hardy. Hardy was astonished by the extraordinary slew of theorems contained in the letter, and quickly realized that he was dealing with a mind of the very highest caliber. The next year, Ramanujan traveled to Britain to begin their historic collaboration, which Hardy would later refer to as "the one romantic incident in my life."

The Hardy-Ramanujan formula

Hardy and Ramanujan complemented each other perfectly as mathematicians, and their combined powers resulted in several major advances in number theory. Their most famous discovery concerned partitions. Instead of aiming for an exact solution to the partition problem, Ramanujan and Hardy discovered a beautiful approximate answer. They showed that the number of partitions of the number n is approximately

$$\frac{1}{4n\sqrt{3}}e^{\pi\sqrt{\frac{2n}{3}}}$$

Although this expression may look alarming on first sight, the answer can be very quickly calculated, especially since the invention of the computer. Although their formula was not an exact value, Hardy and Ramanujan demonstrated that it will provide ever better estimates the larger n itself becomes.

Ramanujan never fully mastered the notion of a rigorous mathematical proof. Despite this, he produced a stream of highly original research, which he claimed was revealed to him in dreams by his family's goddess, Namagiri.

Partitions and modular forms

Hardy and Ramanujan's work was a tremendous breakthrough in its own right. But their approach also provided number theorists with new weaponry to attack the partition problem. In 1937, Hans Rademacher was able to extend their result to produce an exact expression for the number of partitions of any given number. However, Rademacher's result was not a finite formula, but entailed adding together an infinite series. Though technically impressive, it was not of great practical use for counting partitions.

Over the course of the 20th century, the question of partitions was informed by some of the deepest aspects of modern number theory. Indeed, Ramanujan himself had noticed the striking effect of approaching partitions via modular arithmetic (see page 77). He demanded an explanation, and in 2011, one was finally provided by a team of number theorists led by Ken Ono of the University of Wisconsin, drawing on recent analysis of modular forms (see page 393). Their work was a triumph, for the first time yielding a single, finite formula capable of determining the number of partitions of any whole number.

100 Sudoku

BREAKTHROUGH MODERN SUDOKU HAS ITS BASIS IN THE OLDER LATIN SQUARES. BOTH OBJECTS HAVE INTERESTING PROPERTIES. A MAJOR QUESTION ABOUT SUDOKU WAS RESOLVED BY MCGUIRE IN 2012.

DISCOVERER LEONHARD EULER (1707–83), GASTON TARRY (1843–1913), GARY MCGUIRE (1967–).

LEGACY AS WELL AS ENTERTAINING PUZZLE ENTHUSIASTS WORLDWIDE, THE UNIQUE PROPERTIES OF LATIN SQUARES ALSO MAKE THEM VALUABLE IN COMPUTER SCIENCE, WHERE THEY ARE USED AS THE BASIS OF ERROR-CORRECTING CODES.

Sudoku is unquestionably the most popular form of recreational mathematics in the world. But it is based on a far more ancient object, known as a Latin square. These intriguing objects are more than mere puzzles; they have many applications, from considerations of abstract symmetry to safeguarding modern internet communication. The solution to a major theoretical question about Sudoku was announced in 2012.

The appeal of Latin squares is in their pleasingly simple definition. You begin with a 3 × 3 grid. The aim is to fill it with the numbers 1, 2 and 3, so that each number features exactly once in each row and once in each column. One possible solution is this:

1	2	3
2	3	1
3	1	2

Such "Latin squares" are misnamed. So far as we can tell, they first appeared in the medieval Islamic world. The Sufi thinker Ahmad al-Buni studied them in around AD 1225 and formed a 4 × 4 square from the letters of one of the names of Allah, illustrating that numbers are irrelevant: any set of symbols will do. It is possible to build Latin squares of any size, and the number of possibilities grows very quickly: there are two different 2 × 2 Latin squares, and 12 different 3 × 3 squares.

OPPOSITE A detail of a Greco-Latin square at Dartmouth College, USA. The small colored tiles form a Latin square, as do the colored rings, and no combination is repeated.

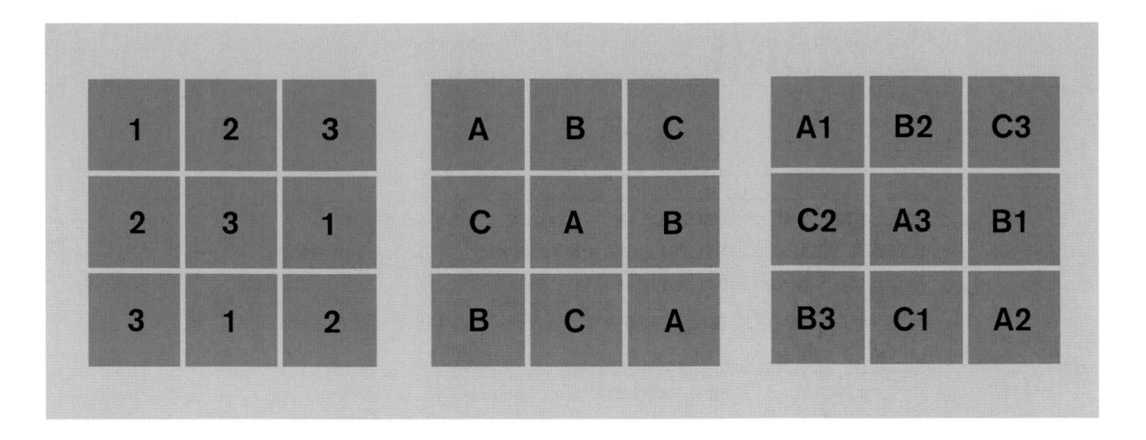

ABOVE Two Latin squares are combined to form a Greco-Latin square. In the result, all 9 pairings from A1 to C3 feature, none repeated or omitted.

Meanwhile, the number of distinct 9 × 9 squares, which are the basis of Sudoku, comes to 5,524,751,496,156,892,842,531,225,600.

The 36 officers problem

In Europe, Latin squares first came to prominence through the work of that master mathematician Leonhard Euler. It was Euler who first noticed the mathematical depths within these squares when he considered the question of how to match two Latin squares together to form a Greco-Latin square (see illustration).

Euler's technique only worked for grids of certain sizes. He was able to build Greco-Latin squares of every dimension, except 2 × 2, 6 × 6, 10 × 10, 14 × 14, 18 × 18, and so on. For these sizes, he conjectured, Greco-Latin squares cannot exist.

Certainly, it is true that there can be no 2 × 2 Greco-Latin square, as there are only two 2 × 2 Latin squares which cannot fit together in the right way. The question of a 6 × 6 square is more troubling, and this became famous as the 36 officers problem: if 36 army officers, of six different ranks, and six different regiments, want to march in a square formation, can they do so in such a way as to represent each rank and each regiment exactly once in each row and in each column?

Euler firmly believed that such an arrangement is "absolutely impossible," but was not able to provide a proof. The definitive answer did not arrive until 1901, when Gaston Tarry analyzed all possible 6 × 6 Latin squares and concluded that Euler was right: no two could match together. When it came to Greco-Latin squares of size 10, 14, 18, and so on, the story was different. In 1959, a group of three researchers (Raj Chandra Bose, Ernest Parker and Sharadchandra Shrikhande, who collectively came to be known as Euler's Spoilers) dramatically revealed that Euler was wrong: Greco-Latin squares of these sizes do exist. It is only those of size 2×2 and 6×6 which cannot.

Sudoku clues

Although Latin squares have long been popular in mathematical circles, in recent years they have taken an unexpected grip on the popular imagination in the guise of Sudoku. First devised by Howard Garns in 1979, this puzzle is a 9×9 Latin square that also satisfies an extra condition. As well as appearing in every row and column, every digit must also appear exactly once in each of the 3×3 blocks into which the grid is divided. In 2005, Bertram Felgenhauer and Frazer Jarvis established that there are exactly 6,670,903,752,021,072,936,960 valid ways to fill in a Sudoku grid.

When a Sudoku puzzle features in magazines and newspapers, it arrives with a few numbers already filled in—usually around 25. Setting these clues is mathematically subtle. The setter cannot fill in numbers haphazardly, but must guarantee two things. Most obviously, there should be some valid way to complete the entire grid, that is to say, the clues must be consistent. But more trickily, they should only allow for one unique solution. The method of solving Sudoku is logical deduction: "there must be a 4 somewhere in this row, and the only place it can go is here." For this type of thinking to work, there cannot be two equally legitimate answers to the puzzle. But of course, a grid containing only one or two clues will be able to be completed in many ways. This raises a delicate question: how many clues are needed to guarantee a unique solution? Puzzlists and mathematicians widely conjectured that the answer was 17, and on January 1, 2012, a team led by Gary McGuire announced that they had established this to be true. Using an ingenious computational search, requiring over 7 million CPU hours, they were able to eliminate the possibility of any 16-clue puzzle being unique.

If 36 army officers, of six different ranks, and six different regiments, want to march in a square formation, can they do so in such a way as to represent each rank and each regiment exactly once in each row and in each column?

That some Sudoku grids can be completed from a set of only 17 clues illustrates a useful property of Latin squares. If data is deleted or even entered incorrectly, it can be put right by examining the rest of the square. For this reason, Latin squares and related objects are commonly employed as error-correcting devices in modern data-transfer.

Glossary

Algorithm
A list of instructions for carrying out some task. All computer programs are algorithms, written in some language.

Arithmetic
Manipulating numbers through addition, subtraction, multiplication and division.

Bases
Methods for writing numbers, dependent on some fixed number. In the usual system, the base is 10. In **binary** it is 2, but bases of any value are possible.

Binary
A way to write numbers, using only the symbols "0" and "1." Binary is **base** 2, while the more familiar **decimal** notation is base 10. Binary is the natural language of computers.

Bit
Short for "binary digit," a bit is a 0 or 1.
Strings of bits are used to represent numbers in **binary**, or to transfer or store information. In computers, bits can be stored by switches.

Calculus
The subject which seeks to analyze changes in geometrical systems. The two branches are differential calculus (**differentiation**), which begins by analyzing the steepness of curves, and integral calculus (**integration**), which calculates the area under a curve.

Cardinal number
A measure of the size of a set of objects. The finite cardinal numbers are the usual counting numbers, 0, 1, 2, 3, ... But there is a whole hierarchy of infinite cardinal numbers beyond that.

Cartesian coordinates
A pair of numbers identifying the position of a point. The coordinates (2,3) represent a point 2 units to the right and 3 units up.

Chaos theory
The study of chaotic systems, which are highly unpredictable and sensitive to slight changes in conditions, even when you know their starting configuration to a high level of accuracy. (This is sometimes known as the butterfly effect.)

Complex number
A complex number is a **real number** added to an **imaginary number**, examples being $3 + 2i$ or $-\pi + \sqrt{6}i$. The system of all complex numbers is the backdrop to most modern mathematics.

Computational complexity
The study of the difficulty of a task, according to how long it takes a computer to carry it out. This is a major part of theoretical computer science.

Conic sections
Ellipses, parabolas and hyperbolas are together known as the "conic sections," because they can be obtained by slicing through a double-cone with a flat plane. These are the simplest curves after straight lines, and feature in physics as the paths of orbiting bodies.

Conjecture
A mathematical assertion which someone believes to be true, but for which no **proof** has yet been found. Examples include the abc conjecture and the Riemann hypothesis. If/when a proof is provided, the statement is elevated to the status of a **theorem**.

Curve
A 1-dimensional geometrical object. When viewed close-up a curve resembles a straight line. Examples include straight lines themselves, circles, and conic sections.

Decimals
A number written in **base** 10, such as 3.14159265358979... The dot is called the decimal point, and the digits following it represent one tenth, four hundredths, one thousandth, five ten-thousandths, and so on.

Differentiation
Analysis of rates of change. Differentiating an expression for a particle's position gives one for its speed, and differentiating that gives an expression for its acceleration.

Diophantine equation
An **equation** describing a condition which a whole number may (or may not) satisfy, in terms of other whole numbers. Famous examples include Fermat's last **theorem** and Catalan's **conjecture**. The study of Diophantine equations is a major topic in **number theory**.

Entropy
A measure of the unpredictability of a stream of data. A fair coin has entropy 1, while a double-headed or double-tailed coin has entropy 0.

Equation
An expression which asserts that two quantities are equal. Examples include $E = mc^2$ and $1 + 1 = 2$.

Exponential function
A rule which takes any input number x to an output of e^x. The formula is $e^x = 1 + x + \frac{x^2}{2} + \frac{x^3}{3 \times 2} + \frac{x^4}{4 \times 3 \times 2} + \cdots$
The exponential function is important in analysis of **complex numbers** and in **calculus**.

Exponentiation
For whole numbers, exponentiation is the same as taking **powers**. To extend this to broader classes of number, more technical work is needed, involving the **exponential function**.

Factor
A factor of a whole number is a smaller number which divides it exactly.
For example, 4 is a factor of 12 (because $4 \times 3 = 12$), while 5 is not.

Factorial
The result of multiplying a positive whole number, such as 6, by all the others less than it. A factorial is represented by exclamation marks, so $6! = 6 \times 5 \times 4 \times 3 \times 2 \times 1 = 720$.

Fractal
A shape or pattern with the property of self-similarity: however far you zoom into the shape, it continues to look the same.

Group
An algebraic structure, consisting of a collection of objects which can be combined in pairs. Examples include the collection of all possible **symmetries** of a chosen shape.

Hindu-Arabic numerals
The way of writing numbers with the symbols 0, 1, 2, 3, 4, 5, 6, 7, 8, 9. Now the worldwide standard, this system began to develop in India, before spreading through the Arabic world.

Imaginary number
The imaginary unit, i, is defined to be the square **root** of -1, $i = \sqrt{-1}$. (This was a fundamentally new type of number, since -1 has no square root among the positive or negative numbers.) Imaginary numbers are multiples of i, such as $4i$ or $\sqrt{6}i$.

Integer
An integer is a whole number, either positive, negative, or zero: …, −3, −2, −1, 0, 1, 2, 3, ….

Integration
The opposite to **differentiation**. Integrating an expression for an object's acceleration determines its speed. Geometrically, integrating a curve calculates the area beneath it.

Irrational number
A number which cannot be expressed precisely as a fraction or **decimal** (those being rational numbers). Famous examples include $\sqrt{2}$, e and π.

Logarithm
Closely related to **powers**, the logarithm of 8, to **base** 2, is the power to which 2 must be raised to reach 8. In this case the answer is 3, since $2^3 = 8$. This is also written as $\log_2 8 = 3$

Matrix
An array of numbers such as $\begin{pmatrix} 1 & 0 \\ 0 & 1 \end{pmatrix}$ (the number of rows and columns may vary). Matrices can be added and multiplied together, and encapsulate many geometric processes such as rotation and reflection.

Nash equilibrium
A situation in which each player in a game has no incentive to change strategy, even when she knows the intentions of every other player.

Negative number
A number which is less than zero, indicated with a minus sign, such as "−4." While a positive number, such as 4, may represent profit, −4 expresses the corresponding debt.

NP-complete
A task is *NP* if it can be verified quickly (technically in **polynomial** time), though not necessarily carried out quickly. *NP*-complete tasks are the hardest tasks in this class.

Number theory
The branch of mathematics studying whole numbers. The two major branches of the subject are the study of **prime numbers**, and **Diophantine equations**.

Paradox
A statement which can logically be neither be true nor false. The archetypal example is "This statement is false." (Some so-called paradoxes are really puzzles or very unexpected facts.)

Perfect number
A whole number whose **factors** sum to itself. An example is 6, since its factors are 1, 2 and 3, and $1 + 2 + 3 = 6$.

π
The circumference of any circle divided by its diameter. The exact value of π can never be expressed as a fraction or **decimal**, since it is an **irrational number**. But its approximate value is around 3.141592653589…

Place-value notation
The method of writing numbers in columns. The value of the "7" in 72 is determined not only by the symbol, but also its place, representing seven tens rather than merely seven.

Platonic solids
The most symmetrical **polyhedra**. There are five: the tetrahedron, cube, octahedron, dodecahedron and icosahedron.

Polygon
A flat shape built from straight lines. Common examples include rectangles and triangles. A regular polygon is one in which the edges are all the same length and the internal angles are all equal. Examples include squares and equilateral triangles.

Polyhedron
A surface built from flat faces, meeting along straight edges. An example is a cube. The most symmetrical polyhedra are the **Platonic solids**.

Polynomial
An expression in which an unknown quantity (often denoted x) is multiplied by itself and the results added together. An example is $x^2 - 2x + 1$. A **root** is a value for which this polynomial is zero, in this case, $x = 1$.

Polytope
A shape built from straight lines and flat faces. Polytope is the general term for these shapes in all dimensions. In two dimensions, they are known as **polygons** (an example is a square). In three dimensions, they are **polyhedra** (e.g. a cube).

Powers
Repeated multiplication. Raising four to the power of five (written 4^5) means multiplying five fours together: $4 \times 4 \times 4 \times 4 \times 4$.

Prime number
A positive whole number, which is only divisible by itself and one. An example is 7, while 8 is not prime, since $2 \times 4 = 8$.

Probability
A measure of the likelihood of an event. Probabilities are measured between 0 and 1. An impossibility has a probability of 0, a certainty has a probability of 1, and the probability of tossing heads on a fair coin is 0.5.

Proof
A logically watertight argument in support of a statement. Proofs are what separate true mathematical **theorems** from **conjectures** (or uninformed guesswork).

Quadratic equation
An **equation** involving an unknown number (often called x), in terms of its square (x^2 or $x \times x$). An example is $x^2 - 6x + 9 = 0$. This has solution $x = 3$, since $3^2 - 6 \times 3 + 9 = 0$.

Quantum mechanics
A major branch of physics in which a classical particle, with its precise location and motion, is replaced by a quantum entity for which location and motion are expressed according to certain **probabilities**.

Real number
A whole number, rational number (i.e. a fraction), or any of the **irrational numbers** in between. Examples include 2, $-3\frac{1}{4}$, π, and $\sqrt{2}$. Any real number can be written as a **decimal** (perhaps with infinitely many decimal places). The collection of real numbers is also known as the real line.

Relativity theory
The physics describing the large-scale geometry of the Universe. It emerged from the fact that the speed of light always appears the same no matter the observer's speed of travel. General relativity additionally considers the effect of gravity.

Right-angled triangle
A triangle in which one of the three angles is a right-angle (90°). The importance of these shapes derives from Pythagoras' theorem.

Ring
An algebraic structure consisting of a collection of objects which can be added, subtracted and multiplied together. The most famous example is the set of **integers**.

Roots
The opposite of **powers**. The square root of the number 16 is the number which when squared (i.e. multiplied by itself) produces 16, in this case 4. This is written $\sqrt{16} = 4$. Higher powers give rise to other roots. For instance the fourth root of 81 is 3, because $3^4 = 81$. This is written $\sqrt[4]{81} = 3$.

Ruler and compass constructions
Ancient Greek geometers were fascinated by the question of which procedures can be accomplished using only a straight unmarked ruler and pair of compasses. A line can be bisected (divided in half) with these tools, but some tasks cannot be completed, most famously the problem of squaring the circle.

Set theory
The branch of mathematics which deal with the properties of sets, meaning abstract collections of objects. Of key importance is the question of when two sets are the same size. Set theory is the setting for a mathematical analysis of infinity, in terms of **cardinal numbers**.

Singularity
A point within a geometric situation at which the usual rules fail to apply. An example is a sharp corner in an otherwise smooth curve.

Spacetime
A 4-dimensional geometry combining three dimensions of space, and one of time. Spacetime is the central object of study in **relativity theory**.

Surface
A 2-dimensional geometrical object, where every small region looks like a patch of flat 2-dimensional plane. Common examples include the sphere, torus, and plane itself.

Syllogism
A logical deduction in which two premises imply a conclusion. A famous example is "All men are mortal; Socrates is a man; therefore Socrates is mortal."

Symmetry
A way of altering a shape which leaves it looking the same, for example rotating a square by 90°. Symmetries may be reflections (i.e. mirror symmetries), rotations, translations (i.e. slides) or any combination of these.

Theorem
A mathematical statement for which a **proof** has been provided. Producing interesting theorems is the primary goal of mathematics.

Topology
An approach to geometry in which two shapes are deemed to be the same if one can be pulled or twisted into the form of the other.

Transcendental number
A number which cannot be turned into a fraction by any process involving only addition, subtraction, multiplication, and division (excluding the trivial move of dividing it by itself to leave 1). Examples include e and π.

Trigonometry
A collection of techniques for analyzing the distances and angles within a triangle. The three main functions are sine, cosine and tangent, each of which relates an angle to the proportions of the triangle's sides.

Turing machine
A theoretical computational device invented by Alan Turing for analyzing **algorithms**. Any computational machine is essentially equivalent to a universal Turing machine.

Waveform
A curve which repeats itself every cycle, used to model physical waves such as sound and light. The mathematician's favorite is the sine wave.

Index

Figures in italics indicate captions.

Quercus
New York

Quercus Publishing Inc.
31 West 57th Street, 6th Floor
New York, NY 10019

ISBN 978-1-62365-054-4

Library of Congress Control Number: 2013937911

Distributed in the United States and Canada by Random House Publisher Services
c/o Random House, 1745 Broadway
New York, NY 10019

Manufactured in China

2 4 6 8 10 9 7 5 3 1

www.quercus.com

123rf.com: /Tramiskincare 300; **akg-images:** 127, /Erich Lessing 55, 71; **Alamy:** /ArcadeImages 338, /Jeffrey Blackler 312, / Gino's Premium Images 159, /B. Hurst 224, /IDREAMSTOCK 368, /Emmanuel Lattes 215, /Chris Mattison 51, /Paul Prescott 228, / Mike P. Shepherd 15, /Mark Uliasz 68; Stuart Errol Anderson, Multimedia Artist, Sydney, Australia: 328; **The Art Archive**/Tate Gallery/ Eileen Tweedy: 194; **Manuel A. Báez** from the Baez Crossing Workshop Suspended Animation Series, Unititled Cellular Construction by Natalia Kukleva, Bamboo dowels and rubber bands, 2004 392; **The Bodleian Libraries**, The University of Oxford. MS. D'Orville 301, fols. 57v-58r: 46; **Bridgeman Art Library**/Royal Geographical Society: 350; Bill Casselman photography, from the Yale Babylon Collection: 31; **Corbis:** 116, 264, /Arcaid/Jeremy Cockayne 28, /Gérard Degorge 222, /Gianni Dagli Orti 16, /Johner Bildbyra AB 8, /Frans Lanting 180, /Ocean 44, /Michael Prince 340, /Roger Ressmeyer 291, /Sylvie Sonnet 200, /George Steinmetz/Menlo Park, California 108, /Steve Terrill 92, /View/Andy Stagg 319, /Philip Wallick 172; **Cortexd:** 48; **Dartmouth College**, photo of the Greco-Latin Square display in Kemeny Hall. Display is property of the Department of Mathematics, and used as the logo for the department: 404; **Sam Derbyshire:** 164; **Dreamstime:** /Hilchtime 304, /Icefields 256, /Icholakov 40, /Msujan 276; © **Foams and Complex Systems Group**, Trinity College, Dublin: 363; Image courtesy of K. Dudley and M. Elliff: 316; **Amy Snyder** © Exploratium, www.exploratorium.edu: 156; **Getty Images:** /Amana Images/Get High Design 124, /Comstock 128, /De Agostini Picture Library 23, / Eightfish 167, /First Light/Simon Gardner 380, /Flickr/Daniel. Candal 192, /Jeff Gross 135, /Oxford Scientific/Tim Oram 284, / Photographer's Choice/Frank Cezus 76, /Photolibrary/David Messent 186, /Markus Reugels 20, /SSPL 43, 111, 212, 383, /Visuals Unlimited/David Fleetham 52, /Workbook Stock/Thomas Barwick 348; **David Greene**, Johnsburg, NY - image generated with CA editor/simulator "Golly" (http://golly.sf.net): 335; **Kai Krog Halse:** 115; **Jan Homann:** 207; **iStockphoto:** /btrenkel 247, /seraficus 32; **Chris King**/dhushara.com/DarkHeart/: 204; Image courtesy of **Hans Lundmark**, Linköping University: 356; **Mary Evans Picture Library:** /Interfoto 94, /The National Archives, London, England 82; MSC Software: 287; **NASA:** /ESA, S. Beckwith (STScI), and the Hubble Heritage Tam (STScI/AURA) 140, /ESA, ESO & Danny LaCrue 216, /ESA, The Hubble Herritage Team, (StScI/AURA). M.Mountain (STScI), P. Puxley (NSF), J. Gallagher (U. of Wisconsin) 234, /ESA, D. Evans (Harvard-Smithsonian Center for Astrophysics) 248, /ESA and Aura/Caltech 384, /JPL/Space Science Institute 120, /N. Benitez (JHU), T. Broadhurst (Racah Institute of Physics/The Hebrew University), H. Ford (JHU), M. Clampin (STScI), G. Hartig (STScI), G. Illingworth (UCO/Lick Observatory), The ACS Science Team and ESA 336; **The Royal Belgian Institute of Natural Sciences**, Brussels/Thierry Hubin: 12; **Mark Newman**, University of Michigan: 292; **Science Photo Library**: /Dr. Jeremy Burgess 360, 372, /Scott Camazine 347, 374, /Maximilien Brice, Cern 271, / Cern, P. Loiez 267, /Isaac Chuang/IBM Almaden Research Center 366, /Stephen Dalton 327, /Wim Van Egmond, Visuals Unlimited 332, /Mark Garlick 123, /GIPHOTOSTOCK 139, /Pascal Goetcheluck 258, /Gustoimages 244, /Richard Kail 364, /Chris Hellier 56, /Matthew Hurst 320, /Bahman Kalantari 84, /James King-Holmes/Bletchley Park Trust 331, 354, /Mehau Kulyk 171, 252, /Laguna Design 168, 176, /Jerry Lodgriguss 239, /Derek Lomas 240, /Bill Longcore 220, /John Mclean 160, /National Institute for Fusion Science Japan 184, /Natural History Museum 36, /NASA 63, /NASA/JPL 196, /Dr. John Nicholls, Visuals Unlimited 152, /David Nunuk 174, /David Parker 88, /Pasieka 268, /Christian Riekoff 148, /Alexis Rosenfeld 188, /Gregory Sams 199, /Science Source 64, /Dr. Gary Settles 182, /Babak Tafreshi, Twan 236 /Sheila Terry 74, /US Department of Energy 154, /Victor Habbick Visions 324, /Dr. Keith Wheeler 344, 400, /Zephyr 130; **Shutterstock:** /Betacam-SP 230, /Dee Golden 315, /Kladej 288, /Oksanika 302, /PHOTOCREO Michael Bednakek 352, yvon52 272; **Steven Snodgrass:** 208; **Solkol**: 311; **suicidebysafetypin.deviantart.com**: 191; **Thinkstock**: /Hemera 132, 308, /iStockphoto 24, 60, 80, 96, 100, 104, 118, 136, 232, 255, 260, 296, 307, 343, 376, 388, 396, /Ron Chapple Studios 280, /Stockbyte 144; **TopFoto**/Luisa Ricciarini: 112; "Topology" by John G. Hocking and Gail S. Young: 211; **Collections artistiques de l'Université de Liège**: 378; **Truan Willis**: 72; **Tony Wills**: 11. **Wikimedia:** 386